AF566974

A MANUAL FOR AMATEUR TELESCOPE MAKERS

With Detailed Plans to Construct Three Different Telescopes

Karine and Jean-Marc Lecleire

Translated and adapted from French by Christopher A. Bechtler

Edited by Harold R. Suiter

With a Foreword by Jean Texereau

Published by

P.O. Box 35025 Richmond, Virginia 23235 USA ☎(804) 320-7016

www.willbell.com

First English Edition

Translated and edited from the French
Réalisez votre Télescope

Printed in the United States of America
First Printing 2003

Library of Congress Cataloging in Publication Data
Lecleire, Karine
[Réalisez votre télescope. English]
A manual for amateur telescope makers: with detailed plans to construct three different telescopes / by Karine and Jean-Marc Lecleire; translated and adapted from French by Christopher A. Bechtler; with a foreword by Jean Texereau. -- 1st English ed.
p. cm.
Includes bibliographical references and index
ISBN 0-943396-79-4
1. Telescopes -- Design and construction -- Amateurs' manuals. I. Lecleire, Jean-Marc. II Title.

QB88.L4313 2003
681'.4123-- cd21 2003044425

03 04 05 06 07 08 09 10 09 08 07 06 05 04 03 02 01

Table of Contents

Part 2: A 130-mm *f*/9.2 Newtonian Telescope

Acknowledgments

We must thank most especially Mr. Jean Texereau for his advice, support and encouragement.

We wish to thank warmly the people who, through their help and their work, have enabled us to bring this project to fruition: Sébastien Debruyne, Simon Fabre, Gino Farroni, Jean Funel, Richard Nowak, Romuald Oumamar, Gauthier Philippon, Michel Sarrazin, David Vernet, Frédéric Auchère, Georges Beaume, Claire Bigot, Nicolas Biver, Mourad Cherfi, Etienne de Foras, Jean-Bernard Emond, Arnaud Gaillon, Renaud Goullioud, Madani Khezam, Daniel Pasquelin, and the Melles Griot Company.

We would like to thank the Société Astronomique de France, in particular Elisabeth Sablé and Michel Laurent, as well as all of the members of its Commission on Instruments.

For this new edition, we also thank all of the amateurs who wrote to us to tell us their experiences with the first edition. Finally, we are most appreciative for the efforts of our English translator, Christopher Bechtler and our technical reviewer, Harold R. Suiter.

KARINE AND JEAN-MARC LECLEIRE

Preface

Amateur telescope-making literature always has been a kind of swamp through which one must hack and trudge to find the occasional sunny clearings. For every good article or book there seem to be ten that teach the reader only enough to be dangerous. There are many manuals giving decent instructions on mechanical assembly, but for mirror making the information is very thin. One good collection was the original *Amateur Telescope Making*, edited by Albert Ingalls. It introduced novice amateur telescope makers—called "ATMs" long before the money-dispensing machines—to the unfamiliar optical world of pitch and grits, together with the unlikely process of achieving perfection by haphazardly rubbing two disks of glass together. However, even the most reverent admirers of this old book would freely admit that its strength is not in testing. For that, English-speaking amateurs had to wait for Jean Texereau and the translation of his landmark *How To Make a Telescope*.

For the first time, the English-speaking amateur was led systematically through the mathematics and—more importantly—the *practical use* of the knife-edge test, including the time-saving technique of the varying reference curve. While this information had been obscurely implied in earlier works directed toward the amateur, it was either presented at an extremely theoretical level or at a cookbook level, with nothing between. Texereau straddled the gap.

Before Texereau, amateurs often struggled for hours trying to correct a perfect mirror. Their entire reduction technique was predicated on the assumption that the hard-to-measure central zone was exactly accurate. A trivial error here resulted in the rest of the wavefront appearing out-of-tolerance. Still other books used no wavefront reduction at all, relying on the raw longitudinal aberration measurements and the relative insensitivity to error of the slow reflectors being made at the time. After Texereau, amateur telescope-making improved a great deal. Aperture ratios came down and the test-reduction techniques that would enable the large, fast reflectors of today became more commonplace in the amateur world.

It seems exceptional that this first comprehensive mirror-making manual in many years (almost the next after Texereau) should also be a contribution of the French, but perhaps it's not so strange. The French have always delighted in exquisite telescopes and other fine astronomical instruments. Perhaps the excellent seeing at Pic du Midi had something to do with it. Maybe the early 20th-century optical geniuses represented by the great names of Lyot, Couder, Duffieux, and Maréchal influenced the French origin of this work.

In any case, this book by Karine and Jean-Marc Lecleire may look superficially like an updated version of Texereau. Nothing could be further from the truth. The basic knife-edge testing method suggested for slower mirrors may resemble Texereau's, true, but that is just reluctance to interfere with something that works. Where this book really shines is in the extensive corrective measures suggested to cure optical difficulties encountered during figuring. It deals more thoroughly than previous books with figuring the kind of mirrors used today with fast Newtonians or Cassegrain primaries. The Lecleires discuss tests less cumbersome than the Foucault test in figuring these very fast instruments, namely the wire and caustic tests.

Actual images using modern physical-optics theory have been added to provide an air of conviction to statements that have in the past been a sort of received wisdom. Most interesting of these to me is the progression of successively more curved diagonals, showing just when this curvature becomes detectable. I recalculated some of these star-test figures to provide more reproducible digital images, but they existed in the French original.

Lately, there has been a movement toward the Strehl ratio as a statement of quality rather than the peak-to-valley Rayleigh tolerance. Although people argue with good reason that the Strehl ratio is a much better way of specifying the wavefront than the somewhat arbitrary Rayleigh limit, a possible degree of duplicity exists in some manufacturers' preference in barely meeting a Strehl ratio = 0.8 criterion. For example, astigmatism can be as large as ⅜-wavelength peak-to-valley and still satisfy the criterion. One can use the Zernike polynomial expansion[1] to show that the ¼-wavelength Rayleigh tolerance is always a tighter tolerance than Strehl ratio = 0.8. In short, using the minimum Strehl ratio allows manufacturers to make sloppier mirrors than heretofore. Commercial makers should bump up their Strehl-ratio criteria to a little higher than 0.8 if they want to make substantially the same mirrors as they did when they met the ¼-wave Rayleigh limit. Several honest makers won't release an optic unless its Strehl ratio exceeds 0.9, a move of which I heartily approve.

Other than a sidebar discussion of just how the Strehl ratio is calculated, this book does not deal much with it, sticking instead with the Rayleigh limit. Some would argue that this omission is a flaw, but I don't think it is much of a defect. We must differentiate between measures useful in specifying finished optics for contractual arrangements, such as a university buying an expensive mirror for an observatory, and measures useful in actually making a high-quality mirror for personal use. Relying on the Strehl ratio to write a purchase agreement prevents the customer from making unreasonable criticisms against a perfectly good mirror, such as complaining about the interior of a tiny pit exceeding the ¼-wave limit. Still other customers might carp about a turned edge that has been pushed right to the perimeter, still theoretically breaking the ¼-wavelength limit but in practice having no effect whatsoever. On the other hand, when making a personal mirror,

[1] M. Born and E. Wolf. *Principles of Optics, 6th ed.*, Pergamon Press, 1980.

there is very little difference between the utility of either method. You can always refocus the best peak-to-valley wavefront to the minimum Strehl ratio wavefront, or vice versa. After all, you don't need to delude yourself using either tolerance; you are your own customer.

In fact, inferring the Strehl ratio from any measure across a diameter or independent diametric measurements is probably a form of speculative fiction. Any test across a diameter leaves vast areas of the disk unsampled, and it is those areas that are most heavily weighted in the Strehl calculation. The Strehl ratio is only reliable when the testing technique has been generated from the whole mirror, such as from interferometry or the two-dimensional Hartmann wavefront test. In a perfect world, it would seem an excellent practice to *require* all tests across the diameter to use peak-to-valley specification, and to only permit the Strehl ratio to be quoted when a whole-mirror test has been used. That way, we would know not only the tolerance but have some idea of the measurement technique employed.

Likewise, the Lecleires will undeservedly attract criticism for using what is known in English—through Texereau—as the Danjon and Couder criterion number 1 (here the authors abbreviate this as the Couder criterion). Various internet pundits have summarily stated that all slope or, equivalently, transverse aberration criteria have been "superseded" by the Strehl ratio. This conclusion is overstated. There is unique information in the transverse aberration of the zones not contained in any one number wavefront-amplitude criterion, including the Strehl ratio and the Rayleigh tolerance. Surfaces can produce acceptable Strehl ratios and still have objectionable features; not until we used measures like the modulation transfer function do these differences manifest themselves numerically. The most common case: a surface with a rolled edge might meet a given Strehl ratio and still distribute an ugly aura around a bright image. Just because transverse aberration does not predict the diffraction shape of the image does not mean mirror slope as a function of radius is not extremely useful in making a superior optic. Again, the Strehl ratio is a convenient and uniform standard of minimum acceptability, but to deliberately throw away the extra information contained in the transverse-aberration perspective is worse than foolish; it is reckless.

The Lecleires also introduce a test for polish quality and high mirror slopes to the American telescope maker. If you replace the lamp on a standard Foucault tester with a super-bright halogen source, widen the slit somewhat, and move its image until it is just covered by the knife, you see only those small-scale errors with enough slope to throw some light beyond the knife. Even though it uses the same apparatus as the knife-edge test, it is actually an inexpensive variant of the Lyot test, which replaces the knife edge with an attenuating phase plate. Because it is not quantitative, this test gains utility with experience, and only by comparison with good optics. Still, it is a handy tool to add to your bag.

Thus, I come to the end of my comments. I wish you joy at reading what is the first thoroughly new amateur mirror-making manual in forty years. My final bit of advice is directed at those who might want to ask me which manual I recommend, Texereau's or the Lecleires'.

Don't be silly; read both!

Harold R. Suiter
Panama City, Florida
March 2003

Foreword

The image of mediocrity usually associated with the work of amateurs is a misconception. On the contrary, amateurs, by the strength of their motivation and dedication, have many chances to stray from the beaten path. Their passion takes them into the realm of productive research and enriches their lives. It is perhaps superfluous to remind those already engaged in astronomy of this; but what is little known, and deserves to be discussed, is that the science of astronomical optics was essentially created and perfected by amateurs.

The English, often more attuned to the concrete than the abstract, have played a dominant role in this development. To begin, we must mention Newton—yes, Isaac Newton. It is not the grave, wigged character, master of the Mint of London, that I wish to evoke here; nor the mathematician of binomials and fluxions; nor the founder of celestial mechanics; and especially not the gawky youth, with his relentless pursuit of mystical subjects like biblical chronology and alchemy. It was Newton, the optician—not only the theorist but, especially, the practitioner—who made the first practical astronomical mirror. This first reflecting telescope, the objective of which measured merely 1⅓ inches in diameter, was considered in 1672 to be superior to long refractors with simple objectives, and it marked the beginning of a fruitful era in the history of astronomical optics.

Artists and professionals, jealous of their secrets, share none of their information with us; but amateurs, sometimes famous ones, make decisive contributions. We know their famous astronomical discoveries very well, remembering here only how they created the indispensable tool, the fruit of patient experiments.

William Herschel was at first an organist and composer. In 1774, helped by his sister Caroline, he began to cast metal mirrors (called *specula*) and to research the best alloys; he eventually made more than 300 mirrors, each one more perfect or larger than the last. He discovered Uranus in 1781 with a simple 16-cm Newtonian *f*/13, and conducted a systematic study of the sky with a variety of instruments, culminating in 1789 with the famous 122-cm telescope that probably required as much effort to use as it did to make.

Lord Rosse was a great landowner who had a large workforce at his disposal in his Irish estate at Parsonstown. Look at the photograph on the next page, of some of Rosse's craftsmen before his colossal 182-cm speculum. Despite the chisel and hammer, he is not about to do any local retouching! The carpenter with the bowler hat has a trying plane under his arm, and the astronomer Boeddicker is holding his portfolio of drawings. This mirror—cast after several attempts, made of a poorly reflective copper/tin alloy, and difficult to polish—was never really

Lord Rosse's 182-cm speculum, photographed at Parsonstown about 1845.

satisfactory; it would have taken a new Herschel to get good results from it. Still, we owe to Rosse the visual discovery of the spiral structure of several galaxies.

William Lassell, a merchant who retired and sought better observing conditions by moving to the island of Malta for three years, constructed the last successful large telescope with a bronze mirror, which he mounted on an equatorial fork. The mirror was 122 cm in diameter and had a focal length of 11.30 m; access to the Newtonian focus required a rather dizzying three-level tower. Lassell was the first to have the idea of supporting the mirror with an astatic lever, one of the best systems still to this day.

The modern era of telescopes employing glass mirrors silvered on the front face opened in 1856 with the work of Léon Foucault. The director of the Paris Observatory at the time was the famous Urbain Leverrier, who was probably less of a pure theoretician than he is normally pictured. He recognized Foucault's genius for experimentation and entrusted the modernization of the observing instruments to him.

Foucault not only introduced glass mirrors, he figured the first 10-cm glass one himself. In 1858, he published all the details, notably the marvellous method of the knife-edge test. This test allows the mirror maker to see the smallest defects, almost to feel them as if he could manipulate them and—for the first time— to retouch them. With the collaboration of two firms, Secrétan and Eichens, he built equatorial fork telescopes of 20- and then 40-cm aperture, and an 80-cm one for the Marseille Observatory.

It is thus the great English-speaking amateurs who first profited from Foucault's contribution. Henry Draper abandoned metal mirrors in 1860; his best instrument was just 39.5 cm in aperture, but he opened the doors to astrophotography, overcoming the disadvantages of his alt-azimuth mounting with a sliding plateholder driven by a water clock to track throughout the shot.

A. A. Common made metal specula, and he was perhaps the last great amateur to do so. After making a good 45.7 cm mirror in 1876, he went on to a 91.5-cm one in 1879. In France, he obtained a plate-glass disk 152.5 cm in diameter and 15.2 cm thick. He tested the surface of this mirror by zonal measurements, and his equatorially mounted telescope went on to become the forerunner of the large photographic telescopes of Mount Wilson that would dominate astronomical research through the first half of the twentieth century.

The origin of the more modest activity of many amateurs can be traced to about 1880 in the journal *English Mechanic,* in which subscribers discussed amongst themselves such diverse topics as the construction of organs, tricycles of unusual design, and telescope mirror making. From 1881 to 1886, a suite of articles by H. A. Wassell brought much information to light; these were reprised in 1909 by C. D. P. Davies and, most significantly, served as the basis for the book by another pastor in 1920: *The Amateur's Telescope* by William F. A. Ellison.

In the United States, the movement grew largely thanks to Albert G. Ingalls, associate editor of the prominent magazine *Scientific American*. He stumbled onto a 1921 telescope-making article by the extraordinary Russell W. Porter, a man of many talents: polar explorer, artist, engineer, optician; a sort of latter-day Leonardo DaVinci. From their collaboration came, beginning in 1926, the famous *Amateur Telescope Making* books, which included the basic articles, magnificently illustrated by Porter, text and figures from Ellison's book, and many other contributions. This first *ATM* was followed in 1937 by *ATM-Advanced* and then by *ATM-Book 3* for the most avid in 1953. From that point on, American amateur astronomy magazines have had regular columns on telescope making.

But in France? Well, in France craftsmanship is not as highly valued as in the English-speaking world, and it is definitely secondary to the contemporary hegemony of literature and the media. In the eighteenth century, the monumental *Encyclopedie* or *Dictionnaire raisonné des Sciences, des Arts et des Métiers* (Encyclopedia or Dictionary of Sciences, Arts and Trades), published between 1751 and 1780 by Diderot and d'Alembert with the assistance of the Chevalier Jaucourt, was a brief success, but it was known more for its theological and philosophical polemics than for its industrial articles, though these were magnificently detailed and illustrated. The Enlightenment remained in darkness when it came to technology. Foucault, despite a popular republication in *Classiques de la Science* in 1905, remained almost unknown by amateurs, although there were already 4,000 members in the Société Astronomique de France (SAF), which was founded in 1887 by Camille Flammarion.

It was not until the 1930s that a new Foucault advanced things: André Couder's 1932 thesis revealed the fine physicist who, along with André Danjon, gave us the admirable *Lunettes et Télescopes* (Refractors and Reflectors) in 1935. The work, of which only 1,800 copies were published, quickly went out of print, and was never reprinted during the authors' lifetimes. Though it did not treat mirror making, it was a bible for the amateur; it answered simply most of the questions asked by its readers.

Things began to progress quickly. At the S.A.F., the idea of making one's own telescope was mentioned in a few isolated articles. The first gathering of the Society's Commission on Instruments was held in November of 1945, and the appearance, in serial form, of *How to Make a Telescope* began in the May 1948 issue of *l'Astronomie*. These articles, once assembled and published, constituted the first edition of *How to Make a Telescope* in 1951, and with additions, its second edition in 1961. It would not have been possible without the gentle but determined pressure of André Danjon—a great astronomer, a great professor, and a great teacher who loved his students as well as the most modest amateurs, because he was always one of us. His death in 1967 left a void, but the seed had been planted and strong roots remained. First, Gauthier Philippon manned the craftsman's workshop. Now, Jean-Marc Lecleire gives us this new book, a modern transposition of my *How to Make a Telescope*.

Jean-Marc, with good reason, presents three separate plans for telescopes of increasing difficulty. I would here underscore the importance of the most modest, the 130-mm Newtonian with a spherical mirror, which many ambitious amateurs would perhaps think they could overlook. Learning first to produce a sphere or a weak parabola, smooth and without zonal defects, is an essential step, the revealing test that allows one to master mirror figuring. One must absolutely resist the premature desire to retouch faulty zones or a turned-down edge; he must find what is wrong with his strokes and, moreover, his polisher, which must be remade when necessary with a mixture of pitch, whose hardness is tempered with turpentine, to yield perfect squares that quickly take the shape of the blank and adjust gradually during the work.

Be careful with the λ/n measurements, because they will be meaningless if the measurements and calculations are not done with great care. Finally, be wary of articles espousing "new methods" of mirror testing; much experience is necessary to understand the sophistication often present in those articles and to arrive at a correct evaluation. In short, bear in mind the spirit of George Bernard Shaw's adage (from *Man and Superman*, 1903): He who can, does. He who cannot, teaches.

Méaux, August 1997 JEAN TEXEREAU

Jean Texereau is the author of the standard reference work on the construction of telescopes and the figuring of astronomical mirrors, How to Make a Telescope.

Part 1: Introduction

1 About this Book

This book details the making of mirrors and the construction of telescopes. It describes the optical and mechanical parts of a telescope and guides you through making a complete instrument. There are complete, step-by-step plans for three telescopes, from the simple to the complex, so that you can make an instrument that suits your desires and needs.

This book also gives important advice on the use and maintenance of telescopes and on the technique of mirror alignment. It is therefore useful to amateurs who are frustrated after having bought an instrument that is complicated and difficult to use.

1.1 Why Make Your Own Telescope?

In our day, when everything is manufactured, what is the use of making your own telescope? Why not just buy one?

Making your own telescope allows you to understand it and its capabilities more fully, and thus get the best performance from it. It is possible to improve the performance of your instrument constantly, to add new accessories to it, and to refine its optical and mechanical alignment.

Making a telescope is exciting. It enables you to discover (or rediscover) manual work and gives you the opportunity to make high-quality optical pieces with your own hands. The mirror is the most important and complex part of a telescope; it is the eye of the instrument. The figuring of mirrors is a very enriching challenge, and it sometimes reveals hidden talents.

Making your own telescope also saves you money. You can make a telescope yourself, without skimping on its quality, for as little as a quarter of the price that you would pay for a commercial instrument. You can make telescopes that are superior to commercial models. The secret to success lies in a good understanding of mirror making and, more importantly, mirror testing.

The qualities most useful for telescope making are patience and carefulness. You must like to do things yourself, have some capacity for doing things with your hands, and—above all—take your time. It is not important for you to know a lot about optics; the formulas you will use in this book are quite simple. Chapters 1 though 5 will give you the background necessary for planning an instrument to

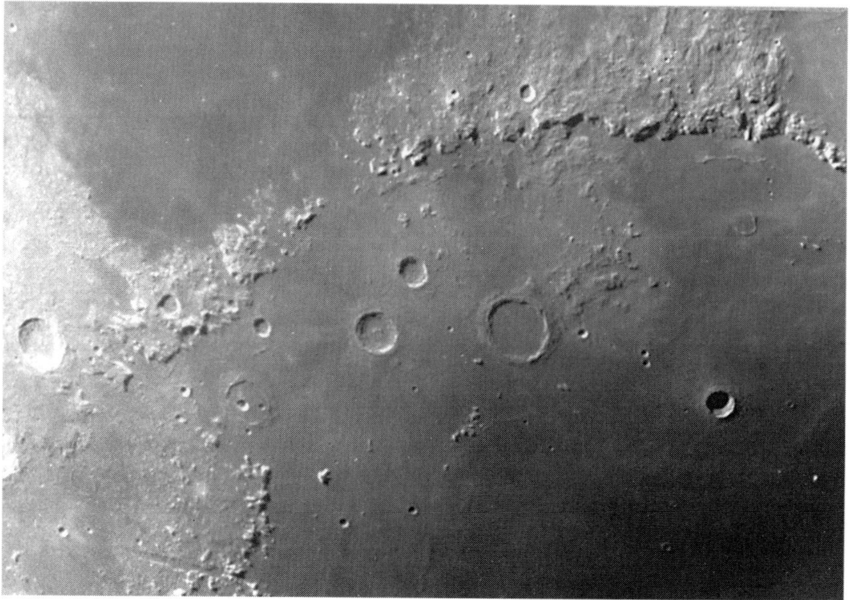

Fig. 1.1 The 130-mm telescope reveals the beauty of the lunar landscape. Here, a photograph of the Archimedes crater region, the Montes Apenninus and Mare Imbrium.

meet your own unique needs and interests. Telescope making is something many people can do; it is not reserved only for scientists. There is no age limit for building a telescope—from children in summer camp to retired persons, anyone can make his or her own telescope.

1.2 What Use Is a Telescope?

A telescope inspires dreams and fantasy. Seeing the stars through a telescope expands the limits of our world and our imagination. A telescope permits you to survey the Moon's mountains like an astronaut, to contemplate the majesty of the rings of Saturn, and to follow the ballet of Jupiter's moons or the phases of beautiful Venus.

To possess a telescope is to have a powerful instrument at your disposal with which to make sketches or take pictures, keep an observation notebook, follow the changing appearance of a comet, detect the presence of sunspots, or observe the polar caps of Mars. A telescope is a window open to the planets, the stars, nebulae and galaxies. Thanks to these instruments, we can examine the universe first hand and try to understand our place in it.

Observing telescopes are tools for research and discovery. In our day, professional astronomers scour the skies in quest of a new supernova, a distant galaxy, or an invisible quasar. They use their giant instruments to concentrate on very small portions of the firmament. Most of the sky remains unexplored, and amateur

Fig. 1.2 Saturn, observed August 26, 1996, with a 250-mm telescope. Drawing © Nicolas Biver.

astronomers can fill this void. Amateurs are building more and more powerful instruments and search for undocumented celestial phenomena like undiscovered asteroids and comets.

1.3 The Instruments Described in this Book

1.3.1 A Beginner's Telescope

The first instrument described in this book is designed for beginners. It is inexpensive and easy to make. It will familiarize you with mirror making and optical testing methods.

The beginner's telescope is a Newtonian with a 130-mm diameter primary mirror and a 1200-mm focal length. It is comprised of a concave, spherical primary mirror and a small, flat secondary mirror placed at the top of the tube.

This instrument can detect stars about 650 times dimmer than you can see with your unaided eye, down to magnitude 13.5—there are over ten million stars brighter than that. It will reveal details on the order of a kilometer and a half on the Moon. You can begin to resolve globular clusters into stars, observe hundreds of diffuse nebulae and galaxies, go on comet hunts, and travel the spirals of the Milky Way. The 130-mm telescope will unveil the rings of Saturn and the cloud belts of Jupiter, and it will distinguish the phases of Mercury and Venus.

1.3.2 The Standard Amateur Astronomer's Telescope

The second instrument in this book is a Newtonian reflector with a 250-mm objective and a focal length of 1000 mm. The primary mirror is paraboloidal (popularly called parabolic); the secondary mirror is flat. This "all purpose" telescope reaches stars of magnitude 14.5 and gives beautiful images. At the eyepiece, the Orion Nebula shines an electric green with pale pink stripes, planetary nebulae appear emerald green, and the bands of Jupiter swarm with details. The Moon unveils the finer details of her topography, revealing valleys, hills and small craters.

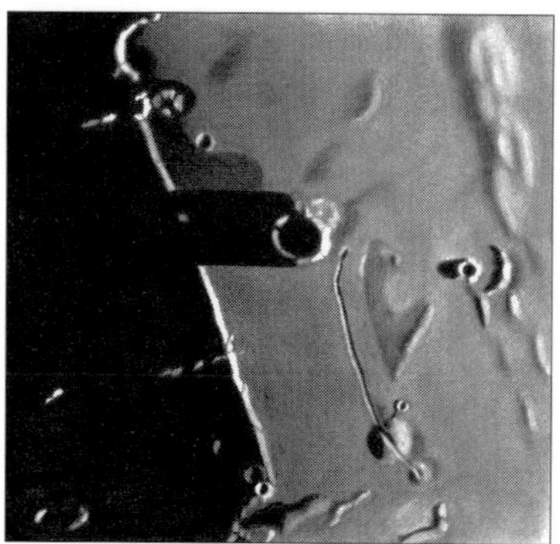

Fig. 1.3 Drawing of the Moon, region of the **Straight Wall**, made August 19, 1995, from 3:40 to 5:05 Universal Time, through the eyepiece of a 200-mm telescope. Magnification 476x. Drawing © Nicolas Biver.

Fig. 1.4 The Orion Nebula, M42, at the focus of a 300-mm telescope. Photo © G. Beaume.

This telescope can be equipped for astrophotography, either at the eyepiece (by projection) for planetary imaging, or directly at the focus to record the dim glow of galaxies. With its focal ratio of 4.3, this high-resolution instrument is particularly well suited for astrophotography; with it, you can obtain beautiful photographs of the sky in color or black and white.

This 250-mm reflector is the standard instrument made by amateurs and is a good springboard to the construction of a larger telescope.

1.3.3 The Telescope for the Experienced Amateur

This is a 300-mm *f*/12 instrument in a Cassegrain-coudé (or Nasmyth) combination. Its round tube fits the equatorial mount for the 250-mm telescope described above.

This project gives you the chance to make several things: a long-focal-length mirror, a convex secondary, and a telescope window. It also introduces you to advanced testing methods such as the wire test or the caustic test.

This long-focal-length telescope reveals the major details of Saturn's rings and will send you to the cores of globular clusters and colorful planetary nebulae. In photography, its versatility lets you capture planets and relatively bright nebulae and galaxies.

The methods detailed in this book are useful for other sizes of instruments. The techniques described for making the mirror and the mechanics of the 130-mm model are usable for smaller mirrors up to that diameter. Those who would like to make a 200-mm telescope can take their inspiration from the method for the 250-mm. Finally, the figuring and testing techniques used for the 300-mm mirror can be applied in the making of 375- to 500-mm mirrors.

2 From Galileo to Hubble: A Short History of the Telescope

2.1 The Pioneers of Astronomical Optics

The first refracting telescopes date from the end of the sixteenth century. Most probably a number of people independently discovered, but did not make known the fact, that two spectacle lenses could be arranged into a refracting telescope but these instruments remained generally unknown until 1608, when Hans Lippershey made and sold several models in Holland. In June of 1609, the Italian astronomer Galileo Galilei (1564–1642) learned of the existence of these Dutch refractors, which were being sold in Paris. He went on to make his own instrument with a 1-inch objective, and began to explore the night sky for the first time on January 7, 1610. He went on to discover Jupiter's satellites, lunar craters, the nature of the Milky Way, and more.

Galileo's telescope consisted of a bi-convex (converging) one-lens objective and a bi-concave (diverging) eyepiece (see Figure 2.1). It had a very small field of view. This optical system is almost never used today except in low-power opera glasses. Johannes Kepler showed in his *Dioptrice* (1611) that the concave eyepiece used by Galileo could be replaced with a convex lens. This converging lens served as a magnifying glass, enlarging the image formed at the focus, and thus provided greater comfort and a larger field of view.

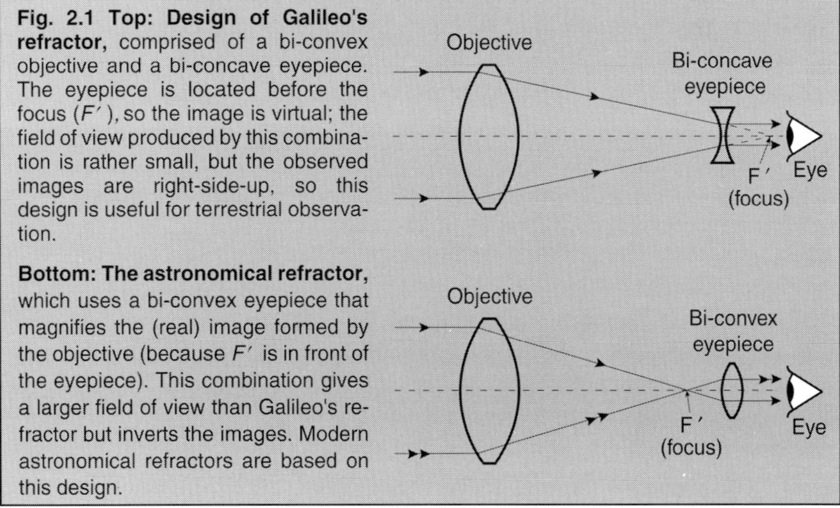

Fig. 2.1 Top: Design of Galileo's refractor, comprised of a bi-convex objective and a bi-concave eyepiece. The eyepiece is located before the focus (*F′*), so the image is virtual; the field of view produced by this combination is rather small, but the observed images are right-side-up, so this design is useful for terrestrial observation.

Bottom: The astronomical refractor, which uses a bi-convex eyepiece that magnifies the (real) image formed by the objective (because *F′* is in front of the eyepiece). This combination gives a larger field of view than Galileo's refractor but inverts the images. Modern astronomical refractors are based on this design.

Fig. 2.2 The first Galilean refractors, preserved at the Institute and Museum of Science History in Florence. It was with these instruments that Galileo discovered the lunar craters, Jupiter's moons, and the rings of Saturn. Drawing © Simon Fabre.

Astronomers such as Christiaan Huygens (1629–1695) made the next improvements in refracting telescopes. Beginning in 1655, he and his brother Constantyn made several refractors of increasing diameters and focal lengths. They observed the planets in detail and discovered Titan, the principal satellite of Saturn. Their instruments, 110 to 130 mm in aperture, had focal lengths over 10 meters and had to be mounted on poles. The large Huygens refractors did not have tubes; their oculars were connected to their objectives with cords or wires. The images obtained were of better quality than those seen in Galileo's smaller telescopes: their long focal lengths allowed them to achieve high magnifications and reduced the severe chromatic aberration inherent in the simple-lens refractors of the time.

In 1729 a British barrister, Chester Moor Hall, devised the first two-element lens, which gave colorless ("achromatic") images, but he never publicized or developed his idea. In the 1750s, Leonhard Euler and Samuel Klingenston published papers theorizing that chromatic aberration could be corrected by combining two different types of glasses—a "crown" and a "flint"—into an objective (see Figure 2.3). John Dollond, a London instrument-maker, built on their work by conducting experiments with crown and flint prisms, and eventually was able to patent and produce commercially an achromatic telescope that was much more compact and performed better than previous refractors.

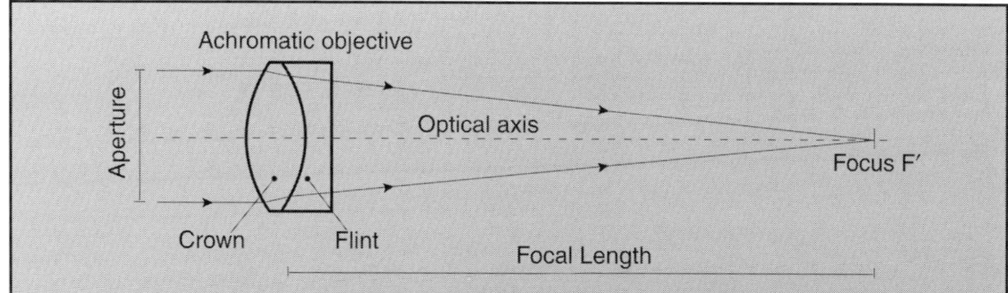

Fig. 2.3 Principle of the achromatic astronomical refractor. The objective is composed of two lenses made of different materials (a "crown" glass and a "flint" glass) whose optical properties are complementary. The doublet thus formed corrects for the chromatic aberration that was present in early simple-lens refractors.

Refractors were not the only instruments capable of forming magnified images. The first studies of telescopes with reflecting optics date from the middle of the sixteenth century, but the technical problems inherent in the figuring of metallic mirrors prevented their construction for another century. In 1663, James Gregory proposed a design for a reflecting telescope that combined two concave mirrors facing each other (see Figure 2.4), but again the technical problems were insurmountable. In February 1672, Isaac Newton (1643–1727) presented, in London, a 37-mm reflecting telescope that magnified 38x. It consisted of a concave primary mirror made of speculum metal alloy (mostly copper and tin) and a flat secondary that reflected the light to the side of the tube (see Figure 2.5). The telescope was achromatic because it formed its images by reflecting light instead of bending it.

In March of 1672, the Frenchman Laurent Cassegrain proposed a telescope equipped with a convex secondary mirror (see Figure 2.6). Since it was more difficult to construct than Newton's telescope, it took several years for this design to gain acceptance. Today, almost all the large professional observatory telescopes are all derived from Cassegrain's design.

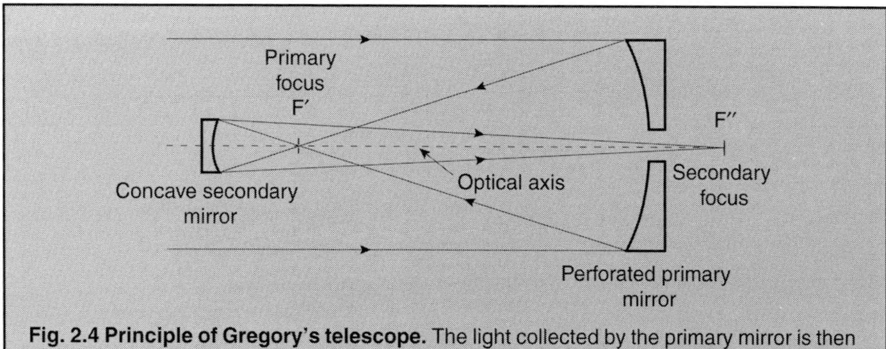

Fig. 2.4 Principle of Gregory's telescope. The light collected by the primary mirror is then reflected by the concave secondary mirror to a focus behind the primary. This combination gives right-side-up images and can be used in terrestrial observation.

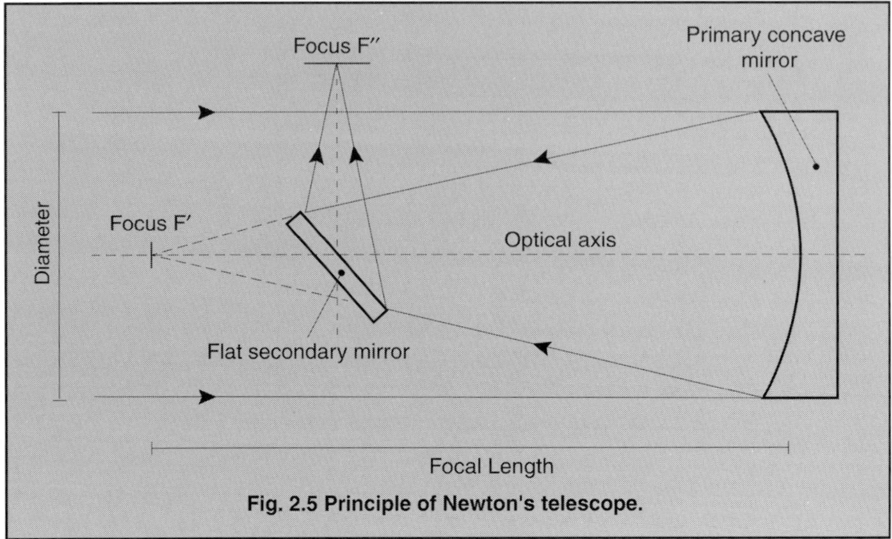

Fig. 2.5 Principle of Newton's telescope.

At the end of the eighteenth century, telescope improvements were made mostly by amateur astronomers. In 1774, the English musician William Herschel (1738–1822) began building telescopes. With a 160 mm speculum mirror of his own make he discovered the planet Uranus in 1781. In 1783, Herschel built his most productive instrument—his "large twenty foot," which had an aperture of just over 480 mm. With this he observed thousands of celestial objects and discovered the general shape of our Milky Way galaxy. Herschel's larger telescopes employed the "front view" design, in which the observer places his head directly at the front of the tube. The mirror is aimed at an angle to the optical axis (see Figure 2.7), which eliminates the need for a secondary mirror and the resultant light loss—which was significant because of speculum metal's tendency to oxidize very quick-

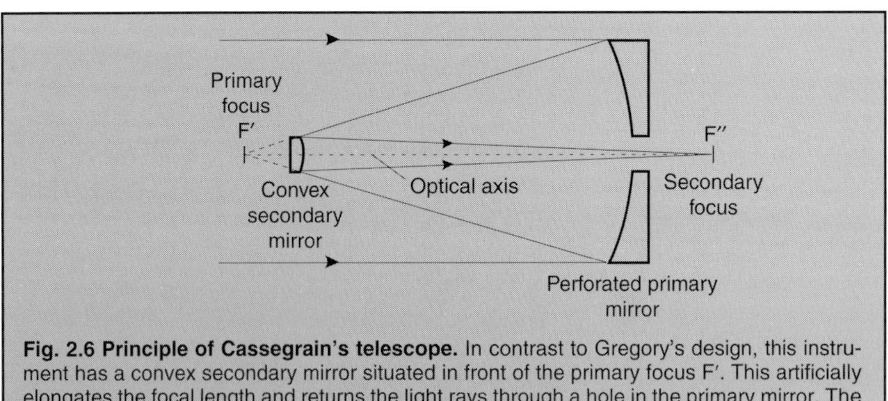

Fig. 2.6 Principle of Cassegrain's telescope. In contrast to Gregory's design, this instrument has a convex secondary mirror situated in front of the primary focus F′. This artificially elongates the focal length and returns the light rays through a hole in the primary mirror. The Cassegrain telescope inverts images.

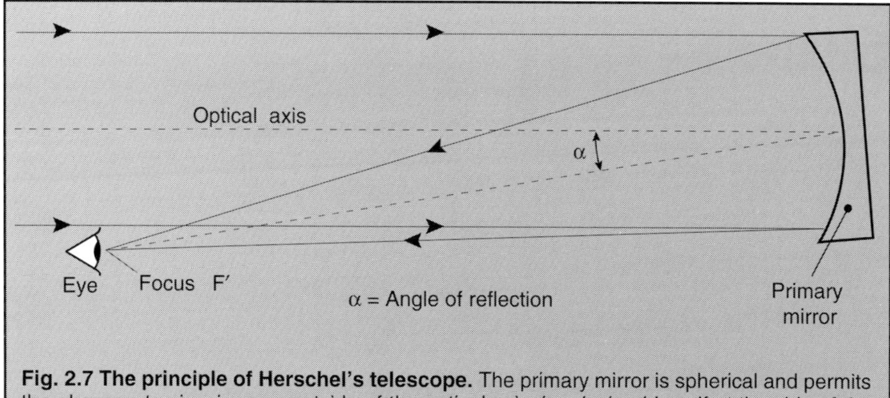

Fig. 2.7 The principle of Herschel's telescope. The primary mirror is spherical and permits the observer to view images outside of the optical axis, by placing himself at the side of the tube. This combination produces essentially perfect images if the focal ratio is greater than 10.

ly and to reflect only forty to fifty percent of incoming light. The use of a speculum secondary mirror would diminish the brightness of an image by a factor of at least two. Herschel's work culminated with the 1789 completion of his giant "forty foot" instrument of 1.22 m diameter, which wasn't surpassed until 1845 by the 1.83 m telescope constructed by the Irish astronomer Lord Rosse.

2.2 The Race for More Powerful Instruments

The nineteenth century saw dramatic changes in telescope making. Astronomical refractors were perfected, thanks to the work of opticians like Alexis Claude Clairaut and, above all, Joseph Fraunhofer. In 1820, Fraunhofer made a 24-cm objective perfectly corrected for a certain wavelength—the largest in the world at the time. Pushed to the extreme, lens-making techniques would permit the construction of instruments of over 1 m in diameter. The great telescope of the Meudon Observatory, the largest refractor in Europe, measuring 83 cm diameter, was made in 1891. In the United States, even larger refractors appeared at the Lick Observatory (91 cm—"36-inch"—1888) and at the Yerkes Observatory (102 cm—"40-inch"—1897).

The large reflecting telescopes of this era were awkward to manage, and their metallic mirrors were very difficult to figure. They were also very heavy, and their reflectance was weak. In 1858, Léon Foucault made the first telescope to employ a glass mirror with a reflective coating of silver. The glass mirror made the instrument lighter, and the mirror was less sensitive to variations in temperature because glass expands and contracts much less than metal. The silver coating was much more reflective than speculum metal, giving images comparable in brightness to a contemporary refractor. Most importantly, the coating could be renewed chemically whenever needed, whereas a speculum mirror had to be repolished and often refigured, a tedious and time-consuming process. The reflector with a glass or ceramic substrate is the basis of almost all modern instruments (although today

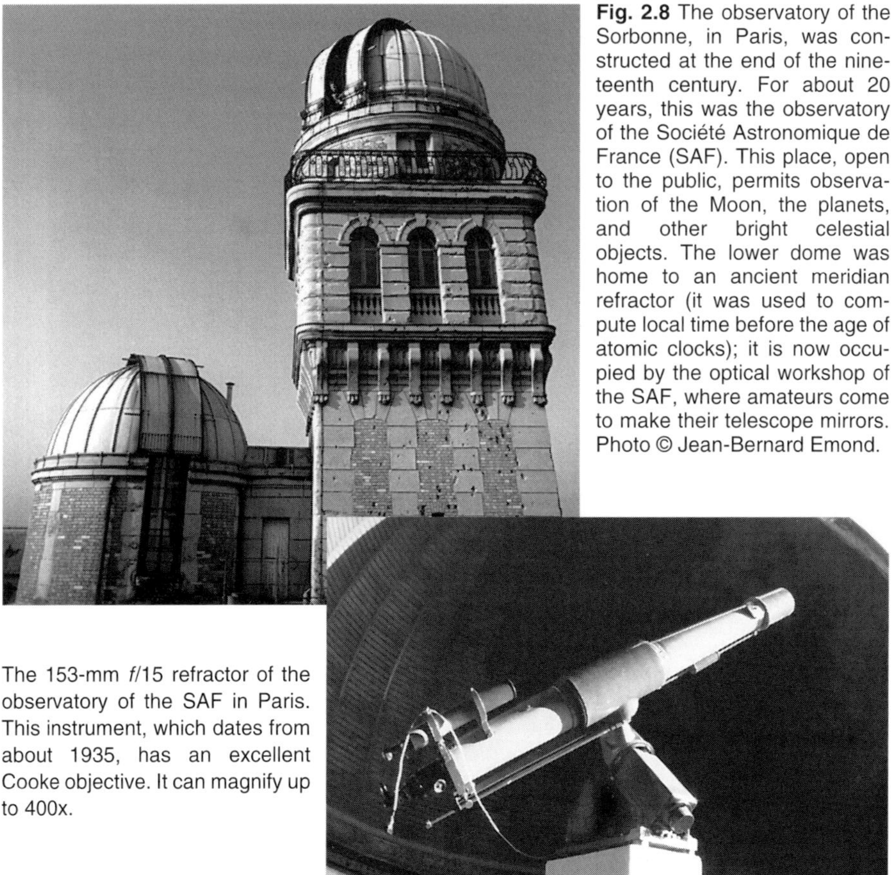

Fig. 2.8 The observatory of the Sorbonne, in Paris, was constructed at the end of the nineteenth century. For about 20 years, this was the observatory of the Société Astronomique de France (SAF). This place, open to the public, permits observation of the Moon, the planets, and other bright celestial objects. The lower dome was home to an ancient meridian refractor (it was used to compute local time before the age of atomic clocks); it is now occupied by the optical workshop of the SAF, where amateurs come to make their telescope mirrors. Photo © Jean-Bernard Emond.

The 153-mm *f*/15 refractor of the observatory of the SAF in Paris. This instrument, which dates from about 1935, has an excellent Cooke objective. It can magnify up to 400x.

mirrors are most often aluminized).

Refracting telescopes, having reached their mechanical and optical limits, were progressively surpassed and replaced by reflectors with glass mirrors. In America in 1908, a 1.52 m reflector, whose mirror was figured by the American optician George Ritchey, was installed at Mt. Wilson in California. This instrument was "copied" and enlarged in 1918 in the construction of the famous Mt. Wilson 2.54 m—"100-inch"— telescope. This instrument remained the most powerful in the world until 1949, when the even more famous 5.08 m—"200-inch"—telescope was installed at Mt. Palomar.

In 1930, Bernhard Schmidt invented a new type of optical system, which bears his name. The instrument is limited to wide-field astrophotography, and has no eyepiece for visual observation. Schmidt's telescope is an extremely fast photographic instrument capable of delivering pinpoint images across a very large field. This type of telescope is called *aplanatic*. Large observatories all have a Schmidt camera. One of the largest in the world, at Mt. Palomar, measures 1.2

Fig. 2.9 Dome of the 200-inch Hale telescope on Mt. Palomar in California. Photo © Jean-Bernard Emond.

cm—"48 inches"—aperture. Most of these telescopes have focal ratios from 1.5 to 3 and are thus photographically "fast." They also have extremely large fields of view, as large as 10 degrees square, or 20 full Moons side by side!

Since early in the nineteenth century, observatories have installed their instruments on equatorial mountings, sometimes enormous ones, in order to compensate for the rotation of the Earth and to produce photographs more easily. A great innovation appeared in 1976 with the development of computers: the former Soviet Union constructed a 6-m telescope on an entirely computer-driven alt-azimuth mounting. Unfortunately the site of the telescope, in the Caucasus at Zelentchukskaya, was rather mediocre, and the instrument, with its slow-cooling monolithic mirror, never performed up to its capacity. However, it was groundbreaking in its computer-controlled mounting.

Fig. 2.10 The large Schmidt telescope of Mt. Palomar. With a diameter of 1.2 cm, this wide-field instrument helped make a complete and very detailed mapping of the sky possible. Photo © Jean-Bernard Emond.

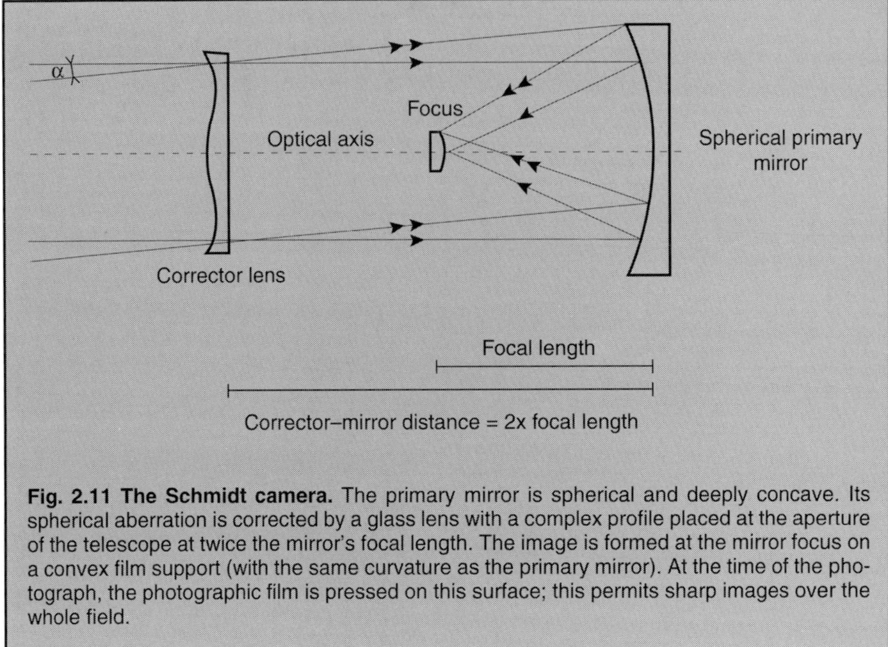

Fig. 2.11 The Schmidt camera. The primary mirror is spherical and deeply concave. Its spherical aberration is corrected by a glass lens with a complex profile placed at the aperture of the telescope at twice the mirror's focal length. The image is formed at the mirror focus on a convex film support (with the same curvature as the primary mirror). At the time of the photograph, the photographic film is pressed on this surface; this permits sharp images over the whole field.

2.3 Modern Instruments

Today, professional astronomers equip their telescopes with electronic cameras and automatic guidance systems to observe the deep skies. The main interest for the moment is in building lightweight telescopes. There are instruments with segmented primary mirrors, like the Keck telescope installed in Hawaii and operational since 1993; it measures 10 meters in diameter and is housed in a dome that's barely larger than that of the 200-inch telescope of Mt. Palomar! Other instruments are equipped with thin mirrors supported by hundreds of remote-controlled hydraulic levers. The 8-meter mirrors of the Very Large Telescope of the European Southern Observatory (ESO), of which the first telescope has functioned successfully since its installation in May of 1998, use this active support process. With this technology, mirrors are easier to make, lighter, and less expensive.

In modern observatories, one no longer sees massive telescopes lodged in enormous domes because instruments are now installed on much less cumbersome alt-azimuth mountings.

To overcome atmospheric problems—turbulence that agitates and degrades the quality of images, humidity that absorbs infrared wavelengths, and dust that diffuses light—telescopes have been sent into space. The Hubble Space Telescope (HST), 2.4 m in diameter and in service since 1990, studies the stars over a wide spectral range and gives astronomers images of remarkable quality. The ISO satellite (Infrared Space Observatory), launched in 1995, studies the infrared universe with a sensitivity unattainable on the ground. Finally, since 1996, the SOHO satellite (Solar and Heliospheric Observatory) examines the Sun in the ultraviolet spectrum and maps the solar corona, which is impossible for Earth-based observers because of our atmosphere and the scattering it causes.

At the professional level, there are systems with adaptive optics that can compensate in real time for the defects caused by atmospheric turbulence. However, space remains the preferred domain of the future, because the absence of atmosphere permits us to expand the limits of our observable universe and to inspect wavelengths filtered-out by the atmosphere.

2.4 The Instruments of the Next Generation

A technique called interferometry, used in radio astronomy, holds promise for the visible spectrum in the decades to come. This method recombines the photons of several instruments pointed toward the same object at a common focus (see Figure 2.12. The study of the light interference thus produced permits the reconstruction of images of that object. Interferometry permits the simulation of very-large-diameter telescopes—thousands of kilometers in the case of radio waves.

Today, about ten interferometers exist worldwide. Future projects envision the placement in orbit of assemblies of telescopes that will synthesize diameters of several kilometers.

2.5 The Evolution of Amateur Astronomers' Instruments

It is difficult to say where we began to distinguish between amateurs and professionals. The vocation of astronomer has existed for a very long time, but it has only been considered a profession since the beginning of the twentieth century.

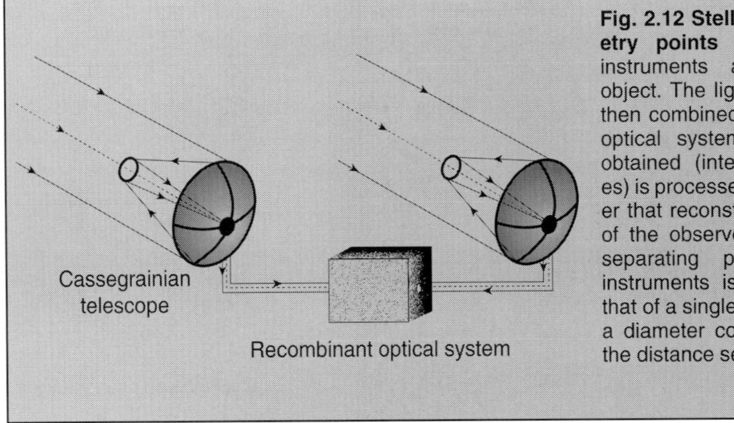

Fig. 2.12 Stellar interferometry points two or more instruments at the same object. The light photons are then combined by a complex optical system. The image obtained (interference fringes) is processed by a computer that reconstitutes the form of the observed object. The separating power of the instruments is equivalent to that of a single telescope with a diameter corresponding to the distance separating them.

Fig. 2.13 The Very Large Array (VLA) near Socorro, New Mexico works at radio wavelengths and is composed of 27 paraboloidal antennae, each about 26 m in diameter. Each is mounted on a railway carriage and movable on tracks arranged in a Y-shape, the maximum length of which is about 27 kilometers. The complete instrument attains, in radio observation, resolution comparable to that of large optical telescopes. Photo © Jean-Bernard Emond.

Since then, instruments have improved considerably. In recent years, the diameters of telescopes have increased: in 1960, an 200 mm diameter telescope would have been considered a large instrument, but today 300-, 500-, and even 800-mm instruments are flourishing. Telescopes are made on assembly lines in factories or by individuals. Certain companies sell, for amateurs with enough money, handmade custom telescopes of 1 meter or more in diameter.

The techniques of mirror and lens making have constantly improved, and today's amateurs use methods that were reserved for professionals only a few years ago. Future amateur telescopes will be large—10 m in diameter, perhaps. Some amateur clubs already possess 100 cm instruments. But it's not sufficient to have a good instrument to make good observations; one must also have a good site. Professional astronomers locate their instruments in the few remaining terrestrial sites

Fig. 2.14 a. M13

b. NGC 3344

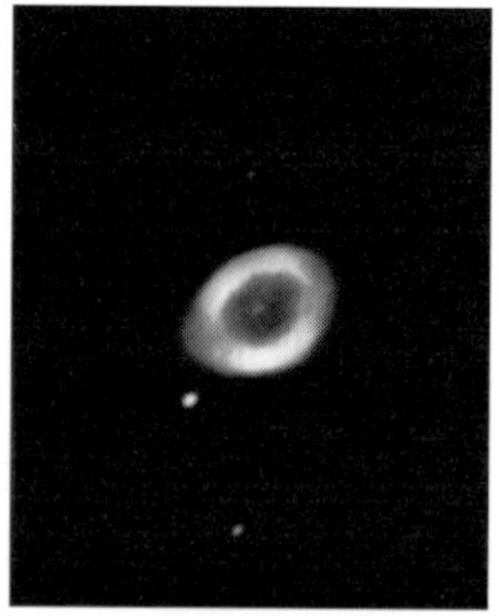

c. M57

still free of light pollution. Amateurs worldwide will surely prefer to build lightweight, transportable telescopes, as is already the fashion in the United States.

Will the average amateur's telescope of the future be a giant transportable instrument, set up on a trailer, with an alt-azimuth mount and a computer drive? A few of these instruments already exist today.

3 How Telescopes Work

3.1 Refractors and Reflectors

Refracting and reflecting telescopes gather light differently. In a refractor, light passes through an objective, and one says that it is "refracted," or bent to a focus; in a reflector, the light is reflected by a system of mirrors to a focus.

Refractors are known for their optical quality and the stability of their images. They have relatively modest diameters compared to reflectors. They are, however, difficult for an amateur to construct and they often present a residual chromatic aberration that introduces false colors around bright images. Their price increases exponentially as their diameters increase, as they require perfectly transparent and homogeneous optical glasses, and have a number of surfaces that must be polished (for even the simplest doublet there are four surfaces to figure). Because of their relatively long focal lengths, refractors are well-suited to observation of the Sun, Moon, and planets, which require the use of high magnification.

Reflecting telescopes offer wider fields of view than refractors. They are easier to make because their mirrors are only polished on one face and they can be made from relatively inexpensive materials, such as porthole glass or Pyrex. The paraboloidal primary mirror remains difficult to make for small *f*-ratios (focus-to-aperture ratios), which require a deep aspherical shape. Reflectors are designed to be shorter than refractors; for a given aperture, they generally have greater light-gathering power and are less cumbersome and lighter. They can have large apertures and still not exhibit chromatic aberration. They are, however, very sensitive to air turbulence unless the tube is capped with an optical window, and the diffraction caused by their secondary mirror and its support reduces image sharpness.

3.2 Parts of the Telescope

A reflector is composed of several parts: the mirrors form the images, the optical tube assembly holds the mirrors in proper position, the mount supports the tube, and a motor drive compensates for the rotation of the Earth so that stars and other celestial objects remain stationary in the eyepiece.

3.2.1 The Mirrors

A telescope mirror is a disk of glass, one face of which is very precisely ground, polished and figured, and on which has been deposited, usually in a vacuum, a thin coating of aluminum, silver, or gold to make it reflective. Unlike household mir-

Fig. 3.1 10-inch mirror, aluminized and readv for use.

rors, where the back of the glass is silvered, the face of a telescope mirror is coated and aimed directly toward the sky. The thin film that covers the polished surface is very fragile, and it is in direct contact with the air, dust and the fingers of clumsy astronomers.

The mirrors are the most difficult part of the telescope to make. In the classical Newtonian and Cassegrain instruments, there are two mirrors: the primary mirror collects the light and concentrates it toward the focal point; the secondary mirror, depending on the type of instrument, reflects the converging light cone either to the side or the back of the tube. The quality of these mirrors is essential and plays an important role in the final performance of the instrument.

3.2.2 The Optical Tube Assembly

In general, the optical tube assembly is a rigid structure with a cell that holds the primary mirror, a spider that supports the secondary, and an eyepiece holder through which you observe. It should be lightweight, so it can be easily transported and manipulated during observing. This book presents three different models of optical tube assembly: an octagonal one for the 130-mm telescope described in Chapter 10 beginning on page 137, a square truss tube for the 250-mm described in Chapter 16 beginning on page 179, and a round one for the 300-mm described in Chapter 26 beginning on page 253.

Fig. 3.2 The optical tube for the 250-mm telescope.

Fig. 3.3 Equatorial fork mount. It can hold telescopes of 250 to 300 mm diameter.

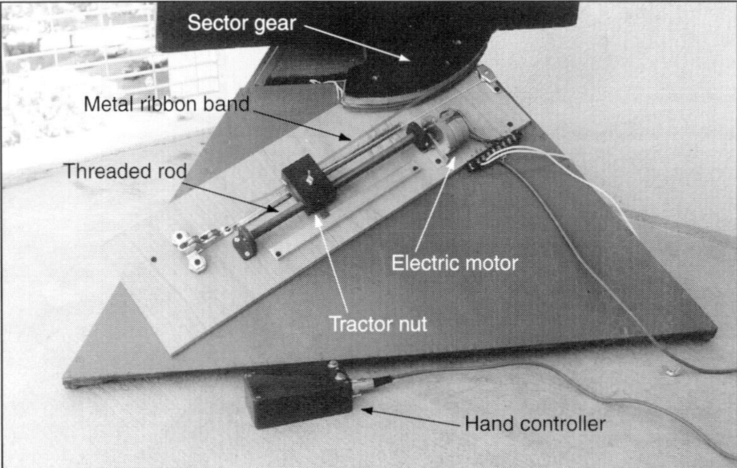

Fig. 3.4 A telescope's clock drive is described in detail in Chapter 17 beginning on page 189. Accurate tracking of the stars is accomplished using a 12-volt "stepper" motor powered by a small battery and an electronic controller. The rotation speed of the motor is reduced by gears placed at the end of a long threaded rod. As this rod turns, a threaded plastic "nut" pulls, along the length of this rod, a flexible steel strap that is attached to the rounded base of a triangular-shaped piece of wood called a sector. This sector drives the polar axis, which is aligned parallel to the rotation of the Earth.

3.2.3 The Mount

The mount is the heaviest and most rigid part of the instrument. It supports the optical tube assembly and rotates on two axes, permitting the tube to be pointed at anything visible in the night sky. It consists of a platform and (in our case) a fork. There are two major types of mounts: equatorials, which are oriented to celestial coordinates (right ascension and declination), and alt-azimuths, which move in altitude and azimuth. These two types of mountings are depicted in Figure 3.5 on page 20. An equatorial mount is composed of two perpendicular axes of rotation that can be easily motorized to "automatically" track the stars with a clock-drive. It is significantly more difficult to motorize an alt-azimuth mount without resorting to a special platform or computer control; even then field rotation complicates photography, and certain areas of the sky may not be observable.

3.2.4 The Telescope Clock Drive

A telescope clock drive, like that shown in Figure 3.4, permits a telescope to compensate for the apparent rotation of celestial objects caused by the rotation of the Earth, a phenomenon that is easily seen in only a minute or two at high magnification. The sky appears to make one revolution about the polar axis in just under twenty-four hours.

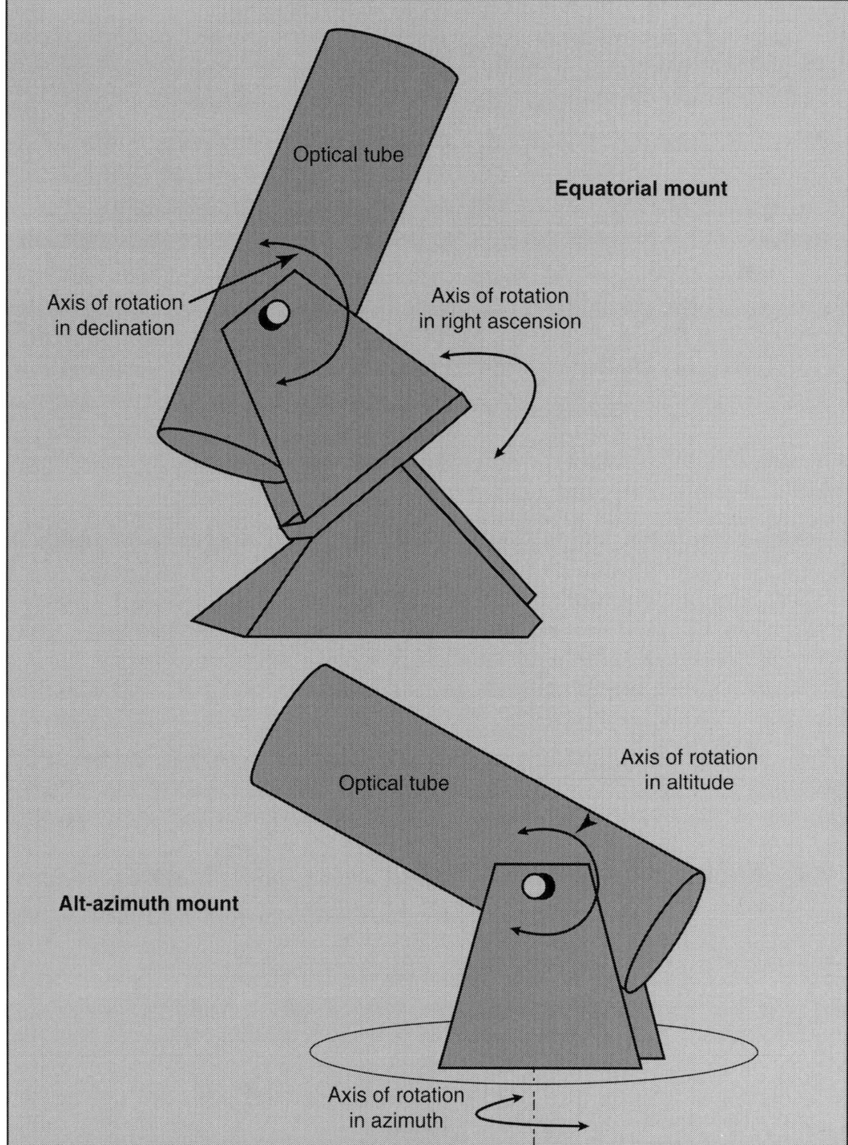

Fig. 3.5 There are two basic types of telescope mountings: equatorial (top figure) and alt-azimuth (bottom figure). The particular examples shown here are called fork mountings. In the top case, the principal axis of rotation of the instrument (right ascension) is aligned parallel to the Earth's rotation axis; and a single, constant movement of the polar axis can exactly compensate for the apparent rotation of the sky. The declination axis lets the observer point the optical tube at objects at various altitudes. The alt-azimuth mounting requires simultaneous drive in both at varying rates (depending on where in the sky the instrument is pointed) to follow a star, but with a computer this is a solvable problem. For similar apertures the alt-azimuth mount is lighter and less cumbersome than a comparable equatorial.

3.3 Accessories

3.3.1 Eyepieces

In order for you to see the small image formed by the telescope objective, both reflecting and refracting telescopes require *eyepieces*, (also called *oculars*). These are actually small microscopes usually composed of two or more lens elements that magnify the images formed at the telescope's focus. There is a great variety of eyepieces, each with different characteristics.

The lens elements of an ocular are usually mounted in a metal barrel; the *barrel diameter* of an eyepiece is the outside diameter of its mounting. Eyepieces and eyepiece holders come in three standard barrel diameters: 24.5 mm (0.965″), 31.75 mm (1¼″) and 50 mm (2″).

Eyepieces are designed for visual observation. At the exit pupil of an eyepiece (see the definition below), light rays emerge in parallel bundles; so, by placing the eye behind a properly focused eyepiece, you can comfortably observe without eyestrain. You can also use eyepieces in astrophotography to project the focal image onto the film to produce negatives of planets or double stars, for example.

3.3.1.1 Choosing an Eyepiece

Eyepieces are designated by their focal length: 40 mm, 25 mm, 12 mm, etc. Which focal length eyepieces you purchase depends upon the degree of magnification you require. For the beginner, a set of two or three eyepieces will usually suffice. For example, you could choose a long-focus ocular to obtain the minimum useful magnification (defined as that for which the exit pupil of the eyepiece is the same diameter as the pupil of the dark-adapted eye, about 7 mm[1]); one that gives max-

What is the Exit Pupil?

An instructive experiment can be conducted in daylight. Place an eyepiece in a telescope, point it at the clear or cloudy sky, and stand about two feet behind the eyepiece. Looking at the rear of the eyepiece lens element you will see a small disk of light, probably hovering somewhat above the rear glass of the eyepiece. This is called the *exit pupil.* This disk is the image of the objective seen through the eyepiece, and its diameter is given by the following relationship:

$$\text{Diameter}_{\text{exit pupil}} = \frac{\text{Diameter}_{\text{objective}}}{\text{Magnification}}.$$

The light rays that enter the telescope exit through this little circle. To obtain the maximum benefit of the light-gathering power of the instrument, use an eyepiece with an exit pupil of 6 or 7 mm, which corresponds to the maximum diameter of the pupil of your eye in the dark (see also footnote 1). All the light collected by the telescope is concentrated in this tiny circle, and you'll get the most out of it when you can see the entire disk.

[1] As the human eye ages the pupil loses some of its ability to fully open to 7 mm. It is a good idea to measure your dark adapted pupil and use that number in your calculations. Few middle aged and older observers will be able to use a fully illuminated 7-mm exit pupil, thus affecting eyepiece choice.

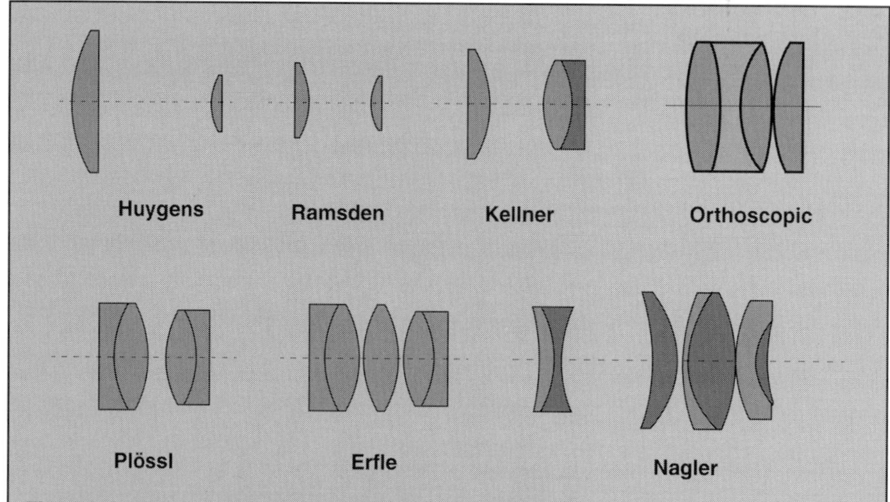

Fig. 3.6 Cross section of the most popular eyepieces. Huygens and Ramsden oculars have the fewest lens elements. The Plössl is composed of two achromatic doublets that give it its optical quality. Wide-field eyepieces (Erfle and Nagler) are very complex and employ seven to eight lens elements to deliver the most perfect image possible.

imum practical magnification (where the magnification is 2 to 2.5 times the diameter of the objective in millimeters), and a third intermediate eyepiece (with an exit pupil of 1 to 2 mm diameter). For a 200-mm diameter telescope at *f*/5, you could thus choose an eyepiece of 35 mm focal length (28x magnification), a 25-mm eyepiece (40x magnification), and an intermediate 10-mm eyepiece (100x magnification).

Because they typically have high *f*-ratios, small refractors usually do not require expensive eyepieces; a Kellner or orthoscopic model with a 24.5-mm (0.965 inch) barrel diameter is sufficient. A lower *f*-ratio 250-mm reflecting telescope using Plössl eyepieces of (31.75 mm or 1¼″ barrel diameter) will give excellent images. A large-aperture telescope (300 mm or larger) will deliver excellent images with wide-field eyepieces.

Fig. 3.7 A selection of eyepieces. The two in the left foreground are Plössls of 25 and 2.5 mm focal length. The two on the right are orthoscopics of 9 and 5 mm focal length. These eyepieces have barrel diameters of 31.75mm. The large eyepieces in the back are wide-field models, 9 and 14 mm focal length (left to right). They have two barrel diameters: 31.75 mm and 50 mm.

Table 3.1
Eyepiece Characteristics

Type	No. of Lenses	App. Field	Barrel Diam. (mm)	Best Use	Avg. Price (2002)	Eye Relief
Huygens	2	30°	24.5	Instruments with *f*-ratios >8	$30	Mediocre
Ramsden	2	30°–40°	24.5	Instruments with *f*-ratios >8	$40	Mediocre
Kellner	3	30°–40°	24.5	All instruments	$35	Good
Orthoscopic	4	40°–50°	24.5, 31.75	Planetary observation	$55	Good
Plössl	4	50°	31.75, 50	All instruments	$140	Very good
Erfle	5–6	60°–80°	31.75, 50	Deep-sky observation	$200	Good
Wide-field	7+	80°	31.75, 50	Wide-field telescopes, deep-sky	$500	Excellent[a]

a. Image sharpness in a wide-field eyepiece is somewhat lower than that of an orthoscopic eyepiece of equivalent focal length.

The quality of the eyepiece is very important and in no case should it be allowed to diminish the performance of the instrument. Beginners buying their first telescope are often disappointed by the flawed, low-contrast images they see. Most often, it is not the objective but the eyepiece that is to blame (those sold with inexpensive telescopes sometimes have plastic lenses; these should be disposed of quickly). One could then invest in an orthoscopic eyepiece with a magnification of D (magnification = diameter of instrument in mm) and experience a significant improvement.

Table 3.1 is a summary of eyepiece characteristics for designs that are commercially available today. They are listed in ascending order of the number of lens elements (and thus price).

3.3.1.2 Eyepiece Prices

A set of eyepieces can add up to a significant fraction of a telescope's total cost. In general, prices vary in relation to their complexity and performance and range from about $30 to more than $600. The least expensive (Huygens, Kellner, Ramsden) are composed of two to three elements and are only usable with small high *f*-ratio instruments (50-mm refractors, 108-mm reflectors) or at very high magnification with faster instruments; they have a small field of view and, if not made by a recognized maker, are often of poor quality. They usually accompany inexpensive consumer telescopes and are difficult to find sold separately.

Higher-quality eyepieces (Erfle, Plössl, orthoscopic) cost between $50 and $250. They usually provide good eye relief and deliver good images over the entire field of view. High-quality wide-field eyepieces (Nagler, wide-field) can easily exceed $300 with a few in the $500 to $600+ range! Their cost may be justified if you own a high-quality telescope.

Why not make your own eyepieces? It's not worth the trouble—eyepieces are

Eyepiece Magnification and Field Size Formulas

An eyepiece is identified by its focal length (indicated in millimeters and engraved on the body of the accessory) and its visual field (in degrees). Modern focal lengths range from 3 to 35 mm and visual fields from 30 to 80 degrees. To calculate the magnification afforded by an eyepiece, divide the focal length F of the instrument by the focal length f of the eyepiece:

$$\text{Magnification} = \frac{F_{\text{instrument}}}{f_{\text{eyepiece}}}.$$

Example 1: A 130-mm telescope, focal length 1200 mm, equipped with an 8-mm eyepiece magnifies 1200/8 = 150 times. The apparent field of view of the telescope is calculated by dividing the visual field of the eyepiece by the magnification. This value is not indicated on the eyepiece, but Table 3.1 (on page 23) will give you the standard values; if you can't find the one you need there, consult an astronomical equipment dealer.

$$\text{Field}_{\text{telescope}} = \frac{\text{Field}_{\text{eyepiece}}}{\text{Magnification}}.$$

Example 2: The eyepiece from Example 1 has an apparent field of 50°; the field of view of the telescope is then 50/150 = 0.33°, or 20 minutes of arc.

not practical projects for the average amateur—it requires knowledge and material very much different from that needed for telescope mirror making. All things considered, it would require a disproportionately large amount of work in comparison to the price of these accessories.

3.3.2 The Barlow Lens

The Barlow lens is a diverging achromatic doublet lens. It is used to increase the focal length of the telescope by a given factor, generally two. Figure 3.8 shows the way this accessory works; the light from the primary mirror has an aperture angle α corresponding to the ratio of diameter to focal length (value in radians). When

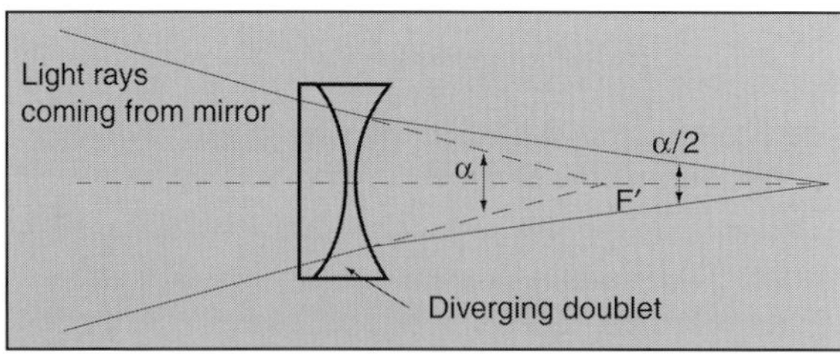

Fig. 3.8 Above: Principle of the Barlow lens. **Right:** Photograph of a Barlow lens with a barrel diameter of 31.75 mm (1¼″).

light passes through the diverging doublet, it is focused behind its original position. The aperture angle is divided in half ($\alpha/2$ in the figure), which translates into a doubling of the effective focal length as it is perceived at the eyepiece.

The Barlow lens can be used both in astrophotography and in visual observation. When used for astrophotography, it doubles the size of the image on the negative (which requires a fourfold increase in the exposure time) and may allow you to achieve sharper photographs with a reduction in field curvature. In visual observing, the Barlow lens doubles the magnification of any eyepiece it's used with, thus increasing the utility of a low-budget instrument. By judicious selection you can often "double" your eyepiece collection. For example, if you purchase 40-, 25-, 18-, and 12-mm eyepieces along with a 2x Barlow, you in effect have gained 20-, 12-, 9-, and 6-mm eyepieces to fill in and expand your magnification range.

There are, however, some inconveniences: a Barlow introduces a small amount of color into the images, and it causes a loss of contrast if the lens is not of the highest quality. You must carefully align your optics in order to obtain the best images. The Barlow lens is often affected by light reflections from the inner metal tube surfaces. Covering the interior of the tube with black flock paper, felt or velvet, can help eliminate these stray reflections.

The price for a Barlow lens ranges from \$40 to \$150 in the standard 1¼-inch barrel diameter. They come in the three standard barrel diameters with the 2-inch models being significantly higher priced.

3.3.3 Focal Reducers

The focal reducer is a lens composed of two to four elements placed slightly in the telescope's normal focal point. It artificially shortens the focal length of the instrument. Use of a focal reducer is justifiable on a Cassegrain telescope—in return for a smaller image and a wider field, it permits you to shorten the duration of a long exposure. For example, you could reduce a one-hour photograph at $f/10$ to a 24-minute one at $f/6.3$. Most commercial focal reducers are designed to be used with Schmidt-Cassegrain instruments; they are not made for Newtonian telescopes with focal ratios of 5 or 6. They also shorten focus in Newtonians so much that the focal plane would move inside the tube. They require significant in-travel.

The focal reducer introduces some aberrations at the edge of the field (stars look like crescents) and a darkening of the edges of the image (vignetting) that can produce quite unpleasant effects in wide-field photographs. The average price of a focal reducer is about \$220.

3.3.4 Field Correctors

Most visual telescopes suffer from a loss of resolution at the edges of their fields of view. In a Newtonian instrument, for example, the image is perfect in the center but degrades progressively as it gets further from the optical axis, making stars look like flying birds or comets (see Figure 3.9 on page 26). This off-axis image

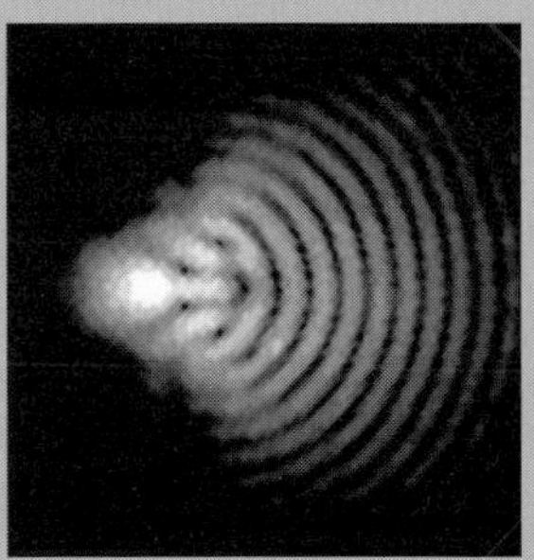

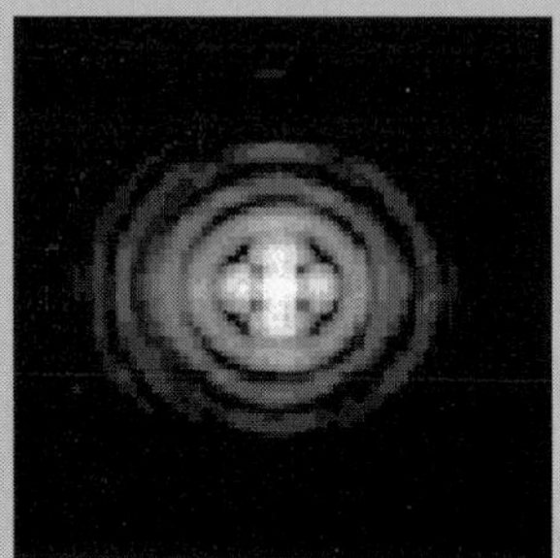

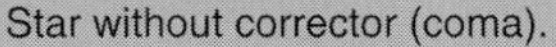

Star without corrector (coma). Same star with corrector.

Fig. 3.9 Appearance of an off-axis star, with and without a field corrector. Newtonians with focal ratios lower than 4 are very susceptible to the problem of coma, in which stars take on the appearance of comets (left photo). Coma is amplified as the distance from the center of the field increases, and it presents a real problem for astrophotography (flawed or stretched images appear at the edges of photographs). To reduce this problem, install a field corrector at the camera's lens. This small lens, comprised of two or three elements, corrects the coma (although it introduces a small amount of spherical aberration—images are less clear at high magnification). In stellar photography, stars remain points all the way to the edge of the field. (Contrast adjusted to show dim details.)

quality reduction is caused by an aberration called *coma*. Its severity is inversely proportional to the cube of the *f*-ratio of the instrument (an *f*/3 telescope is more affected by coma than an *f*/6 telescope); it is not generally distracting in visual observation if the *f*-ratio is greater than 4.

To improve astrophotographs taken through telescopes with small *f*-numbers, a field corrector is placed just in front of the focus of the instrument, at the position of the photographic plate. This accessory, composed of two or three lens elements, corrects the coma and increases the flatness of the field of the telescope, producing sharp images from edge to edge.

We will not discuss the making of field correctors, because their construction requires sophisticated tools and figuring techniques that are beyond the scope of this book.

3.3.5 Finder Scope

The finder scope is an indispensable accessory for your conquest of small and dim stars. You would have to know the sky by heart to find your way among the stars without a finder scope to help you. To be effective, it should provide sharp images of stars down to magnitude 8 or 9, and it should have a wide field (2 to 3 degrees minimum, or six times the apparent diameter of the full Moon).

In place of a finder, you can use a small refractor with low magnification, half of a pair of binoculars, or a device known as a "heads-up" sight, a plan which appears in Chapter 20 beginning on page 211.

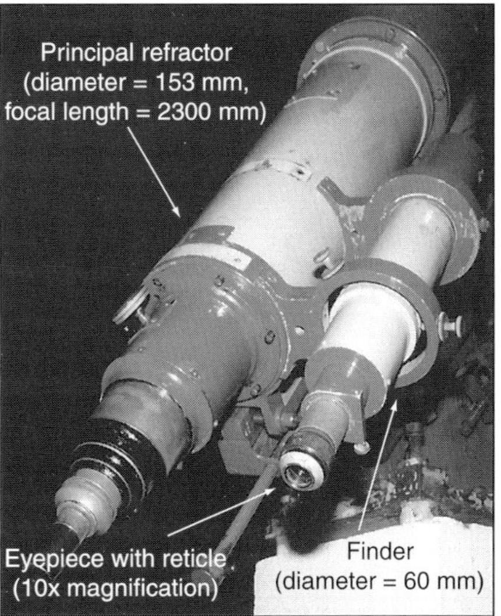

Fig. 3.10 A 60-mm diameter finder scope, 10x magnification, mounted on the 153-mm refractor of the SAF at the Sorbonne. The eyepiece of the finder scope is equipped with a cross-hair reticle that assists in centering of the object in the field of view of the main telescope.

3.4 Important Concepts

3.4.1 The Resolving Power of an Instrument

All telescopes have a theoretical resolving limit that depends on their aperture. Let's consider a perfect circular objective (lens or mirror). At the focus of this instrument, the image of a star appears as a small disk surrounded by concentric rings of diminishing brightness. This image is called the *Airy disk* or *diffraction spot* (see Figure 3.11 on page 28). The linear radius ρ of the first dark ring is calculated:

$$\rho = 1.22\lambda\left(\frac{F}{D}\right),$$

where λ is the wavelength of light (in general, 0.55 μm). The corresponding angle in radians, ρ_{rad}, is determined:

$$\rho_{\text{rad}} = \frac{1.22\lambda}{D}.$$

To convert this to seconds of arc, multiply by 206,265. This gives:

$$\rho'' \cong \frac{140}{D},$$

with D in millimeters. This angle determines the theoretical separating power of an objective.

According to the Dawes criterion, an excellent instrument is one that is capable of resolving two stars if their angular separation is greater than or equal to

0.85ρ. However, the same instrument could show details smaller than ρ in cases of high contrast (stars, bright lines on dark backgrounds). Of course, the image that the objective gives will be spread out over an area equal to ρ.

Examples: The separating power of a 130-mm mirror is about 0.7 arc-second; about 0.46 second for a 250-mm mirror, and about 0.38 second for a 300-mm. The radius of the diffraction spot formed at the focus of an f/5 telescope is 3.3 μm*; the spot formed at the focus of an f/12 Cassegrain measures 8* μm.

3.4.2 How the Eye Works

Astronomical telescopes are optical tools that challenge the capacities of the observer's eye. Let's look at how the human eye works.

The eye is a highly perfected photosensitive receptor, despite its small size and its apparent simplicity. The crystalline lens collects light, and images are formed on the retina, see Figure 3.12. The retina is curved, and its surface is covered with photosensitive cells called *cones* and *rods*.

These cells measure a few microns in diameter. Cones are sensitive to color information and respond mostly to bright light. Rods distinguish only shades of gray and are sensitive to weak light; they allow us to see in the dark and to detect deep-sky objects. Some animals, like chickens, have no rods and are thus incapable of seeing even in twilight.

At night, dim objects appear in black and white (they are perceived by the rods). The sensitivity of the cones is at its highest in the yellow-green (wavelength of 0.55 μm) and that of rods is highest in the blue-green (0.50 μm).

The most sensitive part of the retina is not directly behind the pupil, so through a telescope, observation of dim objects is done by looking to the side of the object. It is important to learn this technique of *averted vision* in order to use your instrument's capabilities to the fullest.

Some substances, like vitamin A (retinol, present in liver, wheat, and butter) and the provitamin A (carotene, present in carrots, apricots, and berries), when consumed in moderation, increase the acuity and sensitivity of the eye.

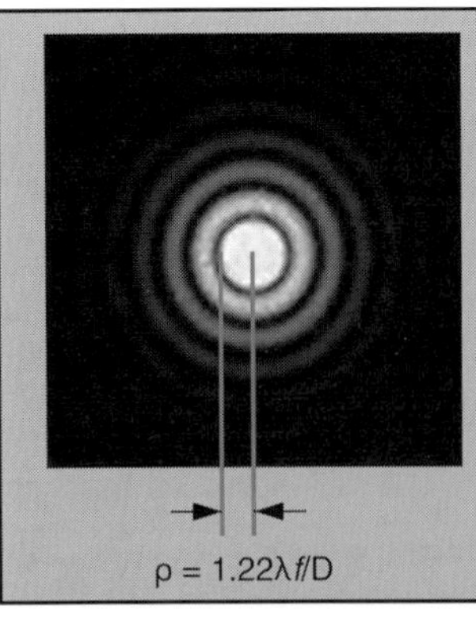

Fig. 3.11 The Airy disk is caused by the wave nature of light. In an instrument with perfect optics (unobstructed, circular aperture), the central disk contains 84% of the light energy, and the remaining 16% is distributed in the diffraction rings (7% in the first bright ring, 3% in the second). Optical flaws, in general, reduce the brightness of the central spot and increase that of the rings. There is a brief study of the effects of optical flaws on the Airy disk in Section 28.1 beginning on page 275.

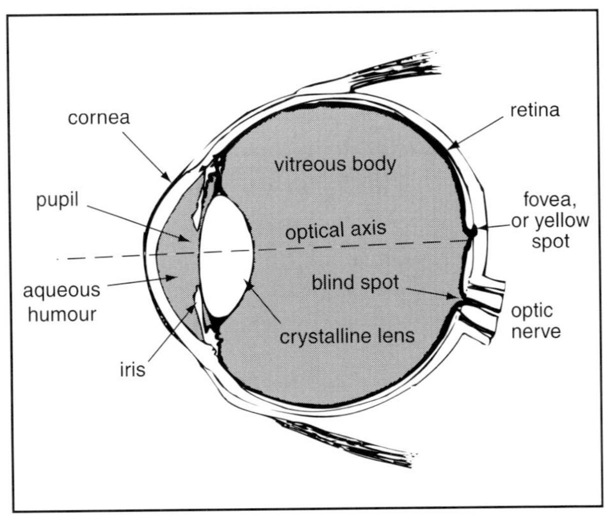

Fig. 3.12 Cross section of the human eye. The crystalline lens is soft and changes shape to focus. The iris opens and closes as a function of light and defines the diameter of the pupil, which can vary between about 1 and 8 mm. The optical axis of the eye hits the retina at a point called the fovea. This region, not very sensitive to weak light, is covered by cone cells. The periphery of the retina is made up of cells called rods (this region is called the retinal purple). There is a blind spot which corresponds to the termination of the optic nerve on the retina.

3.4.3 The Role of the Eye in the Determination of Magnification

The determination of magnifications appropriate for a given instrument depends greatly on the performance of the observer's eye. Let's see how to calculate the *resolving power* and the maximum useful magnification.

To distinguish two very close point-like objects, it is necessary to have at least one cell (cone or rod) between the two images formed on the retina. This capacity to perceive detail in an object is called *resolving power*. It depends not only on the diameter of the eye, but also on the imperfections of the crystalline lens and of the cornea, which become more intrusive as the size of the pupil increases. In theory, on a high-contrast card in daylight, the eye is capable of distinguishing two points ¾ of an arc-minute (or 45″) apart; in practice, the eye distinguishes at best details of 1 to 2 arc-minutes separation.

So that the smallest detail delivered by the objective (the diffraction spot) can be discernible to the eye, one must use an eyepiece that gives *resolving magnification* (see Figure 3.13 on page 30). This magnification measures about ⅔ of the diameter of the primary mirror in millimeters (example: 133x for a 200-mm objective).

How do we determine the maximum useful magnification of a telescope? If the eye was perfect, its angular separating power (ρ) would be only a function of its pupil diameter and would correspond to the relation $\rho = 140/D$. However, as we saw above, the retina imposes, by its structure, a theoretical limit of ¾ arc-minute. Figure 3.13 illustrates the evolution of the separating power of the eye as a function of the diameter of the exit pupil of the instrument. The maximum magnification of an instrument is attained when the theoretical and practical performances of the eye are the same. In the graph this corresponds to the point of intersection of the theoretical curve (dashed line) and the practical curve (solid

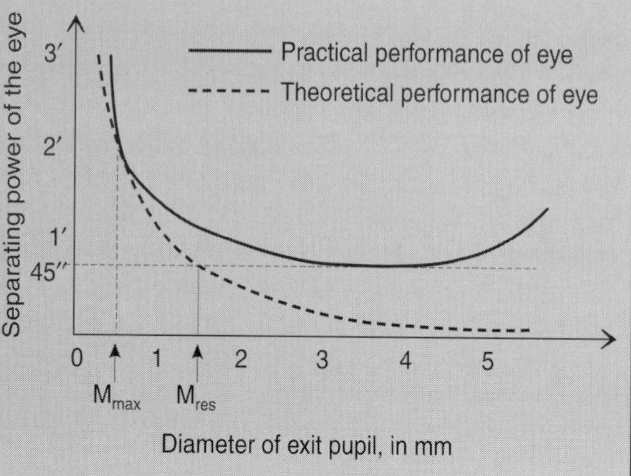

Fig. 3.13 The dashed curve shows the evolution of the theoretical separating power of a "perfect" eye as a function of the diameter of the instrument's exit pupil. The solid-line curve (Arnulf's experimental curve) takes into account the imperfections of the eye and the structure of the retina. Maximum magnification (M_{max}) is obtained with a 0.5-mm exit pupil, and is thus two times the diameter of the mirror. M_{res}, the resolving magnification, corresponds to a 1.5-mm pupil.

line), where the exit pupil's diameter is 0.5 mm, and the magnification is thus two times the diameter of the instrument ($D/0.5$, or $2 \times D$). The exact values vary somewhat from one individual to another. With a smaller exit pupil, the separating power of the eye decreases quickly (the solid-line curve increases as the diameter decreases), and the images in the eyepiece quickly become flawed and unusable.

Thus the maximum magnification of a telescope is equivalent to two times its objective diameter in millimeters—about 300x for a 150-mm objective; about 500x for a 250-mm. In ideal conditions (low turbulence, perfect optics), you can usefully

Formulas for Minimum, Maximum and Resolving Eyepiece Magnification

D is the diameter of the instrument in millimeters; we will consider the diameter of the pupil of a dark-adapted eye to be 7 mm:

Minimum useful magnification: $M_{min} = \frac{D}{7}$

Resolving magnification: $M_{res} \cong \frac{2}{3} \times D$

Maximum useful magnification: $M_{max} \cong 2 \times D$

Remarks:

- All instruments accept a range of magnifications between the minimum magnification, below which the eye effectively stops down the instrument, and the maximum magnification, beyond which the eye must work in extremely poor conditions.
- Too high a magnification results in a lowering of contrast, a diminution in the object's brightness, and a heightened sensitivity to turbulence.
- The choice of an eyepiece is made as a function of the diameter and the focal length of the mirror. Choose a magnification, and then determine the requisite focal length of the eyepiece with the formula:

$$f_{eyepiece} = \frac{F_{objective}}{\text{Magnification}}.$$

employ magnifications of 2.2 to 2.5 times the diameter of the mirror in mm.

3.4.4 Calculating a Telescope's Limiting Magnitude

The magnitude of a star is the number that denotes its brightness. In observation at the eyepiece, the telescope collects light and concentrates it for viewing. The instrument delivers to the eye a quantity of light proportional to the area of the objective. We define the light gathering power *L* of an instrument as the ratio of light perceived by the eye with and without the instrument.

In the case of a star (a point object), the light gathering power is written:

$$L = \tau \left(\frac{D}{D_{eye}} \right)^2$$

or

$$L = \tau \cdot M_{min}^2$$

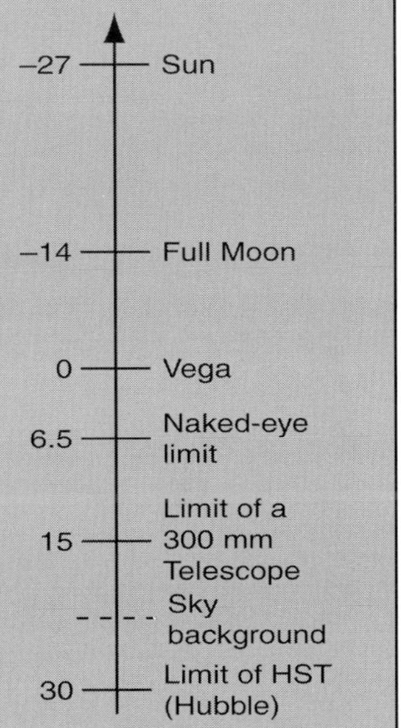

where τ is the coefficient of transmission of the instrument, and M_{min} is the lowest useful magnification; i.e., that for which the exit pupil is the same size as the pupil of the dark-adapted eye.

Example: a 250-mm telescope receives 1,275 times more light than the naked eye (pupil diameter of 7 mm) when τ = 1 (100% transmission).

Fig. 3.14 Magnitude Scale

Magnitude	Object
–27	Sun
–14	Full Moon
0	Vega
6.5	Naked-eye limit
15	Limit of a 300 mm Telescope
- - -	Sky background
30	Limit of HST (Hubble)

Table 3.2 Magnitude Limit of Various Instruments

Diameter	Magnitude Limit
10 cm	12.3
13 cm	12.8
16 cm	13.2
20 cm	13.8
25 cm	14.3
30 cm	14.7
40 cm	15.3

Pogson's formula permits one to calculate the additional number of magnitudes visible through an instrument of light gathering power *L*:

$$-\Delta_{magnitude} = 2.5\log_{10}(L) = 2.5\log_{10}\left(\tau\left(\frac{D}{0.7}\right)^2\right) [D \text{ in cm}] .$$

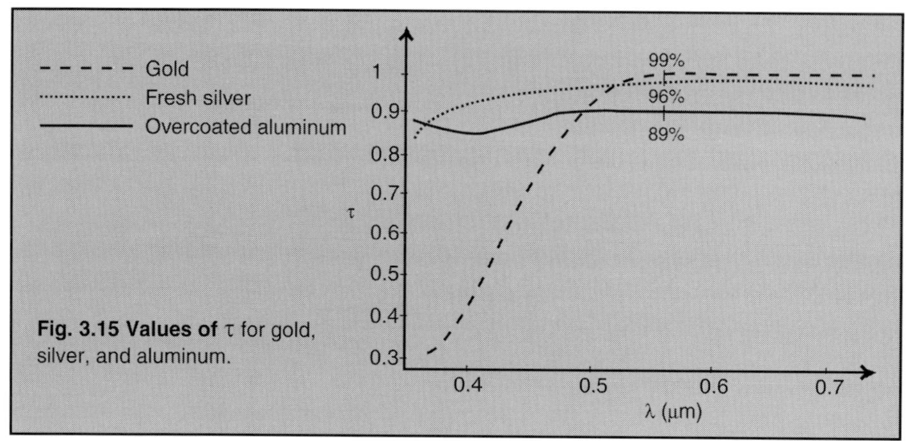

Fig. 3.15 Values of τ for gold, silver, and aluminum.

For a 7-mm pupil and a diameter D in centimeters, use the following formula:

$$-\Delta_{\text{magnitude}} = 0.77 + 5\log_{10}(D) + 2.5\log_{10}(\tau)\ .$$

To find the theoretical limiting magnitude, M_{lim}, attainable by an instrument, add 6.5 (magnitude limit of the naked eye) to the preceding result. Thus, assuming $\tau = 1$:

$$M_{\text{lim}} = 6.5 + 0.77 + 5\log_{10}D\ \text{(in cm)}\quad .$$

In practice, we must take into account the coefficient of transmission of the instrument, which depends on the coatings used on the optics (lenses and mirrors); see Figure 3.15. The average coefficient of reflection for aluminum is 85%, and that of silver is 95%. A uncoated mirror reflects only 4% of the light that reaches it. To calculate the coefficient of transmission, τ, of an instrument, multiply the coefficients τ_n of all the optics between them.

Example: a 12-inch Cassegrain telescope has two silvered mirrors ($\tau_1 = 0.95$ each) and an uncoated telescope window (loss of 4% for each face, or $\tau_2 = 0.96^2 = 0.92$). The instrument transmits

$$\tau = \tau_1^2 \times \tau_2, \quad \text{or} \quad 0.95^2 \times 0.92 = 0.83\,,$$

or 83% of the light that it receives.

The magnitude loss, M_l, caused by this coefficient τ is:

$$M_l = -2.5 \cdot \log\tau$$

or 0.2 magnitude.

One limitation with this formula is the generally increasing blurring visible through larger telescopes when peering through turbulence. However, it is a rough guide.

4 Telescope Making Basics

4.1 Choosing a Telescope

The choice of which telescope to make depends on many things among which are the kind of observing you want to do and the optical and mechanical characteristics that kind of observing program requires, your financial resources, and the amount of time you have available to devote to the project.

4.1.1 Your Motivation

Before undertaking the construction of a telescope, ask yourself some questions. A beginner who wants an initiation into the techniques of mirror-making and wants to have a good-quality instrument is well-advised to start by making a small-diameter telescope—the 130-mm which is described in Chapter 6 beginning on page 85. This inexpensive instrument will enable you to begin your discovery of celestial objects. It is simple to make and does not present any major technical difficulties.[1]

The passionate amateur astronomer who wants to explore the techniques of making paraboloidal, flat, or hyperbolically (convex) mirrors should choose an instrument like the 250-mm (see Chapter 12 beginning on page 149) or 300-mm (see Chapter 19 beginning on page 205) telescopes. These instruments are capable of high-quality performance in visual observation, photography, or CCD imaging. However, their construction demands a sizeable personal investment—several optical pieces are required, some of them very complex. It is almost always not a

Can you observe in the city?

The night sky of large cities is polluted by light, dust and aerosols. In the largest cities, only stars brighter than magnitude 4 are visible to the naked eye (at the zenith). The Milky Way and most deep-sky objects are invisible. One can observe bright objects or use special filters (LPR, OIII) to improve the contrast of certain nebulae. On the other hand, observation of most planets, the Moon and the Sun is possible without difficulty. These images are often better (with regard to turbulence) in cities than in the country. The layer of air over a city can often be quite stable, whereas in the country it is often humid and turbulent.

[1] American amateurs may find 130 mm in between the sizes of available blanks. Either a 4.25-inch (108 mm) or a 6-inch (152 mm) will serve for this simple example.

Table 4.1
Approximate Budgets

Budget	Telescope
< $350	130-mm, alt-azimuth
$750	250-mm motorized equatorial
$1500	300-mm motorized equatorial Cassegrain-Coudé

good idea to begin with the 300-mm telescope; this is an instrument that will tax the skills of even an experienced telescope maker.

4.1.2 Your Resources

An amateur living in an apartment or in a city should opt for an easily transportable telescope. An amateur living in the country or close to clear terrain can install an instrument in a permanent observatory. Someone who can take his telescope to truly outstanding observation sites (mountains or any other place free from light pollution and seeing difficulties) should consider a transportable large aperture, clock drive instrument.

4.1.3 Your Budget

How much does it cost to make a telescope? Table 4.1 gives the approximate budget for various models. These estimates assume that you already have basic tools, such as a drill, a saw, a hammer, etc.

Following are some detailed examples of planned budgets for making the instruments in this book (estimated prices are for the year 2002).

For the 130-mm telescope

The budget is about $350, and breaks down like this:

- Primary mirror, tool, abrasives, polishing compound and pitch: $85
- Secondary mirror: $30
- Vacuum aluminizing or silvering of the mirrors: $60
- Wood and screws: $40
- Eyepiece holder, spider: $60
- Miscellaneous: $40

To this you might want to add one or more Kellner-type eyepieces.

For the 250-mm telescope

The budget, estimated at $750, consists of the following:

- Primary and secondary mirrors, tools, abrasives, polishing compound and pitch: $266
- Vacuum aluminizing or silvering of mirrors: $100
- Wood and screws: $100
- Eyepiece holder, spider: $80
- Motor and electronic control circuit: $100

What Can You See Through a 130-mm, 250-mm and 300-mm Telescope?

In the 130-mm:

- The Moon: all the various formations of our satellite are observable; craters are perceptible down to 1.5 km diameter; grooves; and Copernicus' craterlets.
- Planets: the phases of Mercury and Venus; details and polar caps of Mars; bands and the great red spot of Jupiter; Saturn's rings, including Cassini's division; planetary disks of Uranus and Neptune; four moons of Jupiter and six (below mag 13) for Saturn.
- Stars and nebulae: stars visible to magnitude 13; all the Messier objects: galaxies, nebulae, stars can be resolved in the periphery of the Hercules cluster.

In the 250-mm:

- The Moon: craters a kilometer in diameter; hills and rilles; interiors and ramparts of craters in high detail; evolution of shadows and the color effects caused by varying illumination.
- Planets: details in the clouds of Venus (with a purple filter); dust storms on Mars; storm formations and cyclones on Jupiter; cloud bands on Saturn and details in the rings; Pluto is at the limit of perception.
- Stars and nebulae: stars to magnitude 14.5; resolution of principal star clusters; color in planetary nebulae; distant extensions of the Orion nebula; spiral arms of bright galaxies.

In the 300-mm:

- The Moon: craterlets around impact zones; lunar volcanos; bottoms of valleys.
- Planets: cloud variations on Mars and details in the polar caps; ephemeral formations on Jupiter and spiral structure of the great red spot; cyclones on Saturn and 12 observable satellites; bluish tint of Neptune; a very faint image of Pluto.
- Stars and nebulae: stars to magnitude 15; resolve bright globular clusters to the core; galaxies that are dim in lesser instruments become interesting; details in diffuse and planetary nebulae; quasars and more.

- Miscellaneous: $100

We suggest the purchase of Plössl or orthoscopic eyepieces with a 31.75 mm barrel diameter, and, if you like, a wide-field eyepiece.

For the 300-mm telescope (using the mount described for the 250-mm)

The budget starts at about $1,500 and is divided up as follows:

- Primary, secondary, and tertiary mirrors, tools, abrasives and material used from grinding to polishing: $450
- Telescope window (optional): $300
- Vacuum aluminizing or silvering of mirrors: $120
- Wood, screws, motor and other materials: $350
- Eyepiece holder: $100
- Miscellaneous: $150

4.1.4 Optical Characteristics

The optical characteristics of a telescope (diameter and focal length) should be tailored to the kind of observing that you intend to do. Will you use your instrument primarily for visual observation or for taking photographs or CCD imaging? What

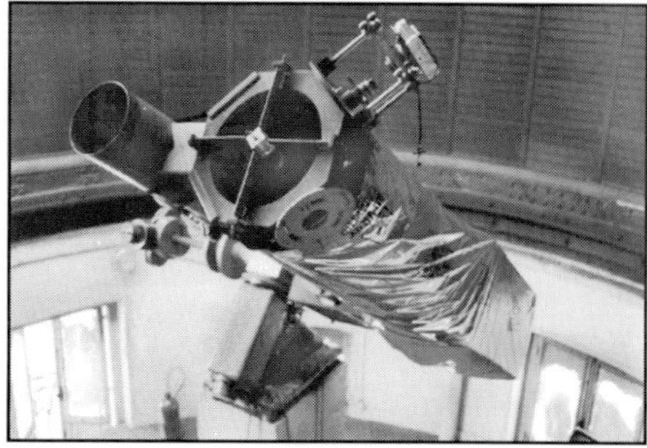

Fig. 4.1 250-mm Newtonian telescope, *f*/8 installed parallel to the 153-mm refractor of the Sorbonne between 1991 and 1995. The mirror of this telescope, made by G. Philippon, has excellent definition.

Fig. 4.2 600-mm alt-azimuth (Dobsonian) telescope made by D. Vernet. The *f*/4 mirror is of excellent quality. This instrument lends itself equally well to deep-sky and planetary observation. It can be completely disassembled and fits into the back of a sport-utility vehicle.

types of objects will you observe with it?

For planetary or stellar observation, a telescope with a clear aperture of between 130 mm and 250 mm and a focal ratio greater than 6 will give good results. These objects are typically viewed under high magnification, and the optics must be of very good quality ($\lambda/10$ minimum).

The handling characteristics of the instrument are also an important consideration; a Newtonian telescope longer than 5 feet can be difficult to handle and transport. A compact tube assembly is advantageous if you require a focal length greater than 2 meters (the 300-mm telescope described in this book is about 825 mm long, despite its focal length of almost 3,600 mm).

For deep-sky objects there is no substitute for aperture. In general a deep-sky telescope should have a clear aperture greater than 300 mm and a relatively small focal ratio (less than *f*/5). The primary mirror can be of average quality ($\lambda/4$ at wavefront or $\lambda/8$ on the glass) if the desired magnifications do not exceed 1x the diameter of the instrument. In general, observation of very faint objects is done at low magnification, and the diffraction disk remains below the eye's limit of resolution. Still, certain deep-sky objects (planetary nebulae, globular clusters) benefit by high power to see detail and increase contrast.

For astrophotography, an instrument with a small *f*-ratio permits shorter exposure times over large fields of view. An instrument with a long focal length will deliver larger, more detailed images, but it will require longer exposures and a high quality clock-driven equatorial mounting.

As for CCD cameras, the telescope's focal length determines the smallest angular detail that a CCD array can record. That detail must cover a minimum of two pixels on the CCD array to be discernible.

The amateur can weigh all of this advice to find the instrument that corresponds to his or her needs. Remember that the "standard" telescope is a 200-mm *f*/6 or a 250-mm *f*/5. This versatile instrument permits high-quality visual and photographic observations.

Fig. 4.3 German equatorial mount supporting a 230-mm refractor at the Triel-sur-Seine Observatory.

4.1.5 Mechanical Characteristics

4.1.5.1 Choosing a Mount

To a high degree, the satisfaction you will receive in using your telescope will depend on how well you match your observing needs to the mount. A simple alt-azimuth mounting (see Figure 4.2) can be both a highly stable and portable mount suitable for visual observing, especially at low power. Photography and drawing at the eyepiece require a clock-driven equatorial or, for ambitious amateurs, a computer-driven alt-azimuth. Other criteria also enter into the choice of a mount; for example, a telescope on a permanent mount can be heavier or larger for maximum stability, good polar alignment and accurate tracking. A transportable instrument, on the other hand, requires a lightweight, but still stable, mount that can be broken down into easily transportable elements. Usually the Dobsonian mount is the most stable transportable mount, but the fork and German mounts (see Figure 4.3) can be adapted for portability and they are more easily clock-driven. Telescopes permanently installed in an observatory building often employ other types of supports like cradles, English, and yoke mounts.

4.1.5.2 Choosing a Tube

A telescope tube can have several forms: round is aesthetic; octagonal is very rigid and less susceptible to flexure; a square is easy to make; a truss tube is open and very light, but although the temperature stabilizes quickly it is sensitive to air turbulence.

4.2 Steps in Construction

There is a logical order for making the different parts of a telescope. The optical elements come first since small departures from the design parameters like focal length can be accommodated if the dimensions of the tube are set after the optics are finished. Likewise, waiting until the tube has been finalized before beginning the mount allows for small but critical variances to be easily accommodated.

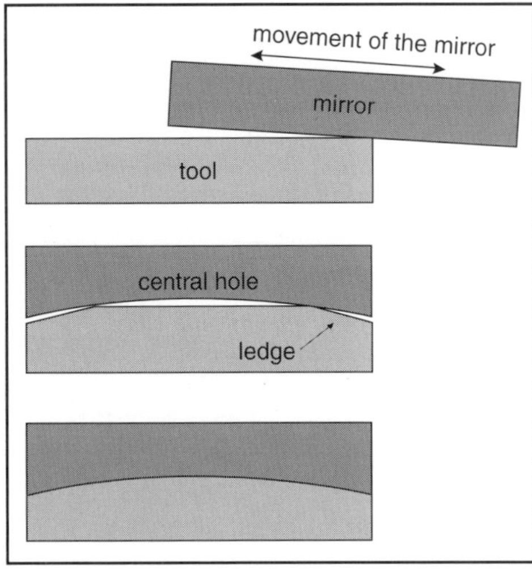

Fig. 4.4 Rough-grinding and fine-grinding a mirror. At first (top), the disks are both flat. The center of the mirror is ground against the edge of the tool. At the end of rough grinding (middle) the disks have defects (a ledge and a central hole) that disappear during fine grinding. Smoothing begins when the mirror and the tool are the same shape (bottom).

4.2.1 Making the Optical Components

4.2.1.1 The Primary Mirror

There are five steps in the making the primary mirror: rough grinding, fine grinding, smoothing, and polishing/figuring. At every stage, the mirror-maker guides the process with specific tests.

Rough grinding (sometimes called *hogging*) consists of systematically working a glass disk (the future mirror) in order to give it a concave form. For this operation, you use a second disk (called the tool) and an abrasive. Placing the flat tool on the workbench, grind the center of the mirror on the outside perimeter of the tool (with the mirror on top), spreading some wet abrasive between them. The center of the mirror and the edge of the disk then wear down simultaneously. The depth of the curve produced by grinding is called the sagitta, and allows you to calculate the mirror's radius of curvature (see Figure 4.4).

Fine grinding eliminates surface irregularities between the mirror and the tool. This phase is complete when the two disks make contact at all points (with defects on the order of 1⁄10 mm).

The smoothing operation allows you to refine the quality of the surfaces of the mirror and tool using a succession of increasingly fine abrasive powders; the smoothed surface (with defects smaller than 1⁄100 mm) is prepared for polishing.

Polishing makes the optical surface reflective and is used to bring the mirror to a spherical form that is precise to within 1⁄10 of a micron. The 130-mm mirror, since it has a relatively long focal length, can be considered finished at this point. For the 250- and 300-mm mirrors, however, polishing is followed by the most challenging and rewarding stage of mirror making—figuring to an aspheric surface (usually a paraboloid).

The finished glass is then aluminized or silvered to improve its reflective power.

Fig. 4.5 The Copernicus region photographed with the SAF 150-mm refractor at the Sorbonne Observatory.

4.2.1.2 Secondary Mirrors

In a Newtonian telescope, the secondary mirror is flat and has no optical power. Its purpose is to direct (usually at a right angle) the converging beam of light from the primary mirror to the outside of the tube near the top of the instrument (see Figure 2.1 on page 5). Since there is no curve and it is often made of polished plate glass to start with, it does not require a rough-grinding step like the primary. The work consists of smoothing and polishing to a plane.

In a Cassegrain telescope, the secondary mirror is convex (see Figure 2.6 on page 8). You must rough-grind it in the same fashion as the primary, but here it's the convex disk that will be used in the telescope. At the figuring stage, the surface is given a hyperboloidal shape.

Like the primary mirror, the secondary mirrors are aluminized or silvered.

4.2.1.3 The Telescope Window (for the 300-mm Telescope)

The telescope window is a thin disk made of optical-quality glass, with parallel, optically flat faces that close the telescope tube to suppress tube currents (see Figure 25.2 on page 246). It is fabricated by alternately smoothing and polishing the two faces until they are plane parallel and flat within predetermined tolerances. Telescope windows usually have an anti-reflective coating applied to both surfaces to increase the light transmission.

4.2.2 Making the Mechanical Parts of the Telescope

4.2.2.1 The Optical Tube Assembly

In general, the optical tube assembly is made when the primary mirror is finished. Since the tube's length is determined by the focal length of the mirror, you must wait until you have polished the mirror to know this dimension precisely.

4.2.2.2 The Mounting

Start work on the mounting after you have made the optical tube assembly. The tube is held in the fork by trunnions and must be able to turn freely without hitting the fork or the mount. It is very important to maintain perpendicularity between the axis of the tube, which will be, in practice, the same as the optical axis, and the declination axis. The length of the fork arms will depend on the tube's center of gravity (the tube with its accessories: finder scope, focusing devices, etc.). The "arms" of the fork can be shortened by moving the tube's center of gravity toward the mirror by adding counterweights at the rear of the tube. This will make for a more compact mount and will also stiffen the arms and reduce vibration.

You will also install on your equatorial mount a clock drive that will track the stars in right ascension (see Figure 3.4 on page 19). This drive is comprised of a motor, a sector drive, and other small parts. A tangent arm will also be installed (see Figure 16.5 on page 183) to fine-tune pointing in declination.

4.3 Preparing Your Workshop and Selecting Materials

4.3.1 Lay Out a Workshop

The ideal optical workshop incorporates three areas. The first is reserved for the processes of grinding and smoothing, the second for polishing, and the third for final testing of the mirror. The latter two areas should have a stable temperature (around 20° C). Air currents must be avoided, and the presence of dirt and abrasives in the polishing area must be completely eliminated.

A complete workshop should have running water, a source of heat, and a clock, as well as the following materials:

- One or two work posts for mirror-making
- Two or three basins in which to immerse the optical pieces
- Sponges or paper towels to clean the optical pieces
- Cloth or paper towels to dry your hands and the optical components
- Newspaper, plastic bags, and a canvas sheet to protect the floor
- Paint brushes to spread the abrasives, jars for abrasive powders and small glass containers for the polishing agents
- Empty jars of abrasive powders and small glass containers for the polishing agents
- Pitch for polishing the mirror

- A hotplate and an old metal sauce pan to heat the pitch
- An electric hair dryer to warm the pitch before pressing
- A large knife and a hammer to trim the pitch facets on the polishing lap
- Plastic water pails in which to rinse the mirror and tool

Fig. 4.6 A polishing stand.

4.3.2 Making a Polishing Stand

To properly grind and polish a mirror you will need a work stand. The one shown in Figure 4.6 has three legs that is very stable and is well adapted to the hard work involved in grinding and polishing. To insure that the tool will not slide off the stand three wooden blocks, spaced at 120°, will hold it (see Figure 4.7, Right). You could also place a thick cloth or several layers of newsprint under the back of the tool. The stand can be further stiffened by loading the base down with heavy weights. An alternative stand is a 55-gallon metal drum filled with water with a wooden platform firmly attached to the top of the drum.

Fig. 4.7 Left, the base of a polishing stand. The triangular structure makes the stand rigid and very stable. **Right,** detail of the top of a stand. The polishing tool is held in place by three wooden blocks spaced at 120°. A little gap is left between the blocks and the tool to allow the disk to be turned in the support.

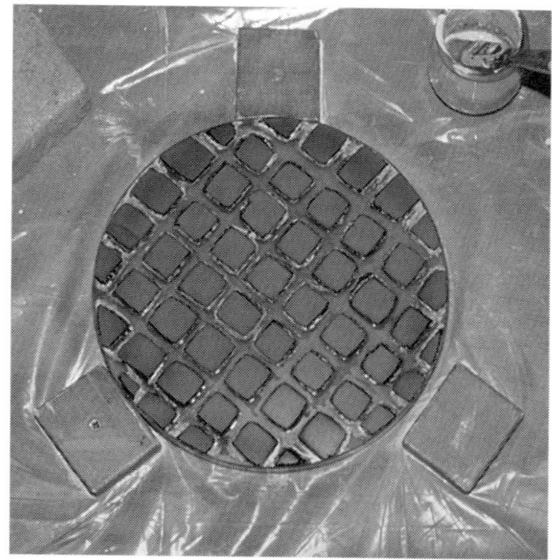

Fig. 4.8 (Top) A 250-mm mirror rests with the optical face down.

Fig. 4.9 (Left) Mirror box equipped with a protective lid and a carrying handle.

4.3.3 Make a Mirror Case

Mirror cases protect the mirrors against shocks and dust during fabrication and later when you send your mirror off to be aluminized.

The case is made of varnished wood. The mirror rests in the interior of the box, face down, on three beveled wooden blocks (see Figure 4.8). The outside edge of the mirror's face rests on wooden brackets. Foam rubber is placed between the mirror's back and the lid of the box to keep the mirror from moving around. The lid can be screwed-on or hinged. The foam rubber is important to make sure that the mirror is well secured in the case. The mirror must not be able to move around freely, but it doesn't have to be bolted down; leave a small amount of play. No object should be able to strike or rub against the optical surface. Double-wrap this case in a larger cardboard box with non-moving packing before shipping.

To protect secondary mirrors (the flat mirror of a 250-mm Newtonian, for example), wrap the optic with soft photographic lens-cleaning tissue. This material will not scratch the surfaces of the optics (even when aluminized), as long as nothing is allowed to rub up against it. If it is wrapped in tissue paper and then with several layers of bubble wrap, it will be well protected from shocks, scratches, or other damage.

To wrap a large secondary (the convex mirror of a Cassegrain telescope, for example), make a rigid foam-rubber-filled wooden box with a cut-out just slightly larger than the dimensions of the mirror. When wrapped with two or three layers of lens-cleaning tissue as described above, the mirror will be held securely in the foam rubber.

Many amateurs do not see the need to carefully wrap their secondary mirror, because it is small. This is an oversight that must be avoided, because these elements will be close to the focus; scratches and chips will have much more impact on image quality than similar defects on the primary mirror.

Before transporting a telescope, you must secure all of its optics, including the secondary mirror. A small jolt could knock it out of its support, resulting in damage to it and other components. These boxes will come in handy when you have to do this.

Fig. 4.10 Jars of different abrasives.

4.3.4 Select the Abrasives

Glass working is always done with an abrasive interposed between the two grinding faces.

4.3.4.1 Grinding and Smoothing Abrasives

In rough grinding, we use a coarse grain (#80) carborundum (silicon carbide), which brings the tool and mirror surfaces in contact very quickly. This abrasive, called "carbo," must be wet to be effective. You can store it dry and mix it with water while spreading it on the glass, or keep it in a jar mixed with water and spread it with a large brush (the latter procedure prevents the abrasive from blowing away at the slightest breeze, possibly preventing contamination of your work area). You must experiment to find the right mixture of water and abrasive: if the abrasive is too dry, the work will be difficult and inefficient; too much water in the mixture pushes the abrasive to the exterior of the disks, cutting its grinding action. The process of working each charge of abrasives, from application to the point at which it no longer cuts glass, is called a "wet."

The abrasive powders for smoothing are finer than those used for grinding but there is a wider range of possibilities. Carbo (grades 280 to 1600), garnet powder (W04 to W8), emery (corundum) (series from 302 to 304), or 25 to 5 micron aluminum oxide. Aluminum oxide is slower acting than carborundum, but it provides a smoother surface.

4.3.4.2 Polishing Compounds

For polishing, we use metallic or rare earth oxides that have been reduced to a fine powder (a few microns in diameter). Pink cerium oxide is the fastest working polisher and is used at the beginning of polishing. Optical rouge (calcined ferric oxide) permits finer polishing than cerium oxide. Zirconium oxide, which is white and gives a good surface, is sometimes used. Aluminum oxide powder, less popular among amateurs, also gives a good surface. It is more often used to work more

Table 4.2
Abrasives Used During Rough and Fine Grinding

Operation	Corundum	Emery	Carborundum	Average grain size (μm)
Rough grinding			60	300
Rough grinding			80	200
Rough grinding	120	W120	120	100
Fine Grinding	180	W180	180	50

Table 4.3
Abrasives Used In Smoothing (extract from G. Kluyskens Co., Inc.)

Average grain size (μm)	Corundum		Emery		Carborundum	
	Grade	Limits (μm)	Grade	Limits (μm)	Grade	Limits (μm)
42					280	22–75
36					320	18–69
34			W 04	12–89		
33			W 03	11–81		
32			W 02	12–74		
31			W 01	9–69		
26			W 1	10–52		
25					400	12–50
24			W 2	11–49		
21	302	10–52	W 3	9–38		
19.5					500	8–42
18			W 4	7–36		
17	302½	8–43			600	7–38
16			W 5	5–37		
14	303	7–49	W 6	7–36		
13	303½	6–40			800	5–35
12			W 7	4–32	1000	4–27
10	304	2–49	W8	3–28	1600	2–21

Shaded boxes indicate the abrasive grains most commonly used for mirror making. It is interesting to note the difference between the size of the largest grain in a given grade to that of the median grain. This explains why tiny pits remain after smoothing.

difficult materials, such as metals or crystal.

Polishing agents can be used alone or can be treated with certain chemicals like caustic soda (dilution = 1⁄1000 or pH = 10). A few drops of acetic acid can eliminate a hard-to-remove thin film that sometimes forms on the surface of a mirror during polishing, especially if the lap is allowed to run dry.

Table 4.4
Principal Types of Glass, Their Optical and Mechanical Characteristics

Type of Glass	Density	Refractive Index	Coefficient of Expansion	Maker
Duran 50	2.23 g/cm^3	1.473 at 587.6 nm	32 x 10^{-7}/°C	Schott
Pyrex	2.23 g/cm^3	1.474 at 589.3 nm	32.5 x10^{-7}/°C	Corning
ZERODUR	2.53 g/cm^3	1.5424 at 587 nm		Schott
CER-VIT	2.5 g/cm^3	1.54 at 587 nm	0.2 to 1 x 10^{-7}/°C	
BK7	2.51 g/cm^3	1.5168 at 587 nm	7.1 x 10^{-6}/°C	Schott
ULE (Silica 797)	2.213 g/cm^3	1.484 at 587 nm	0.2 ±0.3 x 10^{-7}/°C	Corning
Suprax (Boro 8488)	2.31 g/cm^3	1.484 at 587 nm		Schott

4.3.5 Glass Choice for Mirror Blanks

To make a "standard" mirror, choose a glass with a low thermal coefficient of expansion such as Pyrex® or Duran 50®. Changes in the form and focus of the mirror are thus reduced during figuring—and during observation as well. To save money, you could use a disk of ordinary plate glass, but it will be more sensitive to variations in temperature.

There are ceramic glasses with coefficients of expansion of almost zero which are used in high-precision instruments. These glasses are ZERODUR (from Schott), CER-VIT, or ULE (from Corning). These glasses are harder to work with than Pyrex or Duran 50. Their purchase is justified for a large-diameter instrument (larger than 300-mm aperture) or jobs that require extremely high quality (temperature independence such as solar observation).

The material used to make lenses and telescope windows is Bk7 (borosilicate, made by Schott) or B16-64 (the equivalent from Corning).

The rough-grinding tool is a disk of Pyrex or plate glass of the same diameter as the primary (a difference of 3 to 6 mm is tolerable). If necessary, you could get by with a plaster tool (see "Making a Plaster-Backed Polishing Lap" on page 161).

The normal thickness of the glass disks is 20 mm for a 130-mm diameter mirror, 35 mm for a 200-mm, 40 to 45 mm for a 250-mm, and 50 mm for a 300-mm. For even larger mirrors, most builders do not exceed 55 to 60 mm thickness for reasons of weight; the relative thinness of the mirror is compensated for by the use of a specially constructed "mirror flotation" cell.

4.3.6 Preparing the Glass for Rough Grinding

4.3.6.1 Chamfering

The chamfer is a beveled edge, inclined at 45°, ground around the top edge of the mirror blank and the tool (see Figure 4.11 on page 46). The sole purpose of this chamfer is to prevent chipping from abrasive buildup or a bump against something

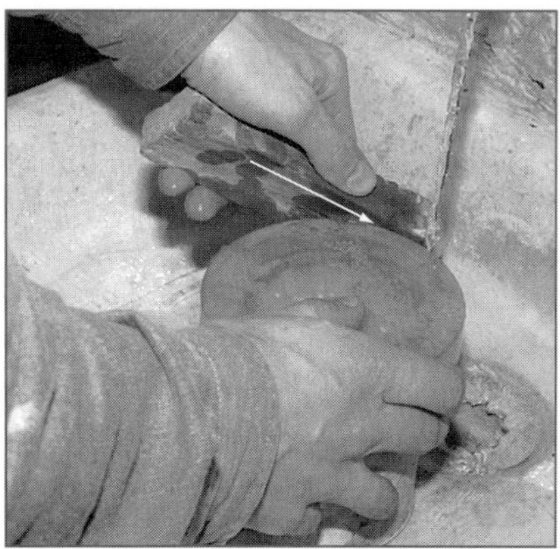

Fig. 4.11 The chamfer is made by grinding the glass disk with a wet carborundum stone. The stone must be inclined at 45° and pressed with the hand solely from the interior to the exterior of the mirror; this way, any chips are made on the side of the disk and not on the optical surface. **Left:** Making the chamfer on a 130-mm mirror; **Right:** On a 300-mm mirror. The arrows indicate the direction of the movement of the stone. Photographs © Arnaud Gaillon.

hard. Under no circumstances should you allow the chamfer to be ground out; it should be renewed when ever it appreciably wears down.

The chamfer is formed by rubbing a whetstone (made of carborundum; sold at hardware stores), inclined at 45°, around the entire circumference of the mirror's face. Press lightly on the whetstone to avoid chipping and moisten it in a washbowl from time to time. You can also make the chamfer with a brass or aluminum block used in the same manner as a whetstone using loose abrasive (#120 or #240 carbo) and water.

The chamfer should be about 3 or 4 mm wide (at 45°) before beginning rough grinding. At the polishing stage, about 1 to 2 mm should remain. A chamfer of at least 5 mm width is necessary for the tool, because its edge wears away more quickly than the mirror's during rough grinding. The interior diameter of the chamfer determines the optical diameter of the component.

If the back of the mirror or tool does not have rounded edges, they should also be chamfered to prevent chipping during handling.

4.3.6.2 Preparing the Backs of the Mirror and Tool

The side opposite the ground face of the mirror and the tool is the back. It is to be made rough and flat, especially if it is thin (under 19 mm thick for a 300 mm mirror, for example). For mirror blanks with a 1:6 thickness ratio smoothing the back can be dispensed with. Roughening the back of the disks prevents hands from slipping during work, an advantage that should not be underestimated.

To roughen the back of the mirror, grind it for about 20 minutes on the back of the tool, using #80 or #120 carbo between them (see Figure 4.13). Use a W-shaped stroke, with a total amplitude of ⅓ the diameter, and refresh the abrasive occasionally. Alternate the mirror and tool positions above and below to prevent the back of the mirror from becoming convex.

The Importance of a Clean Chamfer

On the mirror, the chamfer should be refined by a few strokes at the end of smoothing with a very fine stone (W4 or its equivalent). This gives a cleaner bevel, reduces the risk of chips at the edge of the mirror, and improves polishing quality at the surface's periphery. In effect, the small abrasive grains imprisoned in the faults in the chamfer are freed over the course of polishing (forming very fine scratches around the rim of the mirror) or at the time of vacuum treatment (making the reflective coating more fragile). Making a "clean and neat" bevel thus facilitates better surface quality and less scattering of light from the mirror. You could even polish the chamfer. Simply grind the edge of the mirror for a few minutes with a small polisher covered by felt or wool (pitch is impractical for this exercise) and covered with cerium oxide.

Fig. 4.12 Left: The bevel after smoothing. Several scratches are apparent when viewed under a microscope. **Right:** The bevel is polished, and the surface is clean.

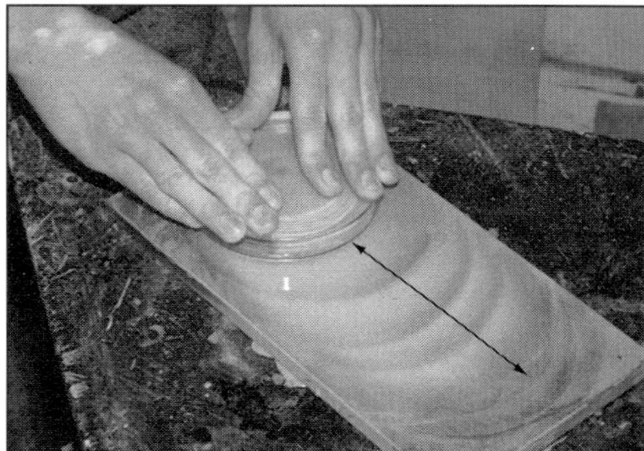

Fig. 4.13 Left: A 250-mm mirror of Duran 50 glass, seen moistened for transparency. Notice the irregularities of the back (clear and unpolished zones). **Right:** Correcting the flatness of the back of a 130-mm mirror by grinding it for a few minutes on a sheet of plate glass. The abrasive used was #120 carbo.

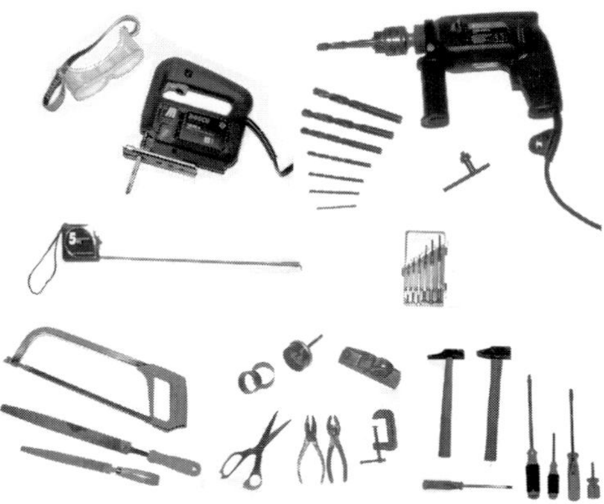

Fig. 4.14 Some useful tools for the telescope maker: a saw (preferably a jigsaw), a drill with the full range of bits from 3 to 12 mm, wood rasps, files, a hammer, screwdrivers (Phillips and flat-head), C-clamps for gluing, chisels, and scissors. Not shown but also useful are drafting tools such as a ruler, pencils, compass, and T-square.

5 Mirror Testing Methods

Read this chapter very carefully, because it contains valuable information on test methods and instruments and the ways they work. An improperly used test method or misinterpreted measurements can have disastrous consequences on how well your telescope turns out. In addition to the information given here, you will find more details on the use of these test methods in the chapters dealing with specific telescopes.

5.1 Finding the Sagitta of a Mirror with Feeler Gauges

Feeler gauges are a set of metal strips scaled and calibrated in thickness (from 0.05 to 1 mm). As the mirror becomes concave, they are used to measure its curvature. The precision of the measurement does not exceed 0.05 mm.

The mirror's radius of curvature, R, is:

$$R = \frac{(D/2)^2}{2s} \quad \text{where } R = 2F \qquad \textbf{(5.1)}$$

Fig. 5.1 Feeler gauges fanned out.

where D is the diameter of the mirror, s the depth of the curve, and F the focal length.

Example: a 250-mm mirror with a curve depth of 4 mm has a radius of curvature of 1,953 mm and a focal length of 976.5 mm.

Fig. 5.2 Measuring the sagitta of a mirror by sliding feeler gauges under a metal ruler. The ruler must be very rigid and must cross the mirror over its full diameter (not a chord of the mirror). To find the curve depth, insert thicker gauges until they no longer slide under the ruler. You can put several gauges together for measuring depths greater than one millimeter.

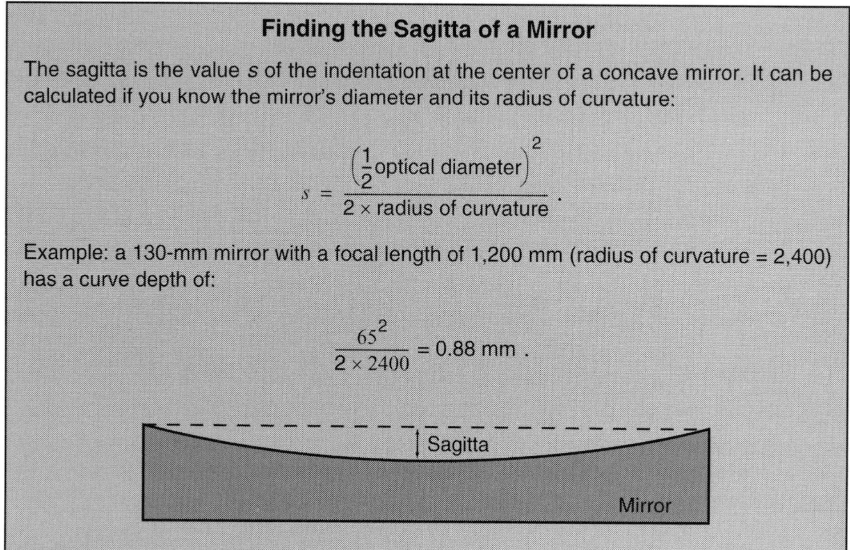

Finding the Sagitta of a Mirror

The sagitta is the value *s* of the indentation at the center of a concave mirror. It can be calculated if you know the mirror's diameter and its radius of curvature:

$$s = \frac{\left(\frac{1}{2}\text{optical diameter}\right)^2}{2 \times \text{radius of curvature}} .$$

Example: a 130-mm mirror with a focal length of 1,200 mm (radius of curvature = 2,400) has a curve depth of:

$$\frac{65^2}{2 \times 2400} = 0.88 \text{ mm} .$$

5.2 Measuring the Curvature of a Mirror with a Spherometer

A spherometer measures the radius of curvature of a mirror more precisely than feeler gauges (see Figure 5.3). It is used in fine grinding and during smoothing.

This apparatus is often smaller than the mirror that is being tested. To obtain the radius of curvature of the mirror, first measure the diameter D_s of the circle that passes under the contact points of the spherometer. Place the spherometer on the mirror and make a measurement (m_1) by turning the micrometer until it contacts the glass. Place the instrument on the convex tool and make a second measurement (m_2). The radius of curvature R is calculated thus:

$$R = \frac{(D_s/2)^2}{2|m_1 - m_2|} .$$

This formula is only accurate for a spherometer with sharp points and involves a slight approximation.

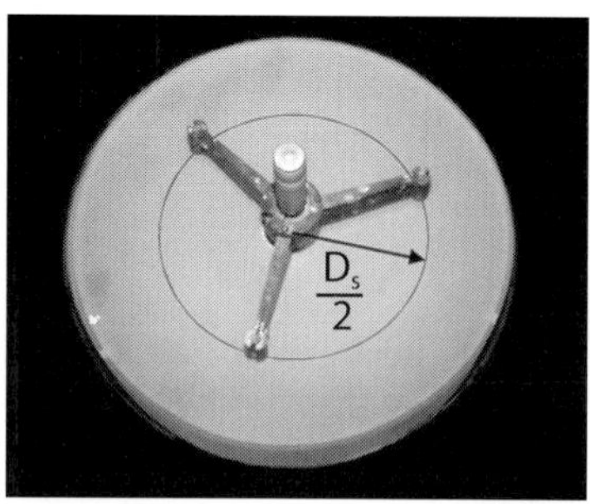

Fig. 5.3 Three-point spherometer on a 300-mm mirror. The measurement is made with a micrometer screw at the center of the device. The diameter of the circle that passes under the three contact points of the spherometer is used to calculate the radius of curvature of the mirror's radius of curvature. See Figure 24.1 on page 240 for a description of a ring spherometer.

Fig. 5.4 Another simple method for checking the curve depth of a mirror during rough grinding is to compare the curvature of the glass disk with that of a metal template that has the desired radius of curvature (the one in the photograph is cut from a sheet of 1-mm-thick aluminum). Here the operator wants a radius of curvature between 2520 and 2530 mm. By placing the template on the face of the mirror, he can tell if the two pieces make contact at every point or if there's a gap, indicating a difference in form.

5.3 Testing a Smoothed Surface with a Loupe

During the process of smoothing, regularly check the surface quality using a microscope or a high-power loupe to examine the quantity and size of the pits caused by the abrasives near the outside edge of the mirror. This test tells you when to move on to the next finer abrasive. You can do this also with an inverted telescope eyepiece. Observe through the field lens, bringing the eye-lens side close to the test surface. A positive eyepiece (orthoscopic or Plössl) with a 12 mm focal length provides about 20x magnification (you will see pits that are only a few microns in diameter). It is also possible to use an inverted photographic lens (Figure 5.5, left). By regularly checking the mirror's edge with one of these optical aids, you will be able to tell when all the large pits from the prior abrasive have been ground out and you are no longer improving the surface. At that point you want to move on to the next finer abrasive.

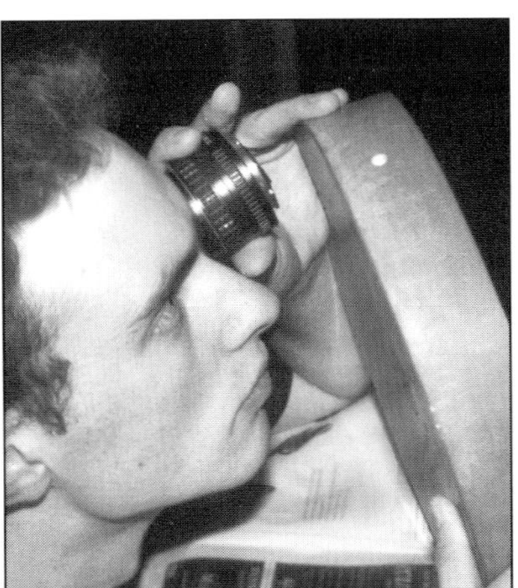

Fig. 5.5 Visually inspecting the smoothed mirror with a jeweler's loupe, inverted camera lens or telescope eyepiece enables you to detect pits and scratches at the mirror's edge. **Below:** a very regular surface, smoothed with W1 emery.

5.4 The Foucault Test

Thus far, we have reviewed only mechanical methods of testing the dimensions and quality of a ground and then smoothed mirror. They have been adequate up to now because the size of the defects measured on the glass was on the order of a 1/10 or a 1/100 of a millimeter. In this section, we will see how to test a polished mirror, for which the measurements are below 1/10 wavelength of light—1/20 of a micron. For a test of this sensitivity, we have the elegantly simple Foucault test. This procedure was invented by Léon Foucault in 1858, and it is within the capabilities of any serious amateur.

5.4.1 What Does the Foucault Test Do?

The Foucault test enables the detection of defects in a telescope mirror, their shape (hills, holes, valleys, etc.), and the calculation of their amplitude in order to gauge what effect they will have on the instrument's images (see Figure 5.6). It is a highly precise method of testing for defects in the form and the smoothness of polished concave mirrors, whether they are spherical, paraboloidal, or otherwise, as long as the focal length of the mirror is not too short.

5.4.2 How Does the Foucault Test Work?

To perform this test, it is necessary to construct a measuring apparatus, of which there are two examples on the following pages. A light source is placed at the center of curvature of the test mirror, the surface of which is then analyzed by the introduction of a bevelled metallic blade (the Foucault knife-edge) into the cone of light reflected by the mirror (see Figure 5.8 on page 54).

Fig. 5.6 The Foucault test shows the shape of the mirror's surface. When the light slit is placed to the left of the eye, and the knife-edge to the right, the mirror looks like it is lit from the left at a low angle. The height of the defects on the mirror's surface appears amplified tens of thousands of times. On a 250-mm mirror, the eye is capable of detecting irregularities on the order of a hundredth of a micron, or $\lambda/50$. By photographing the mirror from behind the knife edge, you can observe defects ten times smaller than that.

The Foucault test is thus a very sensitive method for detecting the presence of irregularities in the polished glass surface (on the other hand, it is more difficult to measure their amplitude).

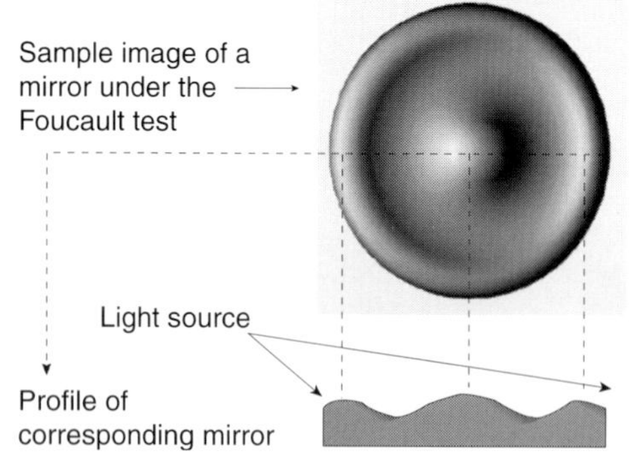

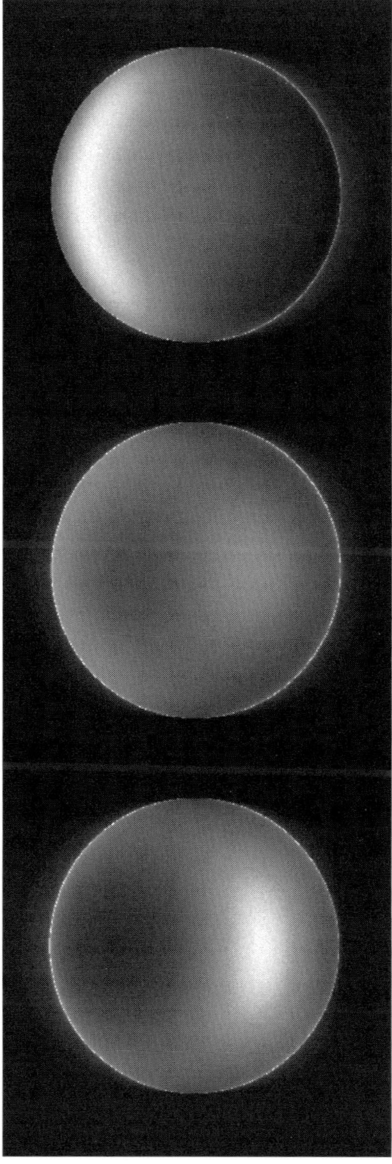

Fig. 5.7 The progression of Foucault-test shapes of a perfect paraboloid is shown as the knife (on the right) is drawn back from a position that nulls (top) the 20-percent radius zone, (middle) the 71-percent zone, and (bottom) the 90-percent zone. The bottom frame is the way the test looks when the knife is farthest from the mirror. This appearance is for a very mild correction, such as for a small, *f*/8 or *f*/9 mirror that a beginner might attempt. Faster and larger mirrors exhibit more contrast in the passage from light to shadow, and narrower gray nulled regions. Also, the edge does not display the solar-eclipse "diamond ring" effect in real tests; the bright ring is smooth. The breakup results from the way these diagrams were calculated. Images courtesy of H.R. Suiter.

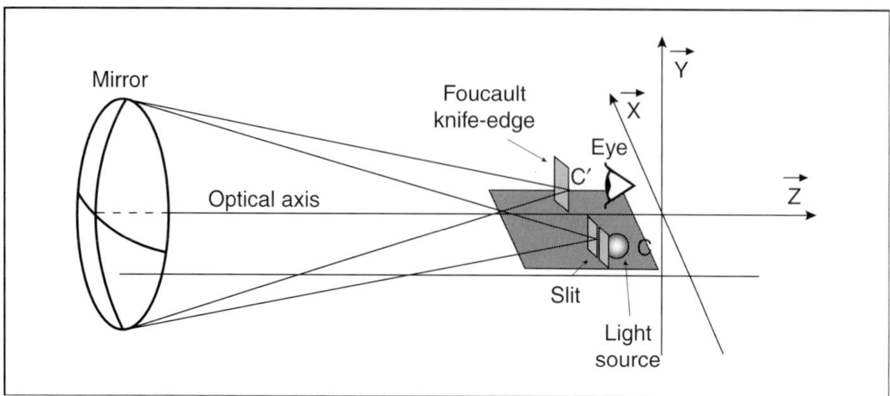

Fig. 5.8 Principle of the Foucault test. The mirror is tested at its center of curvature C. The stage supporting the light slit and the knife-edge is located on the optical axis (Z). The knife-edge is moved along the X axis to intercept the cone of light reflected by the mirror. The eye then observes the mirror lit by the slit at C'.

5.4.3 Spherical Mirrors

Let's consider a spherical mirror of diameter D and radius of curvature R (see Figure 5.9). When a point light source is placed very close to the mirror's center of curvature, C, you see the reflected image of this source formed at C'. Since the mirror surface is a section of a sphere, all the light coming from the source C converges at C' after reflection.

The knife-edge is gradually introduced into the light beam. If the knife is exactly at the center of curvature, and if the mirror is perfectly spherical, then it intercepts all of the light coming from the mirror and the mirror darkens uniformly; this is called a null. If the knife-edge is brought closer to the mirror before being moved into the light-cone, it only obscures one side of it. If the knife-edge is brought back behind C', then the other side of the mirror is darkened.

5.4.4 A Mirror with a Defective Radius of Curvature

This example concerns a mirror with two different centers of curvature, C_1 and C_2 as shown in Figure 5.13 on page 57. C_1 is the center of curvature of the central part of the mirror, and C_2 is that of the outside edge. When the knife-edge is placed at C_1, it intercepts the light reflected by the central zone of the mirror, and this region looks gray (null). Placed at C_2, it intercepts the light reflected by the peripheral zone, which then becomes gray.

The Foucault test permits one not only to visualize this defect but also to measure it. The distance between the points C_1 and C_2 is called the longitudinal aberration near the center of curvature, LA_C.

Figure 5.13 also shows that not all of the light coming from the outer zone of the mirror is not cut at C_1—they converge at C_2. In the following equations, we

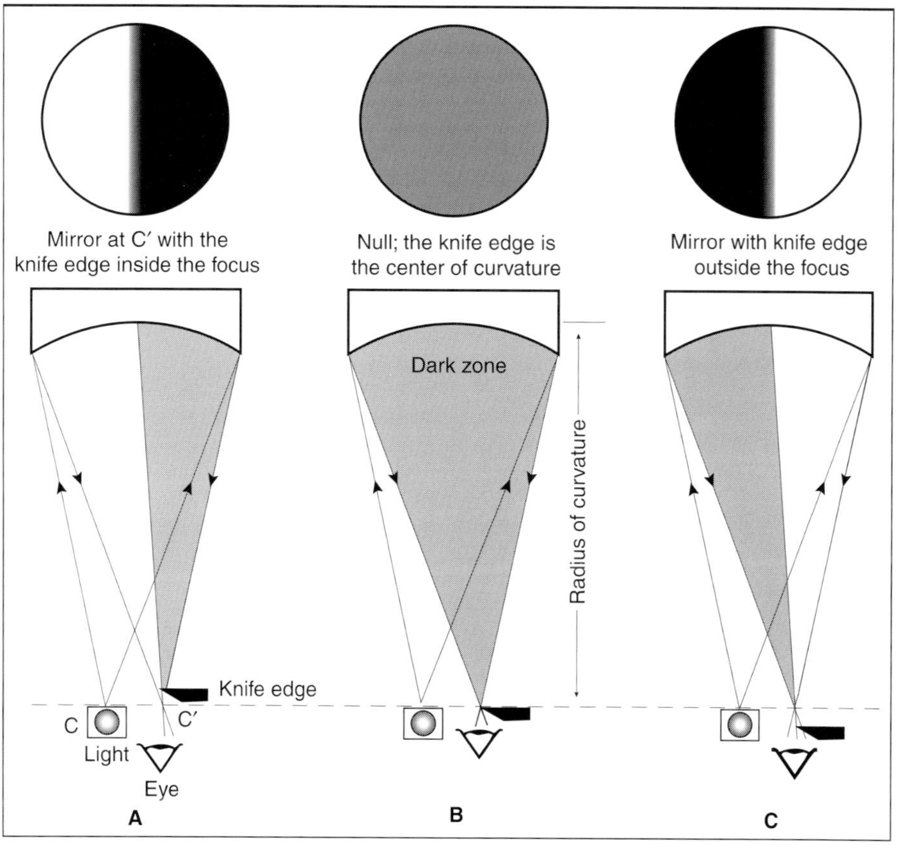

Mirror at C′ with the knife edge inside the focus

Null; the knife edge is the center of curvature

Mirror with knife edge outside the focus

Dark zone

Radius of curvature

Knife edge

C

C′

Light

Eye

A

B

C

Fig. 5.9 The principle of the Foucault test with a spherical mirror. In **A**, the knife-edge intercepts the light rays in front of the mirror's center of curvature. The right side of the mirror is obscured. In **B**, the knife-edge is situated exactly at the radius of curvature of the mirror, and the observer sees a uniform gray tint (called a null) over the whole mirror. In **C**, the knife is placed behind the center of curvature, and intercepts the light reflected by the left side of the mirror. On top, the appearance of the spherical mirror corresponding to each of the cases **A**, **B** and **C**.

Fig. 5.10 Foucault test apparatus used at the Observatory of the Cote d'Azur to test large-diameter mirrors (from 570 to 1,500 mm). The table that supports the knife-slit system can move on two perpendicular axes, which allows the unit also to be used for the caustic test, artificial star tests, and the wire test. The light, provided by a 150-watt halogen lamp, is brought to the slit by optical fibers. Photo by Sebastien Debruyne.

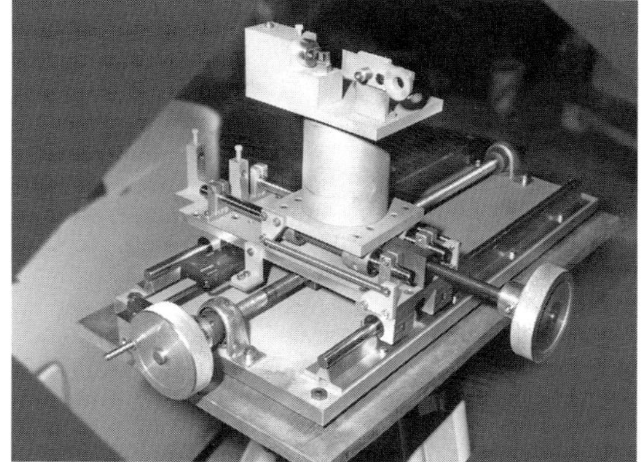

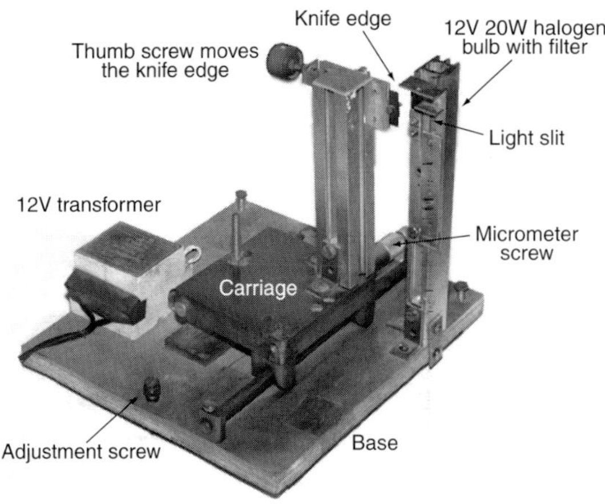

Fig. 5.11 Texereau-style Foucault tester used at the Commission on Instruments of the SAF. The light slit is attached to the wooden base by an aluminum bracket. The carriage moves in only one direction, that of the optical axis of the mirror. It supports the knife-edge, which can be moved left or right by the thumb screw. The movement of the carriage is read off the micrometer screw, graduated by hundredths of a millimeter. This Foucault apparatus has tested dozens of mirrors of 100 to 400 mm diameter and 500 to 2000 mm focal length. It also has a heat absorbing filter to protect the eye from heat and ultraviolet light.

will use R_1 (radius of curvature of the central region of the mirror) as a reference (notated R).

At C_1, the marginal rays illuminate the disc of light with radius TA_C, the transverse aberration near the center of curvature of the mirror. TA_C is calculated with the following equation:

$$TA_C = LA_C \times \frac{h}{R} \tag{5.2}$$

where h is the height of the defect in relation to the optical axis of the mirror and R is the radius of curvature of the central zone of the mirror. However, this mirror will not be used to image at its center of curvature, so TA_C is not of direct interest.

To understand the direct influence of this mirror defect on images, one must know the value of the transverse aberration at the mirror's focus, TA_f, since that's where the images are formed.

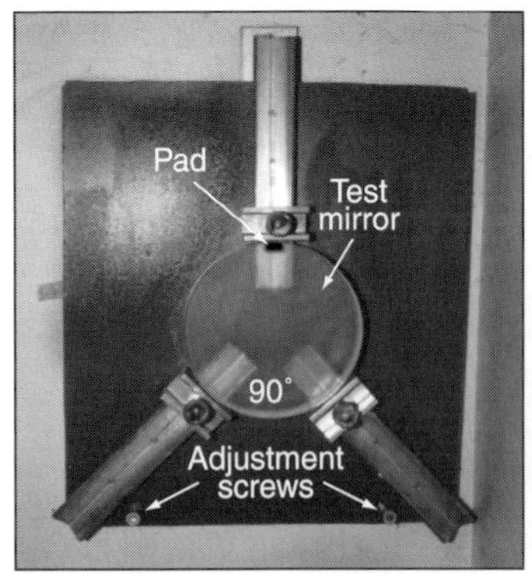

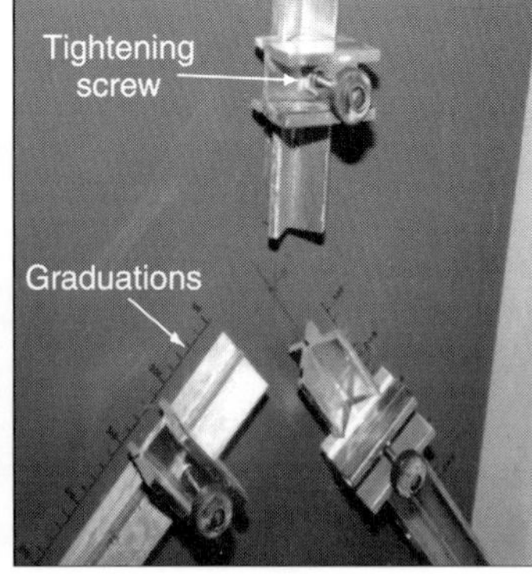

Fig. 5.12 Mirror Test Stand used to test 100–400 mm mirrors at the SAF's Commission on Instruments' optical workshop. **Left,** a 250-mm mirror; **right,** detail of the tightening and adjustment system.

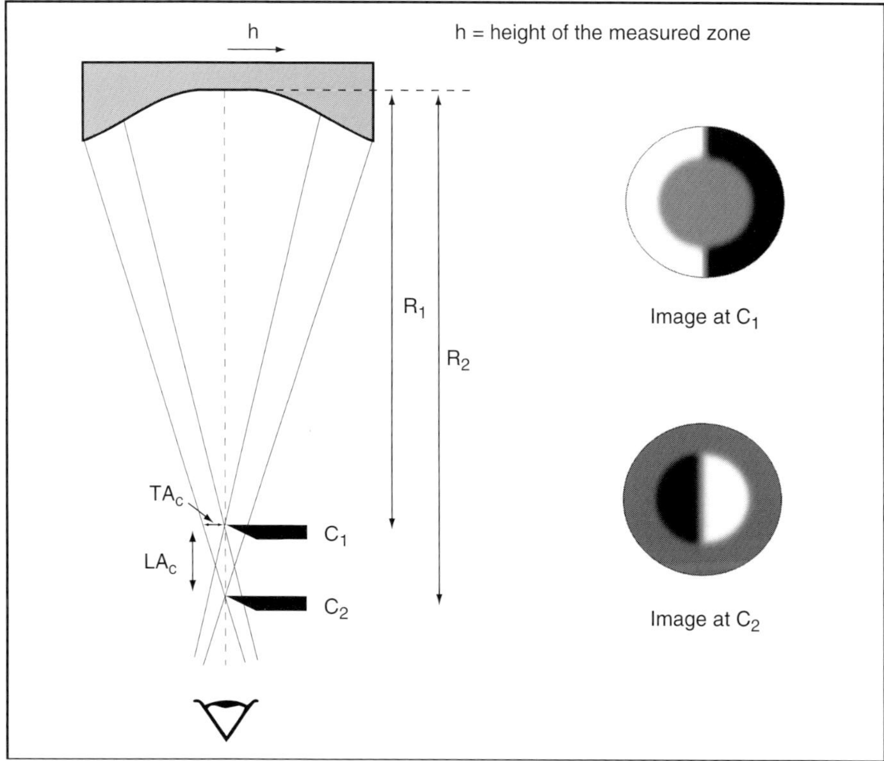

Fig. 5.13 Testing a mirror with different curvatures at the center and edge. (C_1 and C_2 are two centers of curvature.) When the knife-edge is placed at C_1, the central zone of the mirror looks gray. With the knife-edge at C_2, it's the outside edge of the mirror that looks gray. The distance between C_1 and C_2 is LA_C, the longitudinal aberration. TA_C, the transverse aberration, is the semidiameter of the beam of light at C_1.

5.4.5 Paraboloidal Mirrors

To produce a sharp image of an object located at infinity (stars, planets, etc.), a mirror must be paraboloidal, because a spherical mirror cannot focus parallel light rays to a single point (see Figure 5.14 on page 58).

The difference in curvature between a sphere and a parabola is invisible to the naked eye; it corresponds to a few thousandths of a millimeter less glass at the center of the mirror than at the edge, but it makes all the difference in image quality

The radius of curvature of a paraboloidal mirror is different at its center from that at its edge. Thus, the image of the slit source of the Foucault tester is no longer formed at one point (as in the case of a spherical mirror), but is spread out along a segment of its optical axis of length Λ_C (see Figure 5.16 on page 59). This term represents the longitudinal aberration of the paraboloidal mirror at its center of

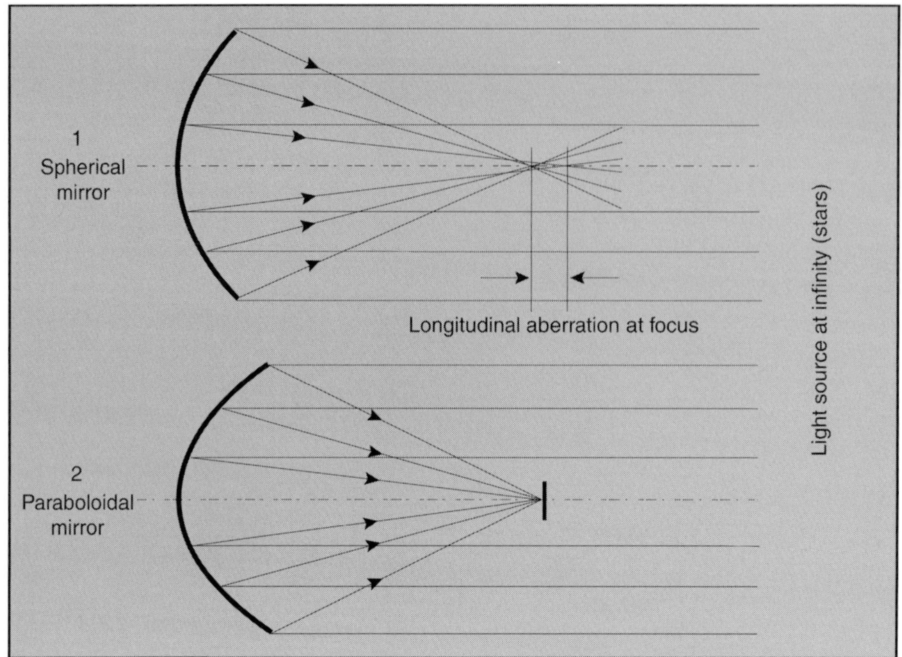

Fig. 5.14 (Above) 1. Path of parallel light rays coming from a source at infinity reflected by a spherical mirror. The image is spread out on the optical axis; the marginal rays converge before the paraxial rays. **2.** To correct this defect, called *spherical aberration*, one gives the mirror a paraboloidal form. The radius of curvature of the outer region of the mirror is increased while that of the inner region is decreased. This way, all the reflected light is concentrated at one point, the focus.

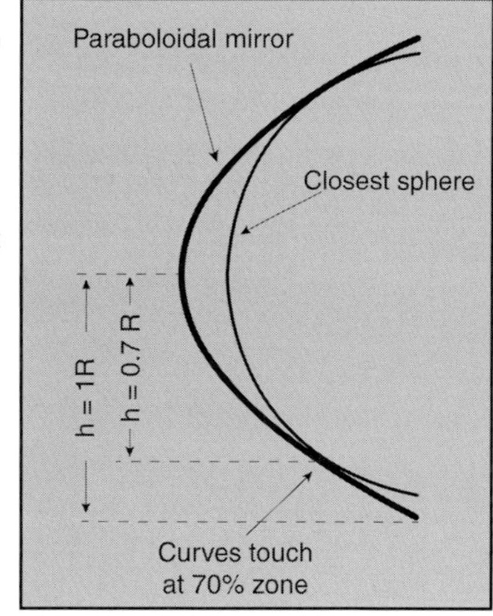

Fig. 5.15 (Right) The deformation of a paraboloidal mirror has been greatly exaggerated to show the difference in curvature (compared to the closest sphere). The radius of curvature at the center of the parabola is shorter than that of the sphere, and that of the periphery is longer.

curvature. Its theoretical value is calculated:

$$\Lambda_C = \frac{h^2}{R} + \frac{h^4}{2R^3} \qquad \textbf{(5.3)}$$

where R is the radius of curvature of the mirror, h is the height of the measured zone (equal to the semidiameter of the piece if you are calculating the total aberration of the mirror). Thus, for most mirrors

$$\Lambda_C \cong \frac{h^2}{R} \ . \qquad (5.4)$$

Important Notes:

1. This formula applies when the source is fixed and the knife-edge is mobile, as in the Texereau-style tester shown in Figure 5.11 on page 56. The term $h^4/2R^3$ can be ignored if its value remains less than 0.01 mm (for standard mirrors, like the 250-mm *f*/4 or the 300-mm *f*/5); however, when testing an *f*/3 or a large-diameter mirror, (>400 mm), use the complete formula to calculate the longitudinal aberration of the paraboloidal mirror. It also only applies for a source slit at the center of curvature of the central zone.[1]
2. With a Foucault tester whose slit and knife-edge are moving together (see Figure 5.10 on page 55), the distance from the slit to the mirror is always equal to the distance from the mirror to the knife-edge and consequently is equal to the radius of curvature of the measured zone (the "null" is obtained in this zone). In this case, the formula to calculate the longitudinal aberration of the paraboloidal mirror is:

$$\Lambda_C = \frac{h^2}{2R} \qquad (5.5)$$

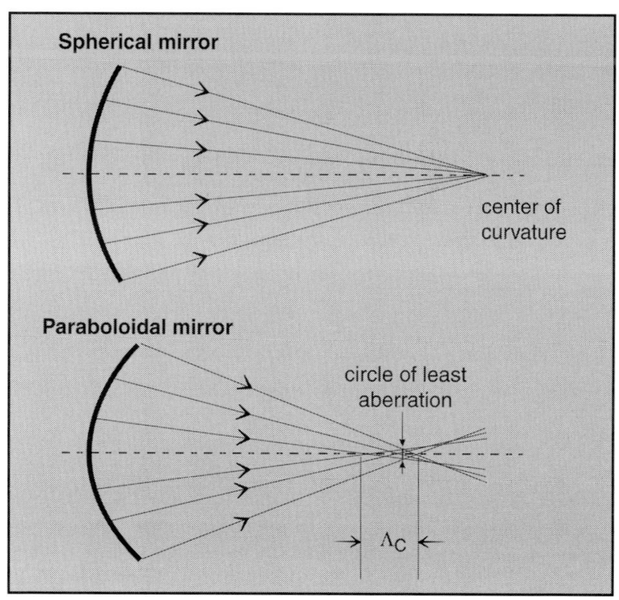

Fig. 5.16 Path of the light rays originating at the center of curvature of, and reflected by, a spherical mirror (top) and a paraboloidal mirror (bottom), in the Foucault test. In the first case, all the light converges at one point. In the second case, the paraxial rays converge before the marginal rays. The theoretical distance between the extremes of this segment is called the longitudinal aberration, Λ_C, of the mirror at its center of curvature.

The circle of least aberration of a paraboloidal mirror observed at its center of curvature is located at ¾ of the segment outlined by the spherical aberration.

[1] See Section B.12.4 of *Amateur Telescope Making Book 1* (2nd edition) p. 314 for detail concerning this tiny correction.

Fig. 5.17 [a] Special support for a large-mirror tester. Here, a 500-mm disk is supported by a strap.
[b] Detail of a 600-mm mirror. Photography permits the detection of small surface irregularities (a small hill in this case) that are invisible with a visual tester.
[c] Severely deformed paraboloidal mirror. The knife-edge intercepts the light reflected by a circular hill close to the center.
[d] Paraboloidal mirror. The knife-edge intercepts the light reflected by the 0.7 zone. This zone separates the light and dark regions of the mirror.
[e] Foucaultgraph of a mirror whose coating was removed with hydrochloric acid. The severely damaged central part shows significant irregularities that require refiguring.

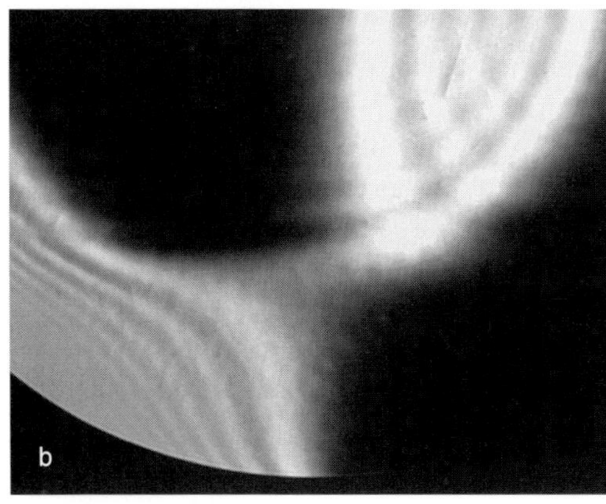

Photo: © S. Debruyne/D. Vernet

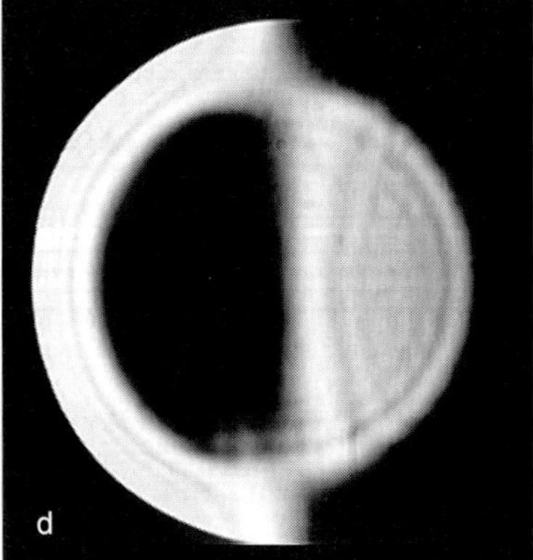

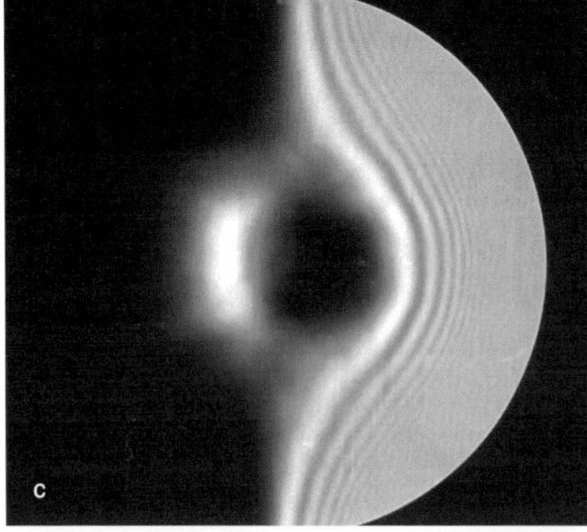

Photo: © S. Debruyne/D. Vernet

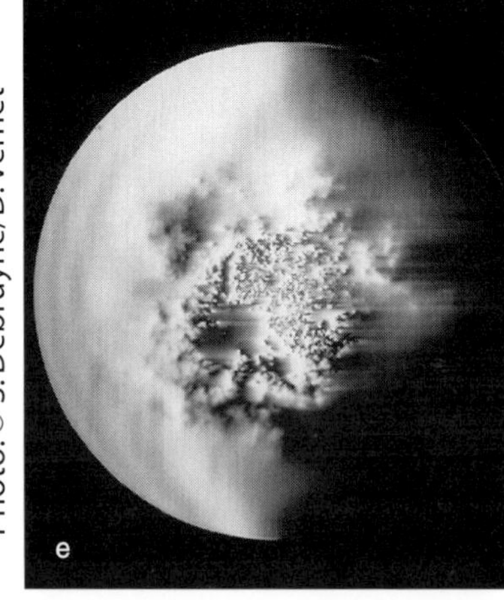

and no correction term exists, as it does for the fixed-source tester.

3. To test a paraboloidal mirror, consider that it is composed of an assemblage of small, concentric rings embedded within each other like an archery target. This division of the parabola into discrete sections allows for the measurement of their respective radii of curvature and thus to reconstruct (piece by piece) the characteristics of the segment of longitudinal aberration and eventually the surface of the mirror. These measured values of LA_C are then compared with the theoretical values (Λ_C) for each zone to derive a difference, ΔLA_C.
4. The focus is located halfway between the mirror's surface and its center of curvature. The transverse aberration at the focus, TA_f, is then equal to half the difference measured at the center of curvature, and is calculated in the equation:

$$TA_f = \Delta LA_C \times \frac{h}{2R} \quad . \qquad \textbf{(5.6)}$$

   The goal is to compare the size of this aberration (calculated for a given defect) with that of the radius, ρ, of the theoretical diffraction image of the mirror. If TA_f is less than or equal to ρ, the defect's effects on images will be negligible. That means that all the light reflected by the mirror will arrive inside the diffraction image (Couder's criterion). In practice, you calculate the value of TA_f/ρ, which should be less than or equal to 1.

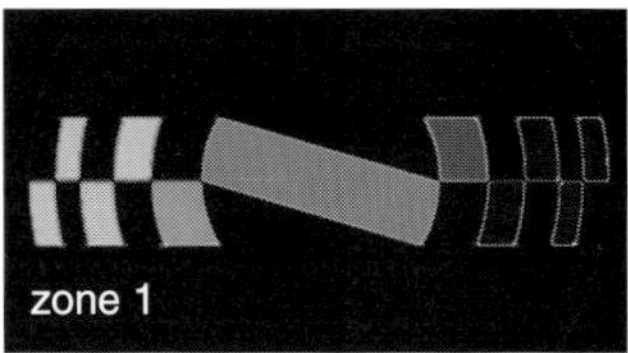

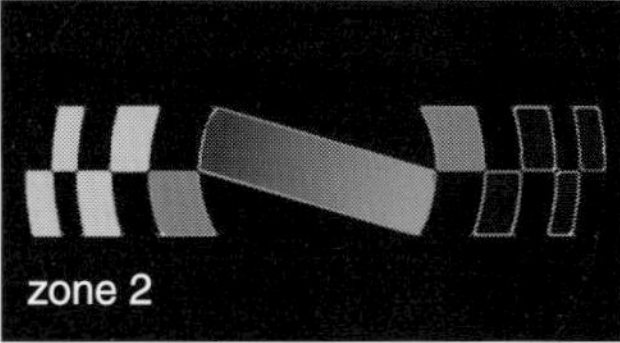

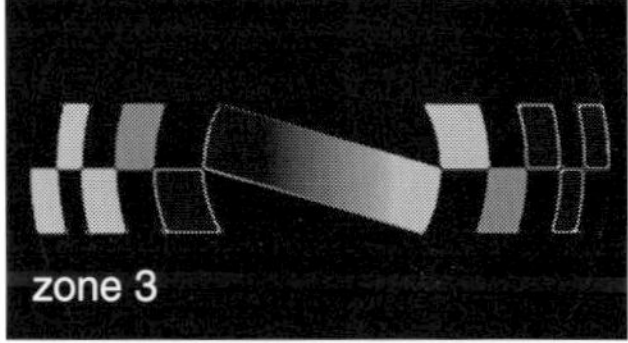

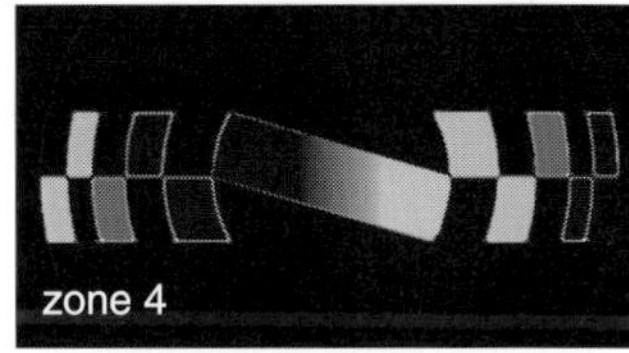

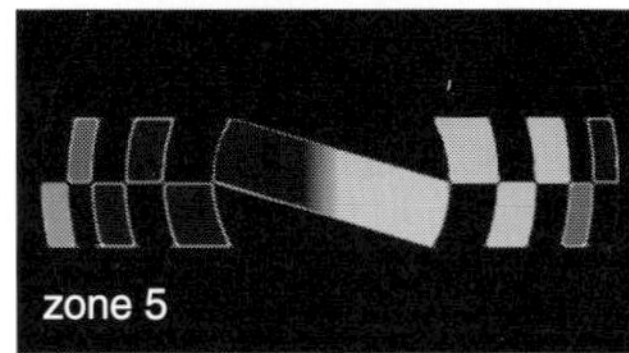

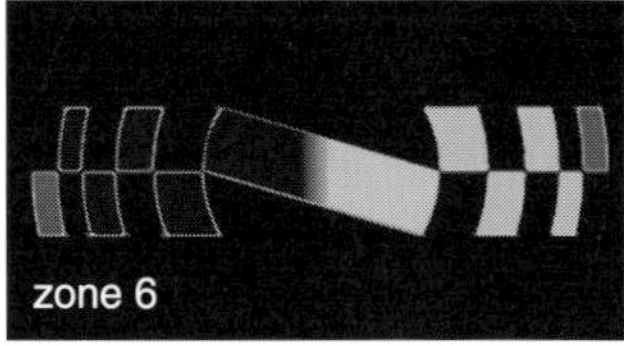

Fig. 5.18 Testing a 245-mm *f*/4.3 mirror with a Couder screen. In each image, a different zone is null. The bright borders around the windows are the effects of diffraction; they frustrate the eye in its attempt to perceive the nulled zones.

5.4.6 The Couder Screen

To help isolate the test zones for measurement, we will make a circular, opaque screen of the same diameter as the mirror in which we will cut "rectangular" windows. This mask, called a Couder screen, is placed in front of the mirror during testing (see Figure 5.19 on page 63).

Dealing with Diffraction Effects

I'm having difficulty using a Couder screen to test my mirror. I'm frustrated by the diffracted and diffused light at the edges of the windows.

There are two ways to make this test easier on your eyes and more precise:
First, the TV-monitor and camera method described on page 223 ("Electronic Foucault Testing") can eliminate all the diffraction caused by the windows.

A second method is to enlarge the windows in the Couder screen as shown below. With larger windows, the test is easier, but a separate screen has to be made for each zone. Great care must be taken not to move the mirror in its support at all during the test. The screens are attached to the edge of the mirror, in succession, by small pieces of tape. The test requires more time, but it's more accurate, as long as the mirror and the tester don't move between measurements.

In these two diagrams, the enlarged windows are shown as shaded. **Left:** A test screen for the central zone of the mirror. A vertical band of opaque tape has been attached to the mirror through the central zone to facilitate measurement. **Right:** Test screen for the second zone of the mirror. The shaded zones are delimited in height by 45° angles originating at the center of the screen. Screens for the rest of the zones are made in the same way.

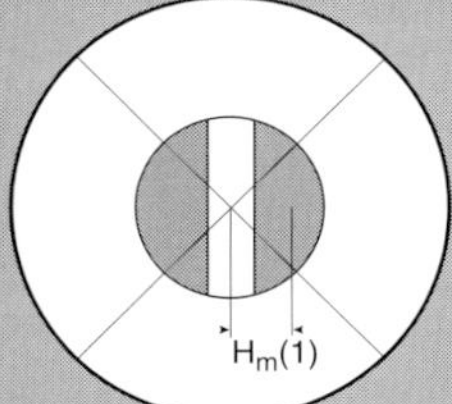

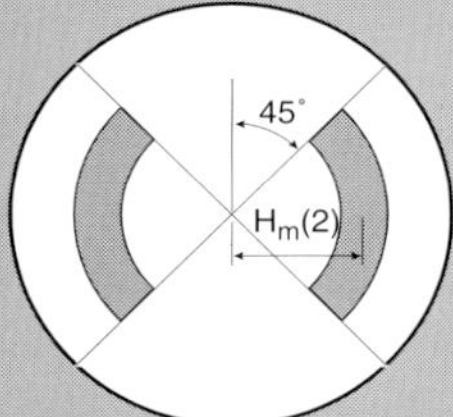

The central zone of my mirror is difficult to measure in the Foucault test because it shows very little contrast. What can I do to make this zone easier to read?

An effective method is to divide the central zone in two with a strip of paper or opaque tape attached to the screen. Positioned vertically on the mirror, this band facilitates the comparison of the left and right parts of the mirror's central zone. An important point: as in the calculation of the test screen for a perforated mirror, you must account for this obscuration when you're calculating the average height, *Hm*, of zone 1.

We will determine by experiment the number of zones to make for the parabola. Practice and experience suggest that you make one zone per millimeter of longitudinal aberration as measured with a fixed-source tester (the aberration is equal to twice the sagitta of the mirror). We will then recreate the "null" successively in each window of the Couder screen—in a parabola, one does not see the null over the entire mirror, but only in the windows of the Couder screen (see Figure 5.19). This way, you can measure the radius of curvature of a given zone when the null is observed simultaneously across (median height, H_m) both the right and left windows. This observation is made more difficult if the windows are narrow and the mirror is far away from the observer. In addition, diffraction effects at the edges of the windows make achieving accurate measurements difficult, especially

Fig. 5.19 Examples of Couder screens used at the workshop of the Astronomical Society of France. **Left**, a test screen for a 250-mm *f*/5 mirror; **Center**, a screen for a 200-mm *f*/5; **Right,** for a 250-mm Cassegrain (the central window is closed).

in the peripheral zones of the mirror.

5.4.6.1 Making the Test Screen[2]

1. Calculate the theoretical exterior diameter of the screen windows, H_x, taken from the example in Figure 5.20 on page 65.
2. Trace the windows on fairly thick, rigid card stock, bristol board, or similar material that is sufficiently large to cover the whole mirror, and then carefully cut out the openings with a sharp knife. It is very important to cut the windows out cleanly—torn paper, a source of extraneous light, is to be avoided.
3. After cutting, re-measure the values of H_x very precisely (half-millimeter precision is required for a pair of windows). You will use these values in the calculations later.
4. According to the values of H_x, calculate the height H_m of the center of each window (see the procedure in the caption of Figure 5.20). Write them in ink on the screen.
5. All that remains is to determine the theoretical longitudinal aberration at the center of each window. This is denoted by the term H_m^2/R (R is the radius of curvature at the center of the mirror). Write each of these values in pencil on the test screen, as R likely to vary over the course of retouching and will certainly vary if the screen is used for another mirror.

5.4.7 The Foucault Test in Practice

From here on, we will use as the example the 250-mm mirror described in Part 3 of this book.

The mirror is set up vertically on the test support. The Foucault tester is aligned with the mirror's optical axis. The knife edge and slit are then aligned (see Figure 8.5 on page 113). The knife-edge is positioned at the center of curvature of zone 1 (at the center of the mirror), and then moved transversely to arrive at the null on this section of the mirror.

Take the reading indicated by the micrometer screw and note this number as

[2] Programs to calculate the values for the Couder mask can be found at: http://www.stellafane.com/atm/atm_mask/atm_zone_calc.htm and http://www.atmsite.org/contrib/Carlin/couder/. As with all software you should independently check the output for correctness before routinely relying on it.

> **Why is a "zone 3" constant being subtracted**
>
> Our tester, when we began this example, had an arbitrary longitudinal screw setting of about 4.2 mm. We could have carefully adjusted the reading off the tester (by scooting the whole tester forward and backward) to be zero at the center of curvature of the central zone. However, we will need to adjust this constant slightly anyway, so there is no point in needless zeroing during the actual data-taking.
>
> So which value of the residual aberration do we take for this constant? Some people advocate the central zone (thus performing in data reduction what would have been achieved mechanically by moving the tester) and others the edge zone, because it is more precisely measured. We take a central path in advocating the zone closest to 70-percent of the radius of the mirror. If your figuring stroke is working well, this will be about the halfway point in the difference in raw measurements between the edge and central zones. It is also the least-deformed position compared to the likely reference wavefront (see Figure 5.15 on page 58).
>
> If you have a Couder screen calculated so that 70 to 75 percent lands on another zone, use that zone's residual. Also, do not despair if you do not have a zone close to 70-percent of the way to the edge. You can, in fact, use any zone's residual value, or use the residual value that is the average of all zones. What you are attempting to do is to remove that gross arbitrary setting. You will search later for a small adjustment constant (called *k*) which you will again use to fine-tune your calculation.

the value for zone 1 (this value has little intrinsic importance; it can be positive, negative or zero).

Next, slowly back off the knife edge, using only the micrometer screw, until you obtain the null in zone 2. Note the reading for this zone, and repeat this measurement for the remaining zones. After recording the value of the last zone (zone 6 in the example), move the knife edge up slightly and repeat the measurements in the other direction (from the exterior zone to the central zone). You will then have two different measurements for each zone. Average the two measurements for each zone, and record the figures as measurement (zone 1), measurement (zone 2), etc.

5.4.8 How to Interpret the Measurements

Measurements alone are not directly useful. It is first necessary to subtract the terms H_m^2/R; the result is called the residual. For each zone, calculate:

$$\text{Residuals} = \Delta LA_C = LA_C - \Lambda_C = \text{Measurements} - \frac{H_m^2}{R} . \tag{5.7}$$

Subtract from each residual the residual of zone 3. These new figures are called the reduced measurements. Why zone 3? See the above box titled "Why is a "zone 3" constant being subtracted". This zone is often referred to as the 70% zone.

$$\text{Reduced measurements} = \text{Residuals} - \text{Residual (zone 3)} . \tag{5.8}$$

Positive values of reduced measurements indicate that the radius of curvature of each measured zone is too large, in comparison to zone 3, and tell you by

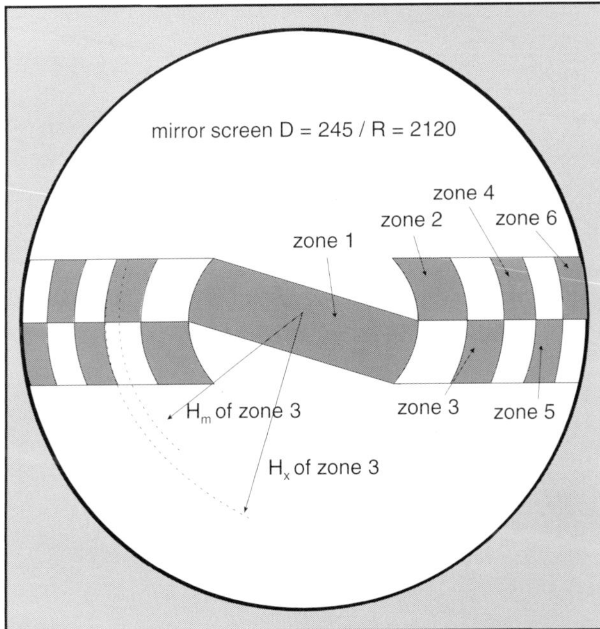

Fig. 5.20 A 6-zone Couder screen calculated for a 245-mm paraboloidal mirror.

Each of these windows is defined by its exterior semidiameter H_x – $H_x(1)$, $H_x(2)$, etc. The H_x's allow the calculation of the median heights H_m of the zones:

$H_m(1) = H_x(1)/2$
$H_m(2) = [H_x(2) + H_x(1)]/2$
$H_m(3) = [H_x(3) + H_x(2)]/2$

and so on.

Finally, calculate the value of the theoretical longitudinal aberration for each zone H_m^2/R.

In our case, $D = 245$ mm and $R = 2120$ mm, which give the values that appear in the table below (in millimeters).

Zone	1	2	3	4	5	6
H_x	49.6	70.1	85.9	99.2	110.91	122.5
H_m	24.8	59.85	78.0	92.55	105.06	116.7
H_m^2/R	**0.29**	**1.69**	**2.87**	**4.04**	**5.21**	**6.42**

Zone	1	2	3	4	5	6
H_x (in mm)	49.60	70.10	85.90	99.20	110.91	122.5
H_m (in mm)	24.8	59.85	78	92.55	105.055	116.705
$\Lambda_C = H_m^2/R$ (in mm)	0.29	1.69	2.87	4.04	5.21	6.42
Measured LA_C (in mm)	**4.59**	**5.91**	**7.10**	**8.31**	**9.50**	**10.60**
	Reduced measurements					
$LA_C - \Lambda_C$ = Residuals (in mm)	4.300	4.220	4.230	4.270	4.294	4.175
Reduced = Residual– Zone 3	0.070	–0.010	0.000	0.039	0.064	–0.055
Constant k to subtract (in mm)	**0.0015**					
Reduced $-k$	**0.068**	**–0.011**	**–0.002**	**0.038**	**0.062**	**–0.056**

Fig. 5.21 Example of measurements made on the 245-mm diameter mirror, with a 2,120 mm radius of curvature, using the Couder screen described in Figure 5.20. All the values are in millimeters. The constant k is 0.0015 mm in this example.

how many millimeters. Negative values indicate too small a radius of curvature (in comparison to zone 3).

The resulting measurements tell us vaguely about the quality of the mirror. It is, however, useless to continue the calculations if these values are larger than 0.1 to 0.15 mm for the peripheral zones; in practice, such differences are generally incompatible with a good paraboloidal mirror. At this stage, one must adequately rework the mirror before continuing.

Next, we must subtract a constant, k, (positive or negative) from the reduced measurements. This constant must be well-chosen to establish the value of the adjusted longitudinal aberration at the mirror's center of curvature.

$$(LA_C - \Lambda_C)_{\text{Adjusted}} = \text{Reduced measurements} - k\,. \quad \textbf{(5.9)}$$

The adjustment constant k may be selected to either balance the maximum excursions of the TA graph or the wavefront. This is best done with a computer-generated spreadsheet. The reduced longitudinal aberration difference allows us to calculate directly the transverse aberration TA_f at the focus of the mirror, or the radius of the smallest circle through which all the reflected light passes:

$$TA_f\ (\text{m}) = (LA_C - \Lambda_C)_{\text{Adjusted}} \times \frac{H_m}{2R}\ , \quad \textbf{(5.10)}$$

where H_m is the median height of the measured zone, and R is the radius of curvature measured at the center of the mirror. At this place, called the circle of least aberration, the extremes of the transverse aberration (TA_f) are equal and of opposite sign (+/–). This condition is only obtained for a single value of the correction constant k (which is obtained through trial and error).

The transverse aberration at the focus gives an estimation of the performance of the mirror. Compare the size of this aberration for each zone with the radius ρ of the diffraction spot at the mirror's focus (ρ = 2.9 μm in this example). We can then draw the histogram of the values TA_f/ρ as a function of the zone (see Figure 5.22). If the values of TA_f/ρ are less than 1, all of the mirror's reflected light fits inside the diffraction spot. This criterion (Couder's criterion) is very important and must be satisfied before proceeding to the next set of calculations. In effect, if the term TA_f/ρ is greater than 1 for even one zone, you must retouch the mirror or it will not give excellent images.

Using the transverse aberration (TA_f) values, we can determine, for each zone, the slopes u of the defects on the wavefront reflected by the mirror:

$$u = -\frac{TA_f}{F}\ . \quad \textbf{(5.11)}$$

Knowing these slopes allows us to plot straight segments on a graph corresponding to the profile of the wavefront for each zone. Starting at the origin (the center of the mirror), we calculate the ordinates $y_{\text{wavefront}}$ for the points of the curve, using for the first zone:

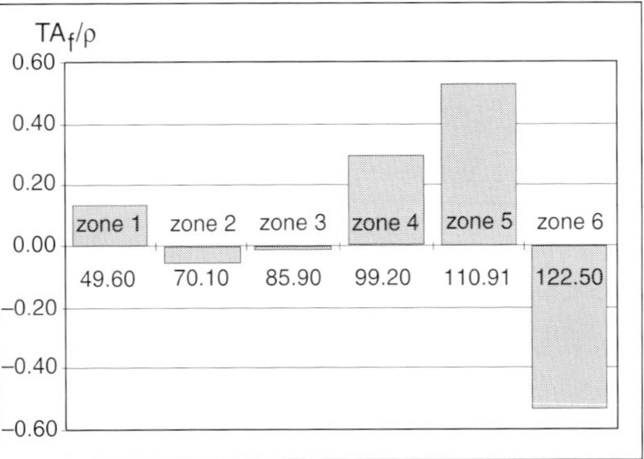

Fig. 5.22 Transverse aberration TA_f compared to the radius of curvature ρ of the diffraction spot. The extremes are equal in absolute value: zones 5 and 6, TA_f/ρ = 0.53 and −0.53; the negative values correspond to light rays that hit the optical axis in front the circle of least aberration and the positive values to rays that hit the optical axis behind it.

Zone	1	2	3	4	5	6
Transverse aberration TA_f at focus (in mm)	0.00040	−0.00016	−0.00003	0.00083	0.00155	−0.00155
TA_f (in microns)	0.40	−0.16	−0.03	0.83	1.55	−1.55
TA_f/ρ	**0.14**	**−0.06**	**−0.01**	**0.29**	**0.53**	**−0.53**

$$y(1)_{\text{wavefront}} = u \times H_x(1)\,. \qquad \textbf{(5.12)}$$

To get the values of y for the other zones:

$$y(2)_{\text{wavefront}} = u \times [H_x(2) - H_x(1)] + y(1)_{\text{wavefront}} \qquad \textbf{(5.13)}$$

$$y(3)_{\text{wavefront}} = u \times [H_x(3) - H_x(2)] + y(2)_{\text{wavefront}} \qquad \textbf{(5.14)}$$

and so on through the last zone of the screen.

Now we know that the coordinates (H_x, y) of the points on the curve represent the profile of the wavefront. The center (A) of the mirror, taken as the origin, has the coordinates (0, 0). The graph shown in Figure 5.23 on page 68 is a plot of this curve taken from Table 5.1 on page 68. The values of y, very small in millimeters, were multiplied by a million. On the graph, the vertical axis is thus graduated in nanometers (1 nanometer = 10^{-6} millimeter).

Figure 5.23 illustrates the profile of the wavefront surface at the circle of least aberration. Experience shows that the sharpest image is not found at exactly this place.[3] It is useful to compare the graph that we have obtained with the closest possible spherical wavefront. The spherical wave will pass through the two extreme points of the wavefront profile, A and G on the graph. The distances between the profile and the best-fit spherical curve vary as a function of the square of the incident height; we will in this way compare the form of the wavefront to the arc of a parabola. This arc can pass above or below our profile, but it must nev-

[3] A. Danjon and A. Couder, *Lunettes et Télescopes*, Editions de la Revue d'optique théorique et instrumentale, Paris, 1935, Section 42.

Fig. 5.23 Profile of the wavefront surface over the semidiameter of the mirror. This curve, measured at the circle of least aberration, passes through the points of the coordinates (H_x, y) for each zone. The point *A* represents the center of the mirror and *G* its right edge.

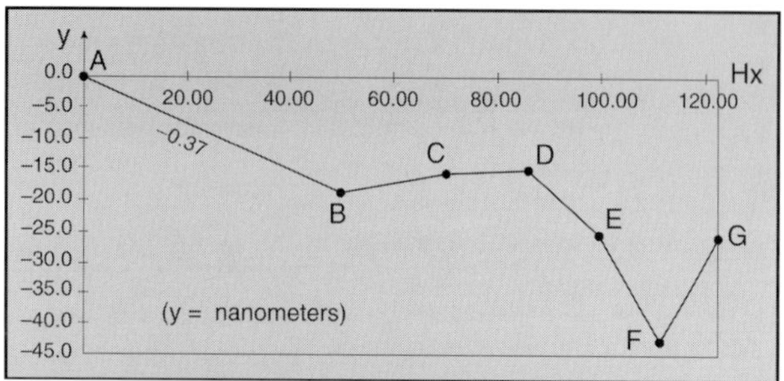

Values of the slopes and coordinates of the points of the wavefront curve.						
Zone	1	2	3	4	5	6
	Graph of wavefront profile					
Slopes $= -TA_f/u$	−0.0000004	0.0000002	0.0000000	−0.0000008	−0.0000015	0.0000015
Slopes × 1 million	−0.3763	0.1508	0.0260	−0.7821	−1.4583	1.4605
y in nanometers	**−18.7**	**−15.6**	**−15.2**	**−25.6**	**−42.6**	**−25.7**

Table 5.1
Foucault Report for 6-Zone Screen.

Mirror Parameters		
Optical Diameter	245.00	mm
Radius of Curvature	2120.00	mm
Wavelength	0.000550	mm
Diffraction spot ρ	0.00290	mm
Focal Length	1060.0	mm

Zone	1	2	3	4	5	6
H_x (in mm)	49.60	70.10	85.90	99.20	110.91	122.5
H_m (in mm)	24.8	59.85	78	92.55	105.055	116.705
$\Lambda_C = H_m^2/R$ (in mm)	**0.29**	**1.69**	**2.87**	**4.04**	**5.21**	**6.42**
Measured LA_C (in mm)	**4.59**	**5.91**	**7.10**	**8.31**	**9.50**	**10.60**
Reduction of Measurements						
$LA_C - \Lambda_C$ = Residuals (in mm)	4.300	4.220	4.230	4.270	4.294	4.175
Reduced = Residual− Zone 3	0.070	−0.010	0.000	0.039	0.064	−0.055
Constant *k* to be subtracted	**0.0015**					
Reduced − *k*	0.068	−0.011	−0.002	0.038	0.062	−0.056
Transverse Aberration TA_f at focus (in mm)	0.00040	−0.00016	−0.00003	0.00083	0.00155	0.00155
TA_f (in microns)	0.40	−0.16	−0.03	0.83	1.55	−1.55
TA_f/ρ	0.14	−0.06	−0.01	0.29	0.53	−0.53
Shape of Wavefront Profile						
Slopes $u = -TA_f/f$	−0.0000004	0.0000002	0.0000000	−0.0000008	−0.0000015	0.0000015
Slopes × 1 million	−0.3763	0.1508	0.0260	−0.7821	−1.4583	1.4605
y in nanometers	−18.7	−15.6	−15.2	−25.6	−42.6	−25.7154
Calculation of the Precise Peak-to-Valley						
PTV on the Wavefront	21.6 nanometers			or	0.039 λ	
	Largest Wave Error:				λ/25.45	

Test sheet: Example of the complete calculation for a 245-mm diameter telescope mirror. The various graphs in this chapter were constructed using the figures from this table. The largest wavefront error, as determined graphically, is at the bottom right of the table (λ/25.45).

Foucault Test Measurement Uncertainties and their Effects

Methodology Uncertainties

The degree of uncertainty of the measurements diminishes as you increase the number of measurements. You will obtain imprecise measurements if you only make the minimum of two measurements per zone (going out and coming back in). In the Foucault test, the average observer will read measurements with a margin of error of ±0.05 mm (this includes the imprecision of the micrometer screw, which is usually less than 0.01 mm). These variations, applied to each of the zones of a 245-mm mirror, can considerably affect the ultimate accuracy of its surface.

The measurements also vary from one observer to another. The perception of the gray shadows is somewhat subjective, and it is advisable to have at least one other experienced person also test the mirror.

Variation of the Radius of Curvature

The radius of curvature can change by several millimeters over the course of figuring, depending on the parabolization method used. For example, to parabolize a 245-mm spherical mirror, the surface must be shaped in such a way as to obtain a difference in the radius of curvature of about 7 mm between the central zone and the edge zone. Thus the simple act of parabolizing the mirror at the center modifies the radius of curvature of that zone by several millimeters. This variation then modifies the value of R used in all of the formulas, most noticeably the H_m^2/R. **Forgetting or neglecting to update the radius of curvature may invalidate all of the calculations.** For example, a variation of ±3 mm in the measurement of the radius of curvature of the 245-mm $f/4.3$ mirror will produce an error of ±0.015λ in the final precision of the wavefront. Thus, a mirror at λ /13 (0.076 λ) would have a variation in precision between λ/10.9 (0.092 λ) and λ/16 (0.062 λ). Moreover, the error caused by measuring the radius of curvature with a tape measure is on the order of ±1 mm. This gives a variation of 0.005 λ for the 245-mm mirror. The precision of the surface (originally λ/13) varies, in this case, between λ/12.2 and λ/14.

In Conclusion

As you add up the uncertainties of the measurements, it becomes difficult to certify a mirror with a great degree of precision using the Foucault test. Beyond λ/16, a mirror can be considered "almost perfect" (see the Françon criterion, page 70). You must also realize that the mirror mounted in the telescope will be, most often, obstructed by the secondary mirror. The effects of this obstruction on the image are more severe than this tiny error.

er cross it. Whether it rises or falls, the parabola must always have a null (i.e., flat) zone at the center of the mirror. It need not intercept the origin at zero height, however.

5.4.8.1 Calculating the Largest Defect on the Reflected Wavefront

Now we compare the wavefront surface of the mirror to the arc of a parabola traced above it (see Figure 5.24). We label the greatest difference between the two curves ε. This value corresponds to the largest error of the reflected wavefront. This error is also called the PTV ("Peak-To-Valley") error. In our graph, this maximum distance is seen at point F, where ε is 21.6 nanometers. We then compare this value to the wavelength of light (λ = 0.55 μm, or 550 nanometers).

The PTV ratio is 21.6/550 = 0.039λ, or λ/25.5 on the wave.

How to Determine the Effects of Mirror Defects on Images

The quality of an image (contrast, resolution) depends on the amount of light energy concentrated in the Airy disk. In a perfect mirror, 84% of this energy is within this disk, and the remaining 16% is distributed through the diffraction rings. Table 5.2 displays the redistribution of the light energy in a mirror that contains defects:

Table 5.2
Relation between the Severity of Defects and Image Quality

PTV Difference (Correction error)	RMS Difference	Strehl Ratio	Percent of Energy In:	
			Airy Disk	Diffraction Rings
0[a]	0	1	84%	16%
$\lambda/16$[b]	$\lambda/54$	0.99	83	17
$\lambda/8$	$\lambda/27$	0.95	80	20
$\lambda/4$[c]	$\lambda/14$	0.8	68	32
$\lambda/2$[d]	$\lambda/7$	0.4	40	60[d]
λ[d]	$\lambda/3$	0.1	10	90[d]

a. Perfect optics.
b. Françon criterion.*
c. Rayleigh limit.
d. At this point, the rings are reinforced at the expense of the energy of the central disk.

* **The Françon criterion** defines the "maximum" precision of a mirror, beyond which further figuring is useless. In the table, it corresponds to a peak-to-valley difference of $\lambda/16$ on the wave. 83% of the energy is then in the disk (as opposed to 84% with a perfect mirror). Beyond this precision, improvements in the image are not visually detectable.

Note: The values in the table refer to an unobstructed image; the central obstruction of a secondary mirror considerably modifies the light distribution in the Airy disk and the rings.

In a large telescope (larger than 1 meter), a defect of λ to $\lambda/2$ is less significant and much less of a problem than it is in a small instrument; the angular diameter of the diffraction spot, even if the mirror has some aberration, remains extremely small and permits the realization of "high resolution" images.

Example: mirror has a defect of $\lambda/4$ *on the wavefront* measured in the Foucault test. It passes the Rayleigh limit, and the Airy disk contains 68% of the light energy (with 32% in the rings).

The Strehl ratio (SR) is defined as the ratio between the intensity of the center of the diffraction disk in an optical system containing an aberration and the corresponding intensity in a perfect optical system. The SR has a direct effect on the distribution of light energy in the diffraction spot. Its value, for small aberrations, is calculated with the following formula:

$$SR \cong \left(1 - \frac{2\pi^2 \Delta \mathrm{RMS}^2}{\lambda^2}\right)^2 .$$

(ΔRMS: see below). In the case of a $\lambda/4$ mirror with low-order spherical aberration, the *SR* is 68/84 = 0.8.[1]

The Strehl Ratio is calculated with a term called ΔRMS (Root Mean Square Difference). It represents the average quadratic difference of the wavefront defects. Let's consider a mirror with a very smooth wave surface; the Peak-To-Valley (PTV) measurement is useful in estimating the quality of the mirror. Imagine now a wavefront with a good general shape but which presents several rough irregularities. In this case, the RMS difference gives a more accurate estimation of the quality of the system because it takes into account all of the defects in relation to the mirror as a whole. Take for example a mirror that is perfect with the exception of one lump. If it covers a small area, this defect has a negligible effect on image quality, even if the corresponding PTV difference is great. The RMS value of a surface is defined as the average value of the amplitude of the surface minus the mean value of that surface. Proper calculation of this quantity must be weighted by the radius, but a Couder screen is already designed to counterbalance this weighting except for perhaps the central zone, which is often of anomalous size. In terms of the zones of a correctly weighted Couder screen, this is the RMS deviation:

$$\Delta \mathrm{RMS} = \sqrt{\frac{1}{N}\sum_{i}^{n}(W_i - \overline{W})^2} = \sqrt{\left(\frac{1}{N}\sum_{i}^{n} W_i^2\right) - \overline{W}^2},$$

where W_i is the edge value of the wave front in the *i*th zone (used for convenience), *n* is the number of zones, and $\overline{W}$ is the average value of the wave front.

However, just any RMS value will not do; you need the *minimum* RMS value to calculate the Strehl ratio. To derive minimum RMS, you must first add a trial reference parabola to the whole wave front of the form

$$W_i^{\mathrm{Trial}} = W_i + Ch^2,$$

where *C* is the trial constant and *h* is H_x. Then you do the RMS calculations above, change *C* and repeat until you have searched the whole vicinity of the minimum RMS value. Clearly, deriving the Strehl ratio from real data requires computer aid. It is also of questionable validity for these sparse, stick-figured, wavefronts taken only along a diameter. For this reason, we stick to the peak-to-valley measures in this book.

Experience shows that the ratio between the RMS difference and the PTV difference for the same surface varies from 2 to 7, depending on the number of measures taken. Low-order spherical aberration gives a 3.5 ratio, which implies the following formula:

$$\Delta \mathrm{RMS} = \frac{\text{PTV difference}}{3.5},$$

which is used as an approximation, but an admixture of higher order terms changes the ratio for mirrors in the Foucault test.

Computerized interferometers sample optical surfaces over thousands of points, and their RMS/PTV ratios most often fall between 4.5 and 6.

For a $\lambda/4$ mirror, the RMS error is $\lambda/14$. For the 245-mm mirror studied in this chapter, we calculate an RMS difference of 7.4 nanometers, or $\lambda/74$ RMS. The PTV difference is $\lambda/25.5$ and the Strehl ratio over 0.99.

[1] "Étude des Effets Combinés de la Diffraction et des Aberrations Géométriques sur L'image d'Un Point Lumineux" André Maréchal. *Revue d'Optique*, Vol 26 (9) pp. 257–277. 1947.

Note: The best point placement of the constant k *may be varied somewhat. Modern computers make it practical for us to use trial-and-error methods to determine the value of* k *that gives the smallest PTV distance without going through preliminary calculations of the transverse aberration. In this example, the best result for PTV is obtained with a value of* k = *0.0169.*

The "curve" of the wavefront is rather jagged because of the small number of measurements. To improve this representation, we can smooth out the curve. If we desire to know the error on the glass, we can determine the profile of the mirror using the following equation:

$$y_{\text{glass}} = \frac{1}{2} \times y_{\text{wave}}$$

since the defects of the glass are half as large as the defects in the wavefront (see Figure 5.25).

5.4.9 When is the Mirror Finished?

Before you can consider your mirror finished, you must first confirm the precision of your measurements by taking a new set of measurements over different diameters of the mirror (for example, by giving the mirror a quarter turn in its support between the first and second set of measurements). You can then graph the profile of the wave surface for each set and find evidence of zonal astigmatism (For more information on astigmatism, see Chapter 23 beginning on 231). To detect the presence of a minor circular defect, use two Couder screens with overlapping windows to measure the same mirror; obviously, the measurements from the two screens should match.

Also, the mirror should simultaneously satisfy two important criteria: the Rayleigh limit and the Couder criterion.

- **The Rayleigh Limit** requires that the surface of the wave coming from the mirror should be comprised entirely within two spheres whose radii differ, at most, $\lambda/4$.
- **The Couder Criterion**, as we have seen, requires that the ratio TA_f/ρ be less than 1. This constant is directly related to the concept of the slope of the de-

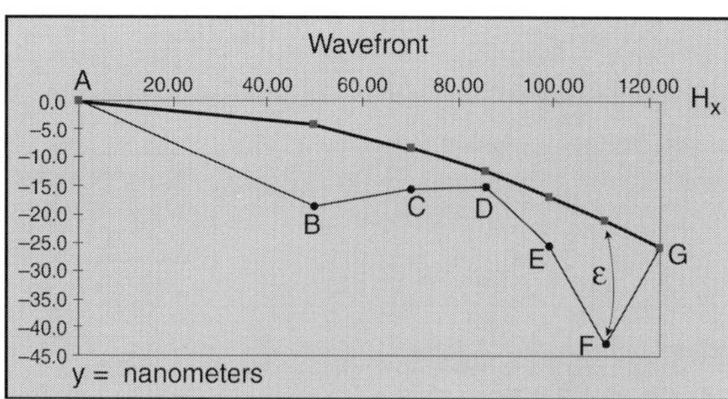

Fig. 5.24 Comparison of the profile of the wavefront as calculated to that of the best reference parabola. This parabola passes through two points of the wavefront at *A* and *G*. It is above the profile, and its slope nulls at the center of the mirror at *A*. The greatest distance ε between the two curves (at *F*) here is 21.6 nanometers, or $\lambda/25.5$ ($\lambda = 550$ nm).

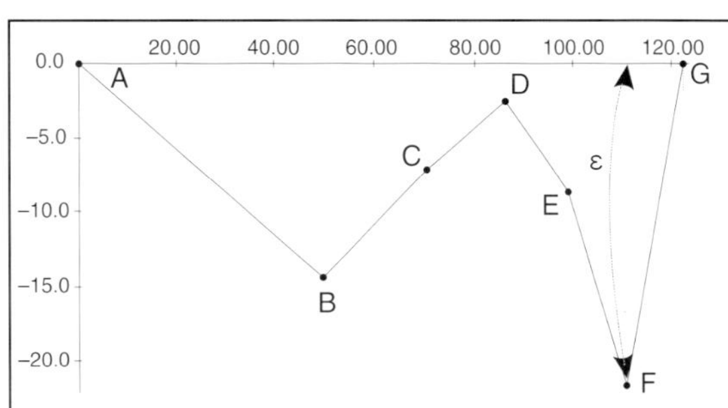

Fig. 5.25 Graph of the distance between the reference parabola and the wavefront profile. The two extreme points are now easily visible at the right and left edges of the graph. This curve can be obtained directly with the calculations using $k = 0.0169$.

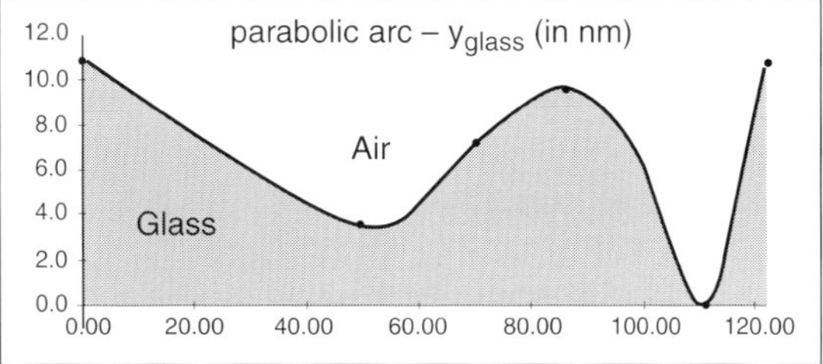

Profile of the smoothed surface of the mirror. The horizontal axis (H_x) represents the ideal reference parabola. The black curve corresponds to the real profile of the mirror compared to this reference parabola (the curve has been smoothed). The gray areas represent the material to be removed from the mirror.

fects; a defect less than $\lambda/4$ can, if it is steep, present a severe slope and scatter light outside of the diffraction spot.

In our example, the two criteria are satisfied. In effect, the largest wavefront error is λ /25.5 and the values of the transverse aberration at the circle of least aberration (adjustment constant k = 0.0015 mm) and at the point of best focus (k = 0.0169 mm) are both less than one. In our example, we recalculated the value of the terms TA_f/ρ for k = 0.0169. We find the maximum in zone 5, with the ratio TA_f/ρ equal to 0.40. The minimum is in zone 6 with the value equal to –0.68. These two extremes each have an absolute value of less than one, so the mirror satisfies the Couder criterion and can be considered finished. (In reality, the analysis of the form of a mirror is not sufficient basis for a definitive judgement of its quality. The condition of the surface is at least as important as the form (see Section 5.6 on page 75).

If the mirror did not meet one of these two criteria, we would have to continue figuring it, following the information given by the mirror's profile. The principal types of defects, and the methods for their correction, are enumerated in the polishing and retouching sections of Chapters 8 (Section 8.3 on page 116) and 14 (Section 14.3 on page 170).

5.5 Fringe Testing Flat Mirrors

We have just described how to test for defects in the form of a concave mirror with a Foucault tester. The method detailed next enables us to evaluate flat or convex surfaces. It consists of comparing the shape of the optical surface, while it is being made, with a reference optic of known quality, and to observe the form of the interference fringes that are produced between the two surfaces. See Figures 5.27 and 5.29.

5.5.1 Some Physical Optics

In physical optics, or wave optics, light is modeled as a sinusoidal vibration (analogous to sound waves) whose period is called its wavelength. White light is a mixture of all of the colors of the visible spectrum, spanning wavelengths from 0.4 to 0.7 μm.

Let's consider a monochromatic light source (which emits only a single wavelength, like a laser) shining a beam in a given direction. In the path of this beam, we place two glass plates with perfectly flat surfaces separated by a thin airspace. When a light vibration arrives perpendicular to the first face of the glass, part of the light is reflected while the rest passes through and creates a second vibration, which is partially reflected by the second face. Before returning to its point of origin, the second light vibration has then traveled twice the thickness of the lens (See Figure 5.26).

The second vibration has traveled a longer distance than the first; this difference is denoted δ. We have the following equation:

$$\delta = (2 \times n \times e) + \frac{\lambda}{2}$$

where e is the thickness of the lens and n its index of refraction ($n = 1.5$ for ordinary glass), and λ is the wavelength of the light beam. The term λ/2 is the amount the vibration is moved out of phase when it is reflected by a slightly inclined surface. If the light beams are not perpendicular to the entrance face of the lens, the angle of incidence is labeled i in the calculation, and the equation is as follows:

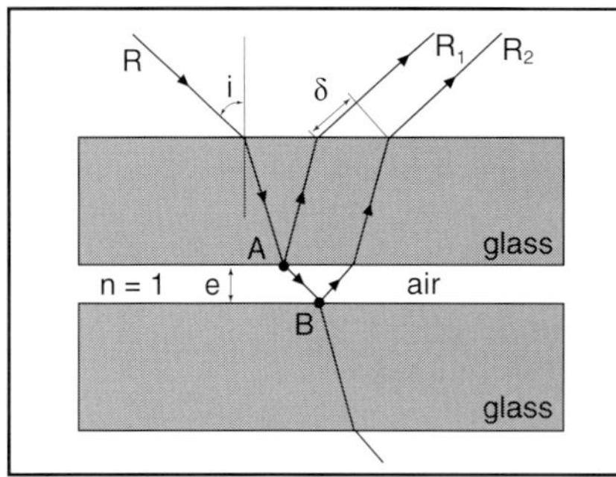

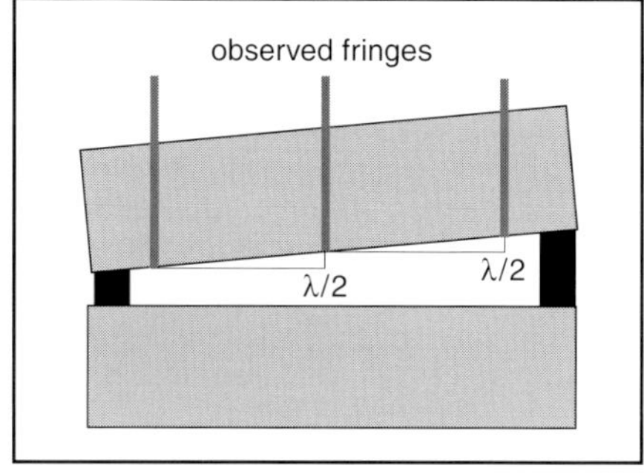

Fig. 5.26 Left: Interference in a layer of air between two flat disks of glass. The portion of the light beam *R*, of incidence angle *i*, which is not transmitted is reflected first at *A* and a second time at *B*. The emerging light rays R_1 and R_2 interfere if their path difference δ is an odd number of half wavelengths. **Right:** Formation of interference fringes in an air wedge. An interference fringe occurs at every place where the thickness of the air wedge increases by λ/2.

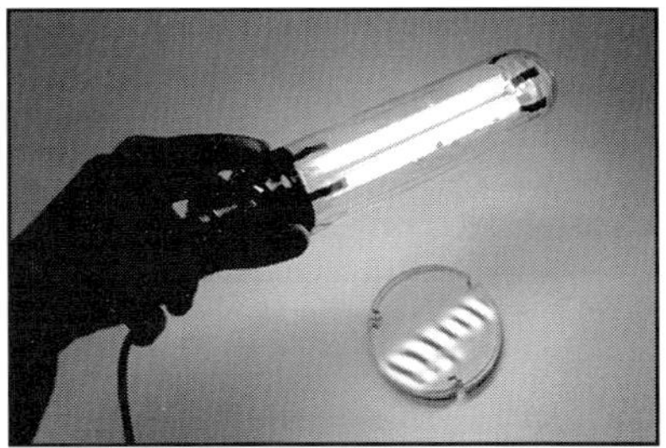

Fig. 5.27 Interference fringes (at high contrast) are visible in the reflection of the sodium-vapor lamp ($\lambda = 0.589\ \mu m$). To obtain uniform illumination, place the lamp behind a diffusing screen or shine it on the support of the test mirror with the lamp.

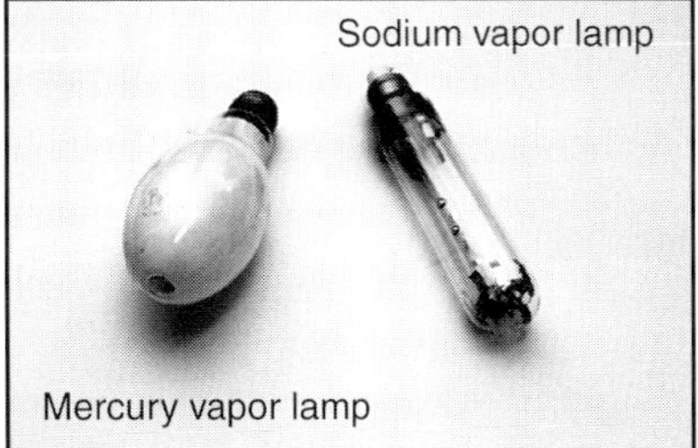

Fig. 5.28 To obtain interference fringes, use a *spectral lamp*. They produce a light composed of only a few wavelengths. You can find these bulbs everywhere: fluorescent lights (mistakenly called neon), energy-saving lights, yellow sodium-vapor lights ($\lambda = 0.546\ \mu m$), and white-green mercury-vapor lights (isolated, the green light is $\lambda = 0.56\ \mu m$).

$$\delta = (2 \times n \times e \times \cos i) + \frac{\lambda}{2}.$$

If the two waves are of the same amplitude, of the same wavelength, and if their path difference δ is equal to an odd number of half wavelengths, the two rays cancel each other out. We say that there is extinction by destructive interference. If the path difference is an even number of wavelengths, the vibrations are added to one another; this is called constructive interference.

In practice, the test consists of placing two glass disks (both considered to be perfectly flat) one on top of the other. The interference is produced within the air space (of refractive index $n = 1$) between the faces of the two disks. For a given incident angle i, the path difference varies as a function of the thickness e (See Figure 5.26 Left).

When the two faces of the disks present a small angle, as occurs when one is tilted into place atop another, they form what we call an air wedge (see Figure 5.26 Right). The thickness of this air wedge increases from one side to the other, and this causes the formation of interference fringes. These fringes look like bright parallel stripes separated by dark lines (called extinction zones), and the distance between two fringes corresponds to an increase in the thickness of the air wedge of $\lambda/2$.

5.6 Testing the Surface Quality of a Polished Mirror

The final quality of an optical piece is defined by two factors: its defects in form (determined in the Foucault test and with interference fringes) and its surface de-

How to Interpret Interference Fringes

The appearance of interference fringes permits us to test the form of flat mirrors (in comparison to a reference flat) or convex mirrors (in comparison to a concave reference of the same curvature). If the test mirror is "perfect," the fringes are straight, parallel, and evenly spaced. Defects in form appear as irregularities in the fringes. The distance between the fringes is known ($\lambda/2$ on the glass), so the amplitude of a defect can be found by examining the inter-fringe spaces. The fringes show a greatly magnified profile of the mirror, and can be likened to contour lines. Occasionally the test piece is curved when compared with the reference; in this case, curved or circular fringes are observed. To determine the type of the defect (concave or convex), press lightly on the mirror with a finger; if the diameter of the fringes diminishes, the test mirror is concave; if it increases, the mirror is convex. Knowledge of the form of the mirror allows the determination of the figuring methods to employ.

Important: 1 fringe = one-half wave on the glass = 1 wave on the wavefront.

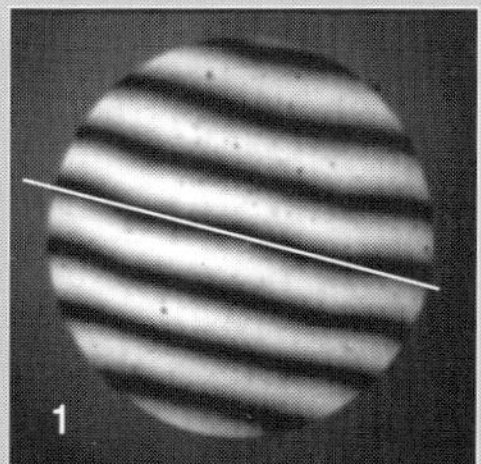

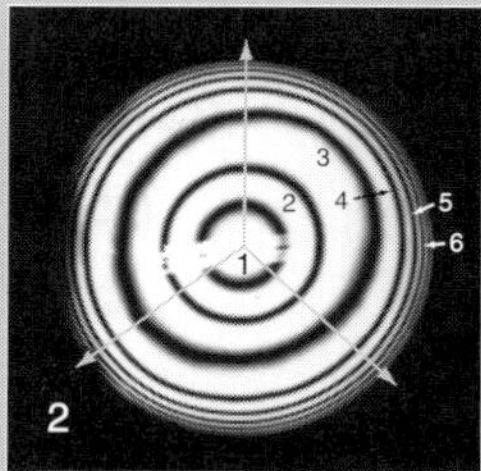

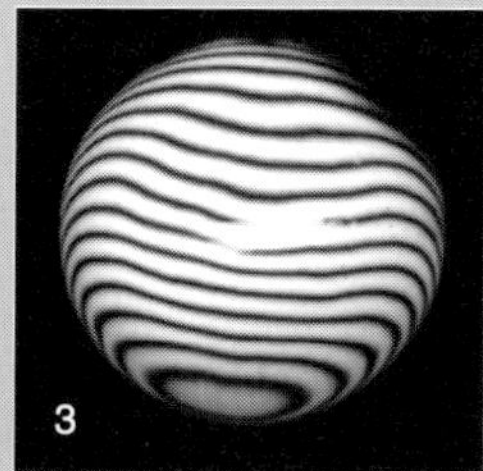

1. The largest defect corresponds to $1/8$ of the distance between two fringes. We say that the mirror is precise to $1/8$ fringe, $\lambda/8$ on the wave, or $\lambda/16$ on the glass.
2. If you press on the center of the mirror, the fringes increase in size. Here, there are 6 circular fringes; the convexity is 6 fringes, so the curvature measures $6 \times \lambda/2 \cong 1.8$ µm ($\lambda = 0.589$ µm) in height.
3. The mirror is concave by 1 fringe and has a turned-down edge of $1\frac{1}{2}$ fringes.

Note on Reference Flats

Telescope making workshops sometimes posses reference flats, either purchased or made. Sufficiently precise, flat, non-aluminized mirrors can be found at reasonable prices. They are available from companies that sell optical components through catalogs (see the list in **Appendix B**); select a disk of a diameter at least as large as that of the test mirror. A precision of at least one-eighth to one-tenth fringe is necessary (get a test certificate for the piece that guarantees its quality).

Fig. 5.29 Photograph of fringes obtained with a reference flat. Defects are smaller than $\lambda/20$.

Some Common Defects on Polished Surfaces

- **Grinding and Polishing Defects**

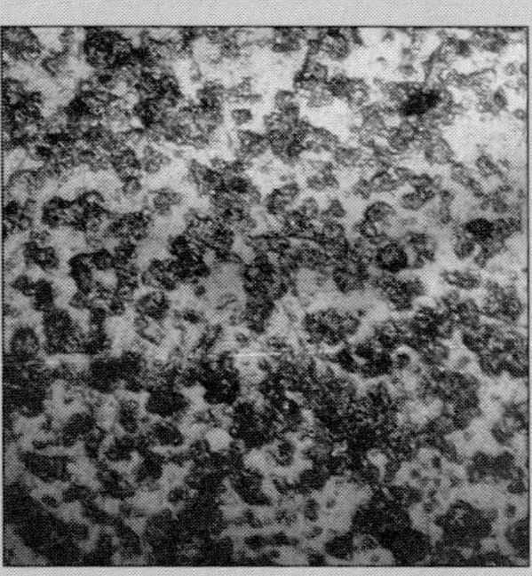

Gray Areas

In the Foucault test or with a powerful light source, a gray area appears as a milky veil of small pits that does not go away when the mirror is cleaned. It betrays an incomplete polishing of the optical surface. Gray areas are a significant cause of contrast loss in an instrument.

Grooves and Fractures

These are straight or curved scratches in the surface of the glass. With magnification, the sides of some scratches are outlined with miniscule chips or bright spots. Some are straight and are not outlined by chips or bright spots. These defects are caused by too large a grain of abrasive between the mirror and the tool over the course of polishing. Grooves can appear during smoothing as well as during polishing. If they are few in number, these defects have little effect on the quality of a telescope's images.

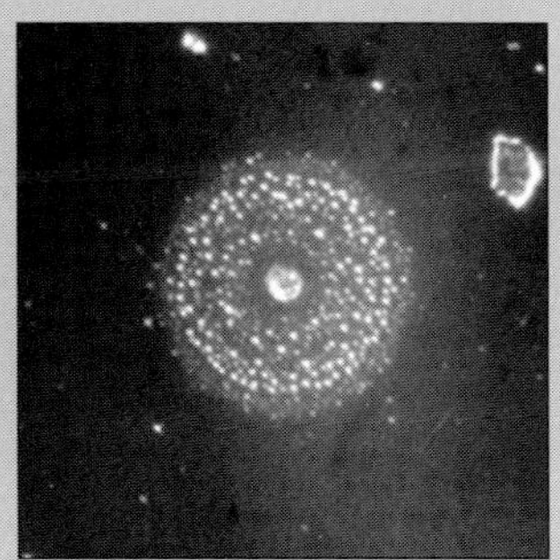

Oxidized Spots

Oxidized spots appear in the surface haze that forms when you breathe on the mirror. These spots do not disappear when the mirror is cleaned with a sponge or a cloth. They are caused by deposits on the surface of the mirror (aluminized or not) by pollution, humidity, or salinity in the air. They occur more frequently if the mirror is left out in the air for a long time without protection. The best way to eliminate them is to rub lightly on the (uncoated) surface of the mirror with a soft 100% cotton cloth soaked in diluted acetic acid.

Sleeks

These are small lines, on the order of one or two microns width and a few millimeters length. They can be observed under a bright light or in the Foucault test in the form of tiny points. Small sleeks are not a problem in observation, as long as there aren't thousands of them on the mirror. These defects appear as a result of polishing that is too rough, too dry, or a wet that is begun with a dirty tool. Certain polishing agents (primarily rouge and cerium oxide) are likely to cause scratches if the lap is allowed to run dry.

- **Material Defects**

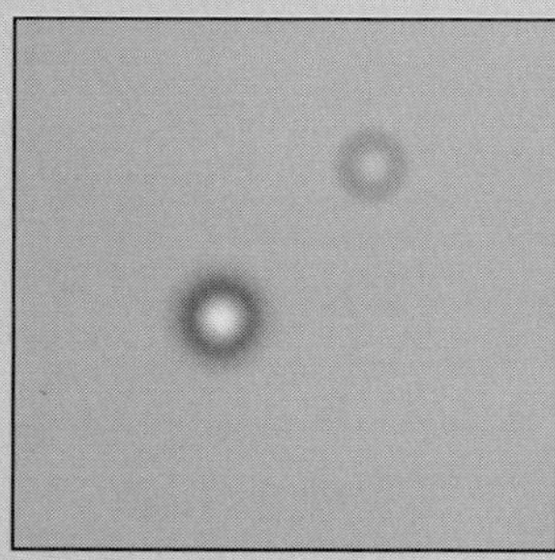

Air Bubbles in the Surface of the Mirror

Air bubbles appear as small opaque spheres imprisoned in the glass. Generally, their diameter is no greater than 2 millimeters. Occasionally, over the course of rough grinding and smoothing, a bubble situated just below the surface opens up to the air and a hole appears in the glass surface. These holes are not aesthetically pleasing, but they have little effect on image quality. In the Foucault test, they appear as very bright points.

Striae in the Glass

The striae that traverse Pyrex disks are caused by local variations in the refractive index of the glass. They are irrelevant in the making of a mirror, but they pose problems in the construction of a lens. You can see them with the naked eye, through the transparency of the glass, if they are pronounced enough. If the striae reach the surface of the mirror, they can be distinguished in the Foucault test (see Figure 25.4 on page 249).

- **Defects Invisible to the Naked Eye**

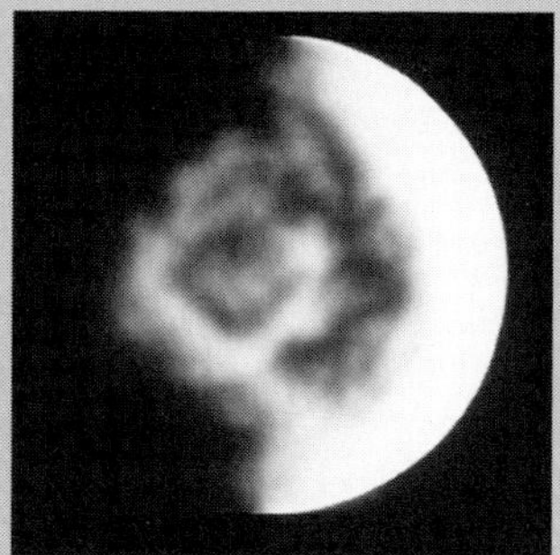

Ripple

In the Foucault test, a rippled surface resembles a wavy sheet. Rippling degrades the diffraction image given by a good mirror (the diffraction rings are absent) and causes a loss of contrast in images observed at high magnification. It is caused during polishing by working too violently or too quickly with an abrasive that is too harsh. An irregular tool or pitch that is too hot or too cold also tend to cause rippling.

Microripple

Much more fine than ripple, microripple often goes unnoticed and requires a very bright light (a halogen lamp). Microripple appears as small ripples on the surface, on the order of one or two millimeters length and a few angstroms in height. In a telescope, this defect causes a slight diffusion of light in the images. Microripple is always present in polished surfaces; it is just more or less apparent. It is caused by the nature of the polishing agent, pitch, and even the structure of the glass. It can best be detected by phase-contrast methods.

fects (bumps, valleys, scratches, and various other traces visible on the polished surface). Surface defects play a smaller role in image formation than do defects of form; still, the surface quality must not be ignored. A well-polished mirror will always give better contrast and more colorful images of celestial objects than will a mirror with multiple scratches, ripples, or gray areas. (A few scratches will have no effect on a well-polished surface; *thousands* of scratches become a problem.)

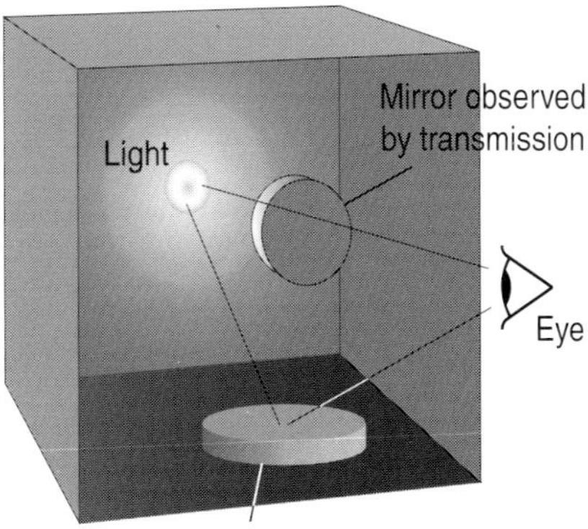

Light
Mirror observed by transmission
Eye
Mirror observed by reflection

Fig. 5.30 Black box test for evaluating local defects on optical pieces. You can evaluate the pieces by reflection or by transmission.

There are several methods for testing the surface quality of a mirror. In this section, we will describe two simple and complementary tests.

The first is the black box test which uses the very simplest of equipment; a filament lamp with a clear (unfrosted) bulb (a 60-Watt bulb does the job nicely) housed in a black box. To check for defects look at the surface near a reflection of the filament. To avoid being blinded and to see the defects clearly, shield your eyes from the direct rays coming from the bulb (hide the reflection with your hand). The test piece must be rotated in order to observe the finest details. This procedure, employed throughout the optical industry, permits the observer to distinguish scratches greater than one micron in size and pits (or digs) on the order of 1/10 of a millimeter diameter. The observer can also detect grayness (an insufficiently polished area) or oxidized hazes (blue or green reflections that do not go away when wiped with a tissue). To test transmission optics (lenses and telescope windows), place them on a block in front of the lamp.

The second method uses the Foucault tester. The mirror is placed in the same position as for the test of a spherical mirror, but with the light slit opened all the way. The Foucault knife-edge is then moved transversely until the null completely disappears. You can then see all of the pits and large scratches on the polished surface. This method, called the strioscopic method, is very sensitive for the detection

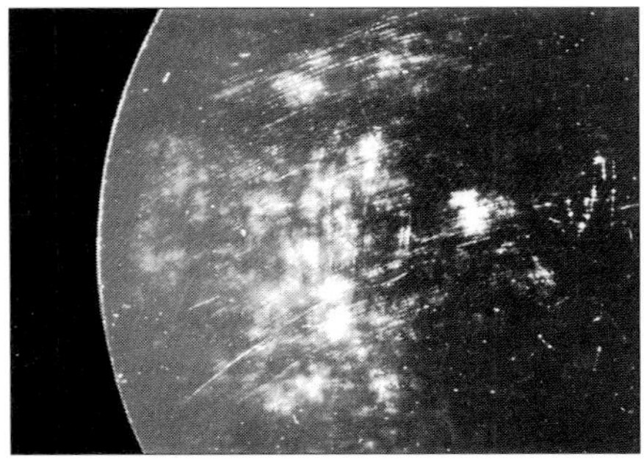

Fig. 5.31 Polishing defects observed in the Foucault test on a 200-mm mirror (detail of the left half of the optic) polished with dried out cerium oxide. The defects measure from 2 to 5 nanometers on the glass. Two-minute exposure on TP 2415 film with image processing to enhance detail.

of gray areas and microripple.

The observable defects on polished glass are classified according to certain standards. The best known are the American scratch/dig standard MIL-0-13830A, the French NFS 10-006 and the German DIN 3140.

In a telescope, the acceptable tolerances of the optical components depend primarily on their placement in the instrument. The surfaces that must have the best cosmetic quality are those that are closest to the focus—eyepieces and filters. Primary and secondary mirrors, on the other hand, are much more forgiving; a large scratch on the primary will have practically no effect on the image at the eyepiece.

5.6.1 How Do Defects Affect Image Quality?

With respect to the telescopes described in this book, the presence of a few scratches and spots on the mirror will not have an effect on the image. However, images begin to noticeably deteriorate if the mirror is very dirty or heavily scratched. You would see their effects as an increase in the diffraction of the light around a bright star as well as a loss of contrast in planetary images.

Grayness is much more serious than scratches. It is caused by thousands of tiny pits, which are signs of incomplete polishing. These little surface blemishes diffuse a significant amount of light outside of the diffraction spot with a large consequential loss of contrast and a dilution of detail.

5.6.2 Photographing Surface Defects with the Foucault Tester

Photography is an excellent way to observe the surface defects of a mirror. When the picture is developed, you can increase the contrast to highlight details that are invisible to the naked eye.

Place a camera with a telephoto lens (135 mm) at the position of the eye behind the Foucault knife-edge. The exact placement is determined by the contour of the mirror. To observe microripple, move the knife-edge longitudinally to in-

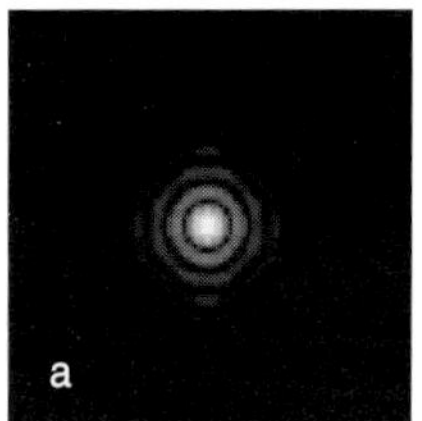

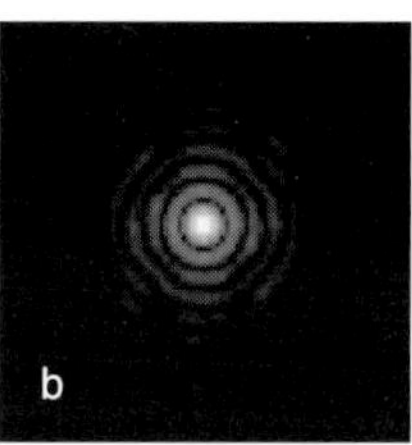

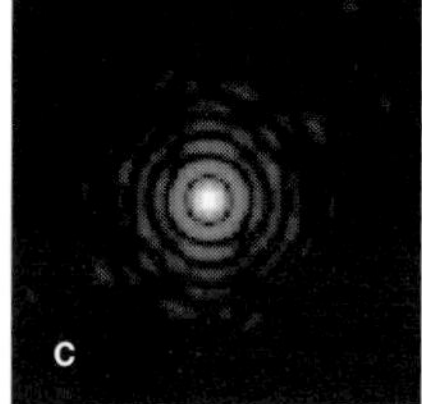

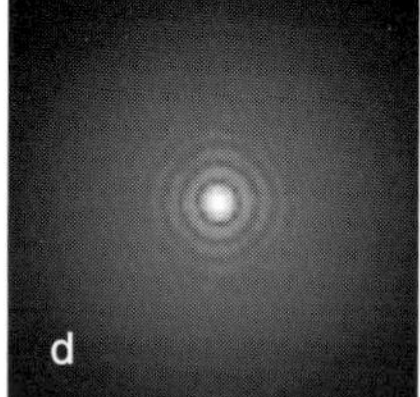

Fig. 5.32 Effects of defects on images. a. Diffraction spot at the focus of a perfect telescope. **b.** The mirror has several large scratches; their total length is 1.5 times the diameter of the mirror. **c.** The total length of the scratches is 4 times the diameter of the mirror. The core of the image is intact but the exterior rings of the Airy disk are discontinuous. **d.** The mirror is gray and excessive light is diffused around the image. The instrument is unusable. Contrast has been adjusted to enhance dim details

The Phase-Contrast Method

The phase-contrast procedure, developed by F. Zernike and B. Lyot, shows the roughness of an optical piece and detects defects of a few angstroms in height. The Foucault tester is used, but the knife-edge is replaced by a phase plate, a transparent plate of glass on which a small square of evaporated aluminum film, of specific dimensions and density (e.g., 0.4 mm by 30 mm with a density of 3), has been deposited in a vacuum chamber. This plate separates the light diffracted by the small surface defects (on the order of a few nanometers) from the rest of the light. As you "null" the mirror, you will see previously invisible structures appear on the surface of the mirror. They correspond to the traces left by the polisher during the last hours of work. The contrast of the observed defects depends on the density of the phase plate.[1]

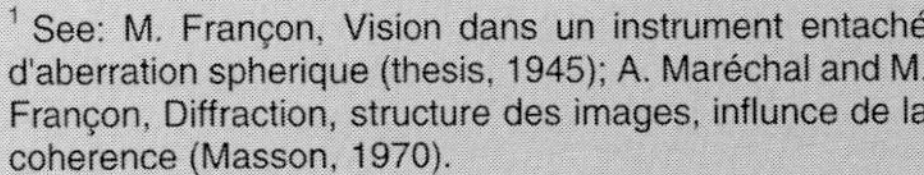

[1] See: M. Françon, Vision dans un instrument entaché d'aberration spherique (thesis, 1945); A. Maréchal and M. Françon, Diffraction, structure des images, influnce de la coherence (Masson, 1970).

Fig. 5.33 Surface detail of a 600-mm mirror presenting a severe microripple, on the order of 30 angstroms. Phase plate density 2.2. Photo © S. Debruyne.

tercept the light from the peripheral zones of the mirror. Then move the knife laterally *almost* to a complete extinction of the mirror. If the room is dark and the source is bright, you can then see the rolling figures (visible in Figure 5.33), which are more or less accentuated by the movement of air in the test area. At the moment that the picture is taken, try to eliminate any extraneous light escaping the Foucault tester. The exposure can last from several seconds to several minutes, and should be taken on a high-contrast black-and-white film such as Kodak TP 2415 or Ektagraphic HC.

Even though this procedure is done with a knife-edge tester, it is not the classic Foucault test, resembling more an amateur's version of the phase-contrast method. The scattered light from tiny defects is viewed without the contrast enhancement effect of interference with a reduced primary beam.

Part 2: A 130-mm *f*/9.2 Newtonian Telescope

Plans for a 130-mm *f*/9.2 spherical mirror telescope on an alt-azimuth mount.

This Newtonian telescope is an ideal beginner's telescope: it is light, inexpensive, and easy to make, yet unlike most commercial instruments, it performs near the theoretical resolution limit for its aperture.

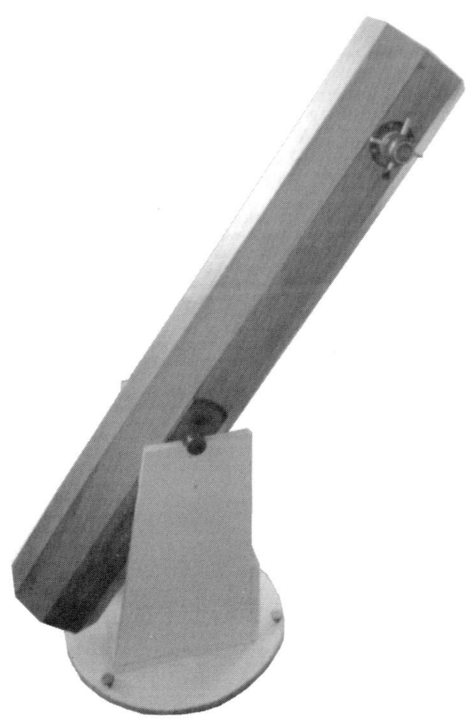

6 Optical Specifications for the 130 mm *f*/9.2 Newtonian

6.1 Optical Characteristics and Theoretical Performance

6.1.1 Optical Characteristics

The primary mirror has a diameter of 130 mm and a focal length of 1,196 mm. Its focal ratio is 9.2. The polished face of the mirror is concave and spherical; its radius of curvature is 2400 mm and its sagitta is 0.88 mm. The secondary mirror has a flat face and an octagonal shape; its minor axis is 25 mm and the major axis is 36 mm. These dimensions provide a fully illuminated field of 0.5°—the diameter of the full Moon (10.5 mm at the focus). The secondary mirror is placed 150 mm in front of the primary mirror's focal point, from which it reflects light at right angles to a point located 85 mm outside of the optical path. The secondary mirror obstructs the primary mirror by 20%, which is within acceptable limits.

6.1.2 Theoretical Performance of the Telescope

The resolving power and the magnitude limit of the telescope are calculated using the formulas given in Section 3.4.1 on page 27 (resolving power) and Section 3.4.4 on page 31 (limiting magnitude).

For the wavelength of light most sensitive to the human eye ($\lambda = 0.55\ \mu m$), the linear radius of the diffraction spot is $\rho = 6\ \mu m$, which represents an angular separating power of $^{140}/_{130} \approx 1$ arc-second. This instrument is thus capable of separating two stars 1 arc-second apart.

The mirror gathers 345 times as much light as the naked eye (for a 7-mm dark-adapted pupil). Under ideal conditions, the magnitude limit of the telescope is about 12.8.

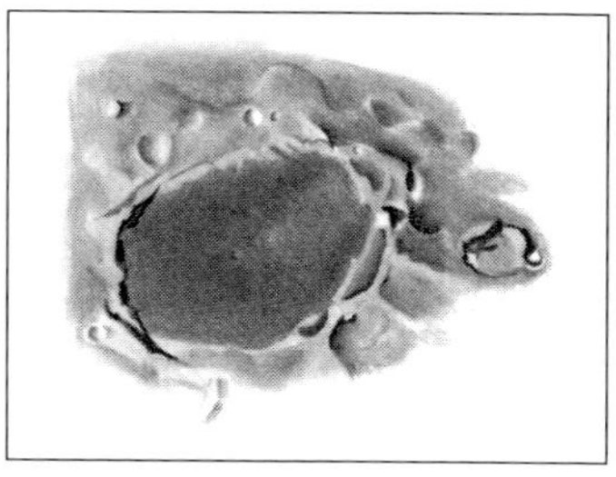

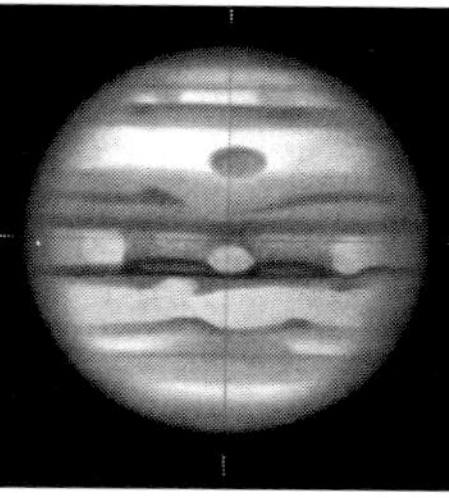

Fig. 6.1 Drawings made at the eyepiece of a 130-mm telescope. **Left,** the lunar crater Plato. **Right,** the planet Jupiter reveals its cloud bands and its famous red spot. Drawings © Simon Fabre.

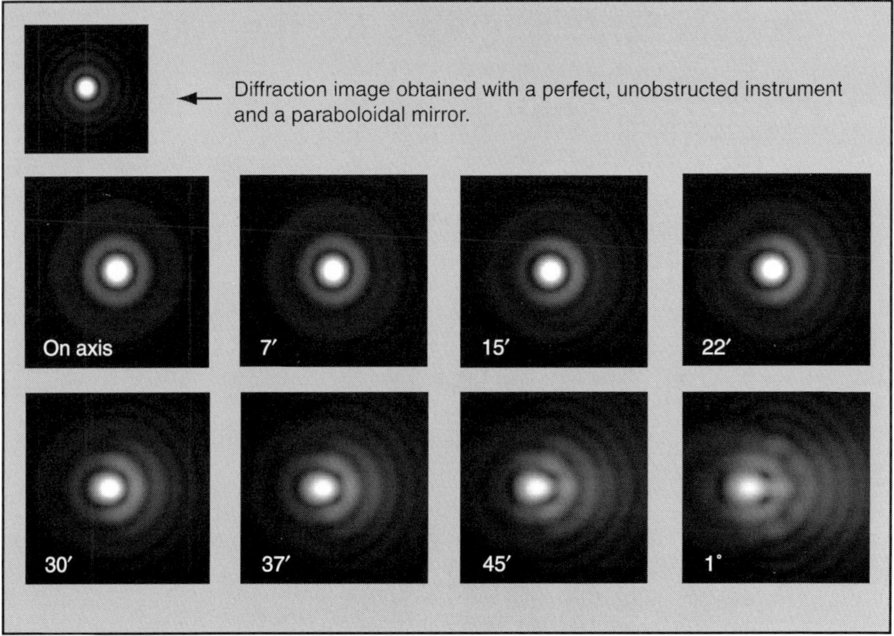

Fig. 6.2 Images observed at the focus of the 130-mm telescope, at increasing distances from the optical axis, taking into account the central obstruction of the mirror (0.2 × D) but not that of the spider arms. Because of its large focal ratio, the mirror is not very sensitive to misalignment. It is instructive to compare these images with those for the 250-mm telescope in Figure 12.2 on page 150 and the 300-mm telescope in Figure 19.3 on page 208. Images courtesy H. R. Suiter.

6.1.3 Magnifications to Use with this Instrument

Equipupilar magnification is on the order of 20x. You would need to have an eyepiece with a 60-mm focal length (a rare find among amateurs) to get to that value. Reasonably, you could observe with a 40-mm eyepiece, giving you a 4-mm pupil and 30x magnification. *Resolving magnification,* obtained with a 14-mm eyepiece, gets up to 85x (⅔ × diameter). Maximum magnification is limited to about 300x (4-mm eyepiece).

6.2 Tolerances for Mirror Making

6.2.1 Primary Mirror

We have seen (in Section 5.4.3 on page 54 through Section 5.4.5 on page 57) that beyond a certain focal length a telescope mirror must be paraboloidal in order to concentrate the reflection of an infinite light source into a point. The 130-mm mirror with a 2,400-mm radius of curvature does not have to meet this condition. Because of its long focal length, the total difference between this sphere and the corresponding parabola is small. The peak-to-valley (PTV) wavefront error refo-

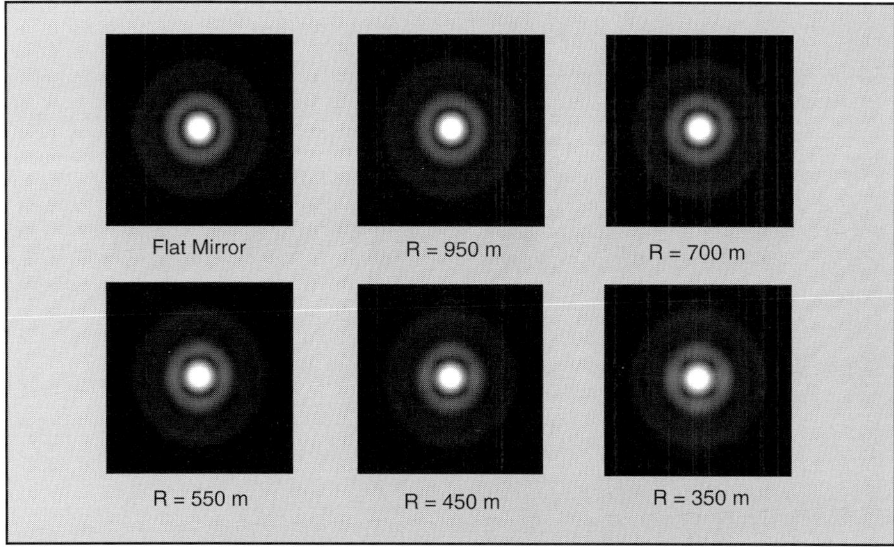

Fig. 6.3 What is the Effect of a Slight Curvature in the Secondary Mirror? At the upper left is the image as it appears with a perfect flat mirror. As the radius decreases from 950 to 350 m in the lower right, it displays the successively stronger characteristic cross pattern of mild astigmatism.

cuses somewhat better than the Rayleigh limit (~λ/6.8), and the transverse aberration at the focus of the mirror just equals the radius of the diffraction spot.

A 130-mm spherical mirror with a 2,400-mm radius of curvature is thus capable of delivering nearly theoretically perfect images when compared to a paraboloidal mirror of the same focal length. Figure 6.2 shows that this instrument is also not very sensitive to coma so that images will for the most part be sharp across a 1° field.

6.2.2 Secondary Mirror

There is no such thing as a perfectly flat mirror. At the end of polishing, there are always residual curvature or local defects similar to those observed in the primary mirror (rings, scratches, pits, ripples). A flat mirror is considered adequate if its defects do not reduce the quality of the images formed by the primary mirror (see Figure 6.3).

For the 130-mm spherical mirror telescope, the radius of curvature of the 25-mm minor axis secondary mirror should be greater than 950 m to keep the induced astigmatism negligible compared to the spherical aberration already contained in the optics. This curvature corresponds to a sagitta of 0.09 μm, or 0.156 λ (≈ ⅓ of a fringe) on the minor axis of the mirror. On the 36-mm major axis, the defect is 0.6 fringe or λ/3 on the glass.

In practice, not all of the diagonal is illuminated from a single object spot (see Figure 6.4 on page 88), so a minimal radius of curvature of 550 m is tolerable for the flat mirror. The resulting sagitta is 0.14 μm on the minor axis of the sec-

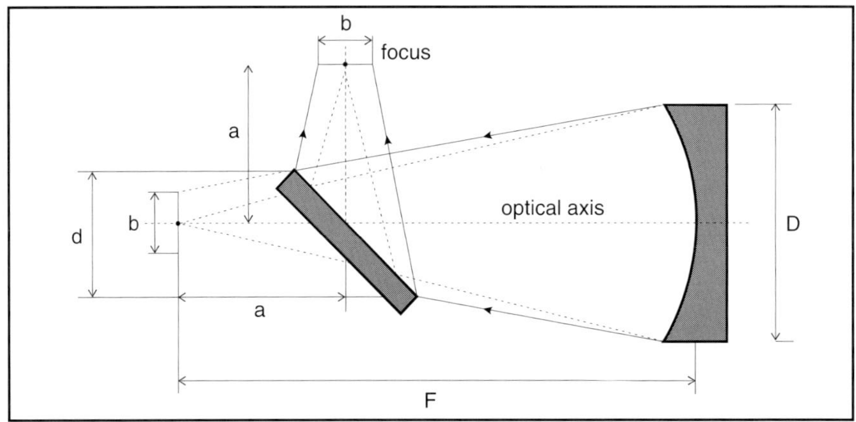

Fig. 6.4 Calculating secondary mirror dimensions.

ondary or on λ/4 the glass (λ/2 along the major axis). Of course, these numbers change if the deformation is non-spherical.[1]

6.3 The Dimensions of the Secondary Mirror

The flat secondary mirror of a Newtonian telescope is inclined at a 45° angle to the optical axis of the primary mirror. It can be rectangular, elliptical or octagonal, and its major axis (length) is the square root of 2 (or 1.414) times its minor axis (width). To define its dimensions as shown in Figure 6.4, calculate the value of its minor axis with the following very close approximation:

$$d = \left[a \times \frac{(D-b)}{F}\right] + b$$

where a is the distance between the secondary and the focus, D is the diameter of the primary mirror, F its focal length, and b the linear diameter of the field to be covered.

Inversely, the dimension b of the fully illuminated field as a function of these other parameters is calculated with the formula:

$$b = \frac{d - \dfrac{a \times D}{F}}{1 - \left(\dfrac{a}{F}\right)}.$$

[1] This is much larger than the error on the wavefront. Error on the glass does not take into account re-focusing.

7 Making the Primary Mirror

The steps for making a spherical telescope mirror can be broken down as follows:

1. **Rough grinding** creates the basic curve on the mirror's surface using a glass tool and coarse abrasives. This process creates a concave surface on the mirror, the shape necessary to focus incoming light.
2. **Fine grinding** eliminates the large pits and irregularities caused by rough grinding using intermediate sized abrasives.
3. **Smoothing** begins when the mirror surface is a regular sphere. This operation continues to reduce pits and prepares it to be polished with increasingly fine abrasives.
4. **Polishing** makes the mirror reflective through the use of a pitch-polishing lap and special polishing compounds. At this stage, you are correcting the remaining small-scale mirror defects to arrive at a spherical surface.

7.1 Using Abrasives

7.1.1 From Rough Grinding to Smoothing ...

You will grind the mirror against a tool, using loose abrasives mixed with water (see Table 7.1 on 90 for a listing of the abrasives to use). Using your hand or a brush, begin by spreading a little moistened abrasive on the lower disk (tool). The right mixture of water and abrasive is found through trial and error; too much water will push the abrasives to the sides of the mirror and halt grinding; too much abrasive makes a thick paste that will not spread out evenly between the disks.

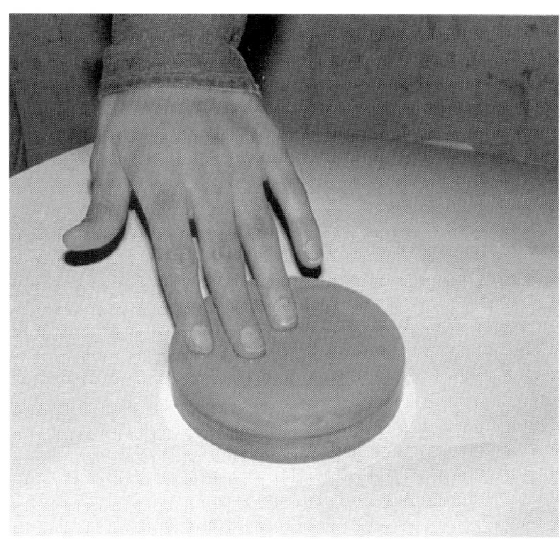

Fig. 7.1 At the beginning of a wet, use your hand to spread the abrasive over the entire surface of the mirror.

Next, gently lay the second disk (which will become the mirror) on top of the abrasive coated tool. Grinding can now begin by using various back and forth strokes which will be discussed in detail in the next section. The

Table 7.1
Estimation of work times and abrasive quantities for making the primary mirror

Operation	Amount of Abrasive	Type of Abrasive	Other Materials	Approximate Time
Rough grinding	400 g	#80 Carbo		1 hour
Fine grinding	200 g 150 g	#120 & 220 Carbo		2 hours
Smoothing	100 g 100 g	#320 & 600		10 hours
Polishing	100 g 50 g	Cerium Oxide, Zirconium Oxide or Rouge	Pitch, 500 g	10 hours

sounds you hear during grinding will tell you how the work is progressing: they will be loud at the beginning (especially with #80 carbo), and then as the abrasives break down, the noise will diminish. Once the grinding action appreciably slows, separate the disks by carefully sliding the mirror (top disk) off the tool (lower disk), then clean the spent abrasives off the mirror and tool with a sponge or by alternately dipping the mirror and tool—one-at-a time—into a bucket of clean water. You have just completed a process that opticians call a *wet.*

The duration of a wet primarily depends on the abrasive used. During rough grinding, #80 carbo gets used up very quickly, and a wet lasts only a few minutes. During final smoothing, wets can last up to twenty minutes if water is judiciously added to prevent the binding or seizing together of the disks as the abrasives dry out.

7.1.2 To Polishing

In polishing, you begin with a pitch lap covered with polishing compound. As the polishing progresses, the mirror is occasionally offset while still on the lap, and polishing compound added to the exposed area of the lap using a brush. A polishing wet can thus be quite long and is usually limited only by the operator's endurance.

7.2 Rough Grinding

7.2.1 Strokes for Rough Grinding

There are four different strokes you can use for roughing in the basic curve for a 130-mm mirror with #80 carbo. These four strokes compliment each other and let you quickly attain the correct concave shape of the primary mirror. In each case, the mirror is on the top and the tool is on the bottom.

No matter what stroke you use during grinding, you must rotate the mirror under your hands about a ¼ turn after every three or four strokes, while you slowly

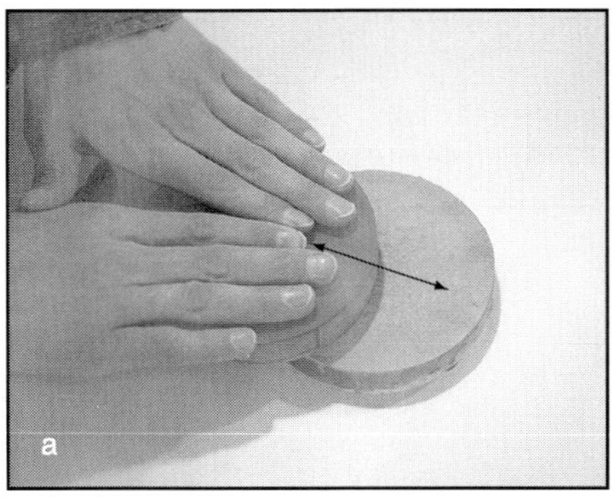

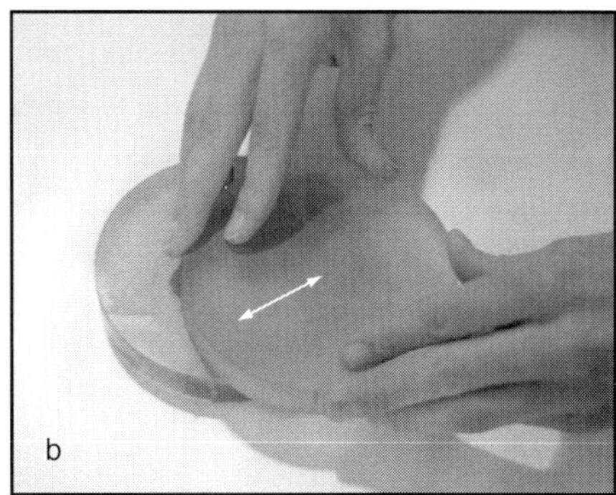

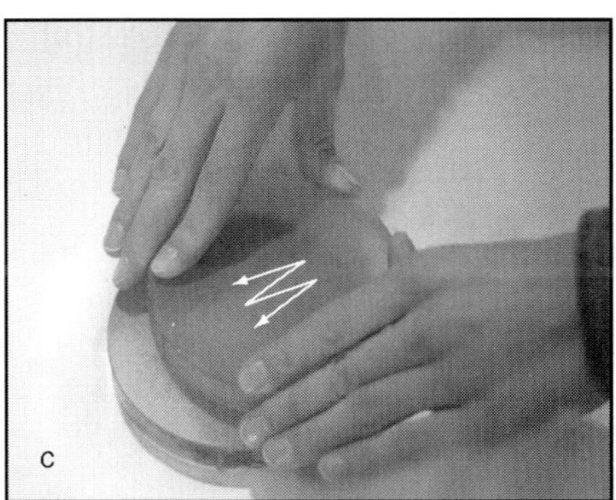

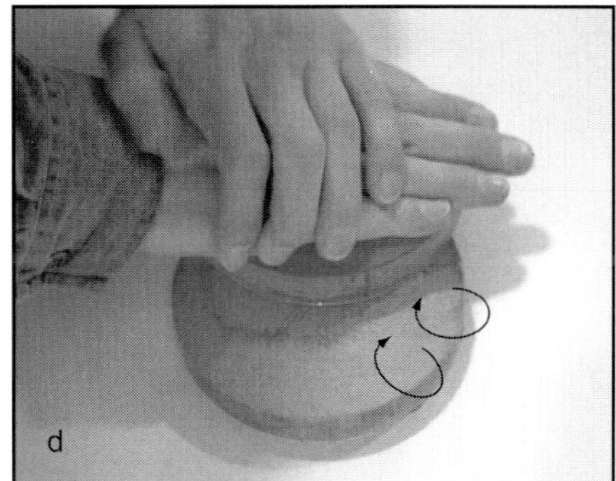

Fig. 7.2 The different polishing strokes for rough grinding. For all of these strokes, the mirror is shown here on top, and the tool is on the bottom.

walk around the work post in the opposite direction. This keeps you from creating periodic defects on the surface of the mirror and is the "secret" to creating a sphere. You should also occasionally turn the tool ¼ turn in its support (keep the directions of these movements constant).

Figure 7.2 illustrates these four strokes. In each case, work using both hands as shown in the pictures.

a. **The Center-over-Center Stroke** describes straight lines of ¾ the mirror's diameter, passing the center of the mirror over the center of the tool. Firm pressure is exerted on the mirror throughout the entire stroke. This stroke deepens the mirror's surface slowly and regularly, at the rate of about ½ mm per one hour. This is a good stroke to increase the mirror's depth (sagitta) by a fraction of a millimeter relatively quickly.

b. **The Tangent Stroke** makes a back-and-forth motion of ⅓-diameter amplitude, grinding the center of the mirror *around* the edge of the tool. This stroke deepens the center of the mirror quickly at the rate of about 1 mm per hour. Alternate the tangent stroke with the center-over-center stroke to keep the surface roughly spherical.

c. **The W Stroke** describes a W with an amplitude of ⅓ of the mirror's di-

Rough Grinding Tricks

Here's how to mix and use carborundum abrasives for faster rough grinding with less wasted abrasive. In a jar, dissolve a small amount of dishwashing detergent in water and stir or shake the mixture to make some suds. Moisten the tool and sprinkle some carbo on it. Spread some of the water and suds from the jar on top of the carbo-coated tool before laying the mirror on top of it. The suds facilitate an even distribution of the carbo between the disks and enhance the performance of the abrasive. The carbo is spread by the suds but does not leak out of the sides of the disks. After a few strokes, the gray sludge (the color is a result of the mixture of the black carbo and the white glass powder removed during the work) runs out of the sides, leaving fresh carbo between the surfaces, allowing the wet to last longer. After a wet, clean the sludge off with a sponge before recommencing the work.

ameter and ⅛- to ⅙-diameter width over the entire surface of the tool. This stroke does not work very quickly, but it creates a spherical surface and will reduce the time spent in fine grinding. Use this stroke at the end of rough grinding. Speed is about ¼ mm per hour at the center of the mirror.

d. **The Elliptical Stroke** causes the mirror to describe small ellipses while positioned over the edge of the tool. This stroke deepens the center of the mirror quickly at the rate of about 1 mm per hour.

At the beginning of each grinding wet you can also lightly run the center of the mirror, in large circles, around the edge of the tool two or three times. This motion allows for a better distribution of the abrasive between the two disks. This should be done using light pressure.

Note: Rough grinding can create zones. To show which zones are high or low, make hatch marks on both the dried mirror and tool faces with a permanent marker. Then as the next grinding session progresses, you will be able to see where the lines remain (low areas) and where they are ground off (high areas).

7.2.2 Finding the Sagitta during Rough Grinding

Measuring the sagitta from time to time throughout rough grinding keeps you appraised of your progress and lets you know when to move on to the fine grinding. The desired sagitta for the 130-mm mirror is about 0.88 mm (measuring the sagitta with feeler gauges or a spherometer is covered in Sections 5.1 starting on page 49 and 5.2 starting on page 50). Stop rough grinding and proceed to fine grinding when the sagitta is within 1⁄10 or 2⁄10 of a millimeter of 0.88 mm, say about 0.70 mm.

7.3 Fine Grinding

The large abrasive grains used in rough grinding seldom create a truly spherical surface. Fine grinding smooths out the gross defects resulting from rough grinding and gives both the tool and mirror complimentary surfaces of the same radius of curvature. In general, fine grinding is begun with the mirror on top, using the same abrasive used in rough grinding (#80 carbo). Use strokes *b* and *c* shown in Figure 7.2 on page 91. When the surfaces become as regular as you can get them, move to a finer abrasive, #120 carbo, and begin inverting the mirror and tool between each wet. However, before changing to a finer abrasive, do a thorough clean-up of the work area, tool, mirror and yourself (use plenty of water and even a vacuum). The work you will be doing at this point demands that you and the mirror be free of larger abrasives to avoid scratching the mirror. Repeat this clean up process each time you move to a finer abrasive.

7.3.1 Strokes for Fine Grinding

Fine grinding requires more finesse than rough grinding. As in rough grinding, you must make sure you turn the mirror and the tool frequently, and walk yourself around the workpost; but the form and the amplitude of your strokes are now more important to success. Here is a list of problems (and their solutions) that are often encountered during fine grinding.

- *A ledge (exterior zone of the mirror that was not fully ground by the abrasive) and a central zone (center of the disk that was ground too much) remain at the beginning of fine grinding (see Figure 4.4 on page 38).*

When the correct sagitta is reached, there may still be a flat exterior zone of 1 or 2 cm width around the outer edge of the mirror. Begin correcting this defect with the mirror on the bottom, using light pressure and making straight strokes of ⅓-diameter amplitude (see Figure 7.3a). Invert the mirror and tool every two or three wets. Re-measure the sagitta occasionally with a spherometer. After about an hour of work, begin making W-shaped or figure-eight strokes of half the amplitude of the straight stroke (or ⅓ of the mirror's radius). When the ledge and flat spot are mostly gone, the mirror should begin to slide over the tool with less resistance or grabbing. You can also check for flat spots by viewing low-angled light reflected off the mirror or tool (Figure 7.7 on page 99).

- *There's a thick bubble of air between the two disks.*

At the center of the mirror, you may see a round bubble of air. This means that the mirror and the tool do not have the same radius of curvature and that there is a "hole" at the center of the mirror or perhaps a flat spot at the center of the tool. You must invert the mirror and tool between each wet and make straight or figure-eight strokes of a total amplitude between ⅓ and ⅔ of the mirror diameter (Figure 7.3*b*). The greater the mirror's *f*-ratio (*f*/3 or *f*/4, for example), the smaller the amplitude of the strokes. Do not work too quickly because the resulting greater momentum will tend to create longer strokes, and thus rendering your retouching ineffective. You can measure your progress by changing the shape of the bubble: from a circle to a square or a pincushion. It should finally disappear after about 30

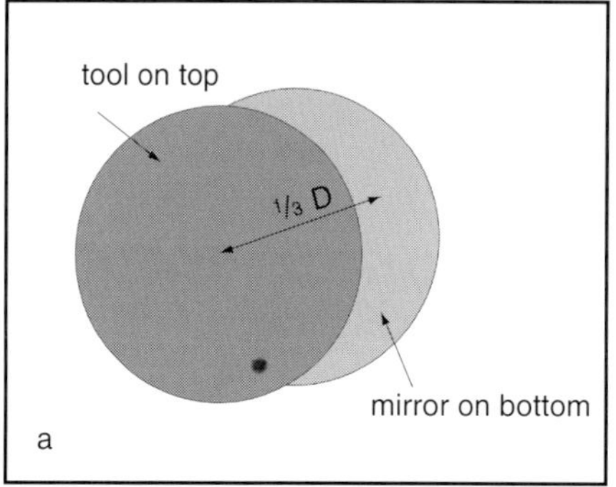

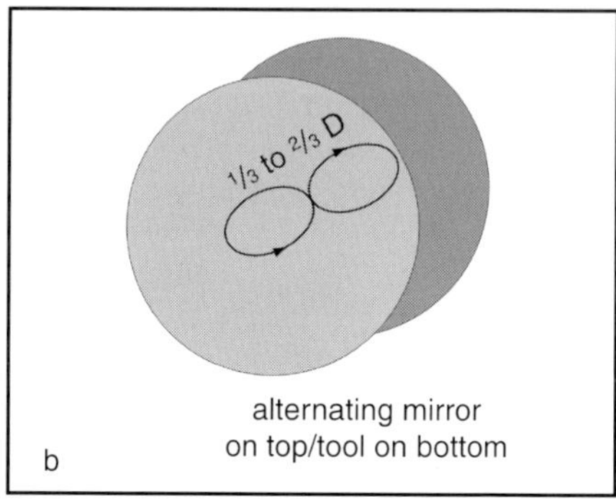

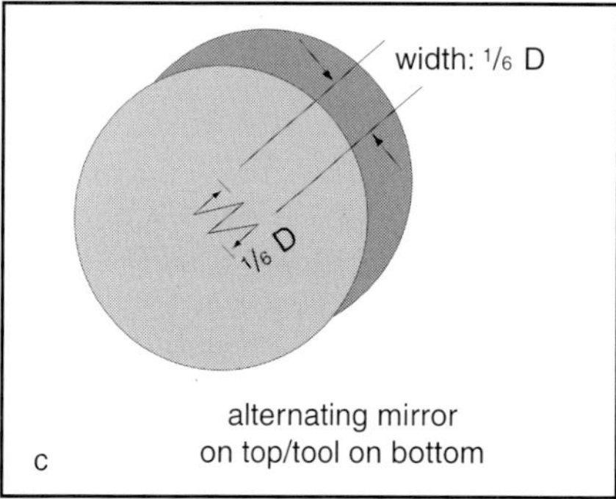

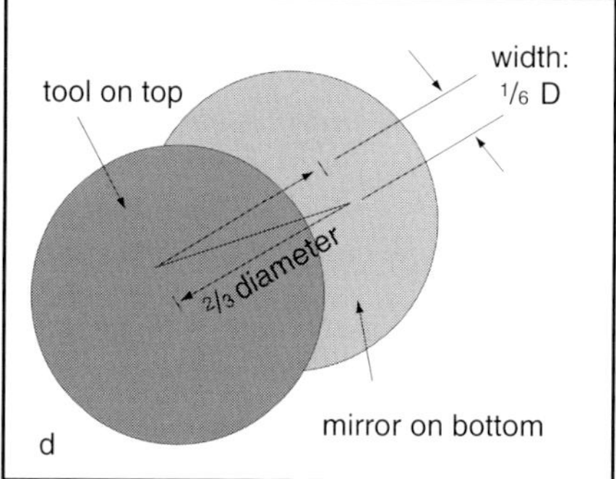

Fig. 7.3 Different strokes for fine grinding.

to 60 minutes.

- *Small air bubbles are trapped between the disks.*

Small bubbles grouped at the center of the disks indicate a minor difference in curvature between the two surfaces. The bubbles will slowly move away from the center toward the sides of the mirror with a stroke amplitude between ⅓ and ⅔ of the mirror's diameter.

- *The abrasive does not flow out to the edge of the disks.*

During a wet, you may find that the abrasive does not spread out evenly to the edge of both the tool and mirror. First, to make sure that the abrasive is not too dry, offset the disks and add a few drops of water. The abrasive should then spread out a little more toward the edges. If that does not happen, or if the mirror sticks to the tool at certain points, the outside zone of the mirror and/or the tool is too flat. In this case, make ⅙-diameter W-strokes, alternating the mirror and tool on top (see Figure 7.3c).

- *The mirror is too concave.*

After rough grinding, the sagitta may be several tenths of a millimeter too large

Rough Grinding and Bubbles

I've finished rough-grinding my primary mirror and I'm still using #80 carbo. I can't get rid of a huge bubble between the mirror and the tool. What should I do?

First, you need to change to the next finer grade of abrasive. Stop using #80 carbo and move on to #120. Effectively, #80 carbo is made of large grains (0.2 mm on the average) that let water circulate freely between the mirror and the tool.

There are two phenomena that can occur here. Working with the mirror on top, a large air bubble can appear at the center of the mirror, because the water is pushed to the edge of the tool, bringing with it a large quantity of abrasive. Conversely, when working with the tool on top, the movement of the two disks can cause a sludge, made of pulverized abrasive and glass mixed with the water, near the center of the mirror. This accumulation of abrasive and water lets the central hole in the mirror persist (see Figure 4.4 on page 38) so the bubble won't go away, although sometimes it will fill up with water and look like it has disappeared.

This movement of the water doesn't happen with a finer abrasive grain, like #120 or #180 carbo. Smaller and closer together, the grains hold the water and the abrasive between the two disks—not only at the center of the mirror—and let you work more efficiently. Make sure that you don't use too much abrasive or too much water; either way, it will make the wet much less effective.

The best stroke now is the W-stroke with a total amplitude of ⅙ of the mirror's diameter (see Figure 7.3c) or figure-eight stroke over ⅓ of the diameter. For *f*/6 to *f*/12 mirrors, smaller strokes than these are more likely to produce a bubble.

(the radius of curvature is thus too short). To lengthen the radius of curvature, work with the mirror on the bottom with straight or figure-eight strokes of a total amplitude of ⅔ the mirror's diameter (Figure 7.3d). Every third wet, use ⅙-diameter W-strokes (Figure 7.3c). Remember to check the sagitta of the mirror about every five wets.

- *The mirror is not concave enough.*

Work with the mirror on top with ⅔-diameter strokes, as shown in Figure 7.3d.

7.3.2 Finishing Fine Grinding

Fine grinding is finished when you can no longer see a central air bubble between the tool and mirror, and there is no point where they stick or grab during a wet. At the end of fine grinding, the abrasive will spread out uniformly between the mirror and the tool during the wet; a few bubbles might get trapped in the abrasive, but they rapidly move to the outside and disappear. After changing sponges and thoroughly cleaning the workpost and the mirror, finish fine grinding with the next

What to Do if the Mirror has a Chip

It happens—at any stage in mirror making, including polishing, chips can appear on the mirror or the tool resulting from a shock between the two disks. If a chip appears on the optical face of the mirror, there are several ways to eliminate it. If it's not very deep (less than 1 mm), you can work the mirror with #80 carbo using straight ⅓-diameter strokes, with the mirror on top and on the bottom, pressing strongly; this is equivalent to recommencing rough grinding. You reduce the thickness of the mirror without changing its form (for example, with #80 carbo on a 130-mm mirror, you remove about 1 mm of thickness in one hour of work). If the defect is situated as in the photograph, round the sharp edges with a fine carborundum stone or with a large diameter brass rod covered in #120 or #240 carbo. This operation is called "making-up" or "camouflage."

If the chip is very deep (more than 2 mm), it is better not to waste your time trying to eliminate it; simply continue smoothing—you'll have to live with the chip. This kind of defect is mostly cosmetic, and alters the quality of the images very little.

If a very large chip is situated on the periphery of the mirror, and if you absolutely must get rid of it, the last solution is to reduce the diameter of the mirror using a grinding stone until the defect disappears.

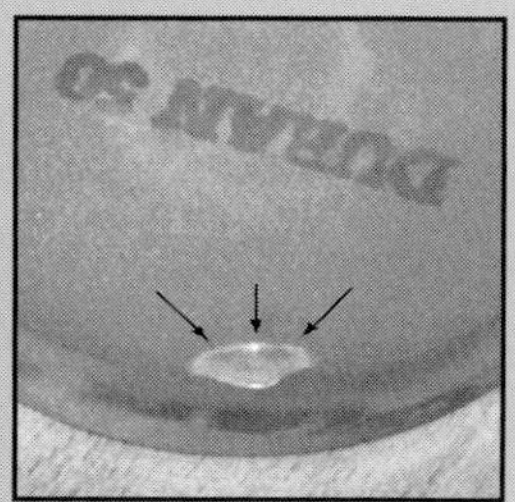

Fig. 7.4 Be careful of chips during grinding and smoothing. When the chipped glass is still attached to the mirror, you must remove it with a grinding stone.

abrasive stage (#220).

From here on, do an extra-thorough clean up before moving to a finer abrasive. Use new sponges, rinse the sink, and clean the workpost, mirror, tool and yourself. This will reduce the risk of scratching the mirror.

7.4 Smoothing

At the beginning of smoothing, the surfaces of the mirror and the tool are regular and have the same radius of curvature. During the course of smoothing, you remove pits with increasingly finer abrasive powders. You will maintain the desired radius of curvature by using well-defined strokes.

Smoothing begins with #320 carbo. It finishes with 5-micron aluminum oxide, when the surface is perfectly smooth and devoid of pits. The surface will have a uniform, nearly transparent, gray cast. Check the edge of the mirror carefully for pits. This is especially important because the edge is the last part of the mirror to be ground by each grade of abrasive, and pits tend to persist here. It is a common failing of amateur mirrors to show pits or a gray outer ring when the mirror is aluminized because the mirror was not thoroughly smoothed and fully polished right

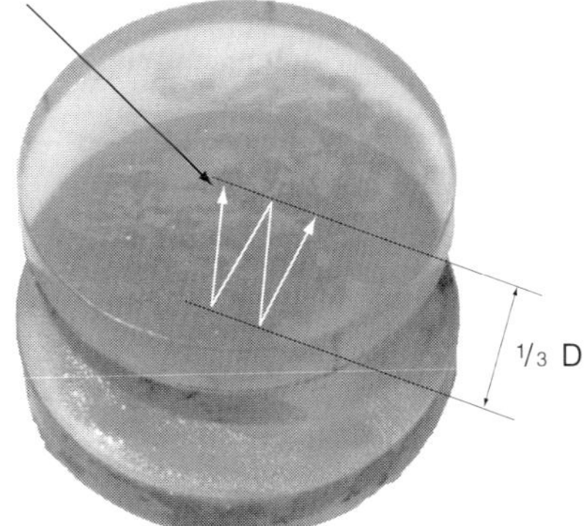

Fig. 7.5 Normal smoothing stroke. Invert the mirror and tool every other wet. The center of the disk describes a W of ⅓-diameter amplitude and ⅙-diameter lateral width. After two or three series of W-strokes, rotate the top disk about ⅛ turn clockwise. Simultaneously, move yourself around the grinding post about ⅒ turn counterclockwise every few strokes. This way, you avoid repeating the same exact stroke over the same part of the mirror and creating irregularities on the surfaces.

out to the edge.

7.4.1 Smoothing Strokes

The standard stroke for smoothing is the now-familiar W stroke, with a total amplitude of $\frac{1}{3}D$ and a lateral width of $\frac{1}{6}D$ (see Figure 7.5). Invert the mirror and tool every other wet in order to maintain the radius of curvature. Continue to rotate the top disk regularly beneath your hands, and move around the workpost several times per wet.

The wetted smoothing abrasives should spread out and uniformly break down between the mirror and lap. If an air bubble forms, this is a sign of trouble; check the amplitude and lateral width of your strokes; in effect, strokes that are too large with the mirror on top tend to shorten the focal length, and conversely with the mirror on the bottom, tend to widen the curvature and lead to irregular zones. It is thus important to find a happy medium in order to maintain the same radius of curvature and a smooth figure of revolution. The prolonged presence of an air bubble between the disks results in poor smoothing at the center of the mirror. When you finish smoothing, the surface quality from the center to the extreme edge of the mirror must be identical. Polishing cannot eliminate the large surface defects left after smoothing, especially at the outside edge; this step is important for the success of your project and should be done with great care. Unlike rough grinding, when smoothing, you must press moderately on the glass disks and position your hands in a regular manner in order to exert a uniform pressure. Avoid pressing more on one side than on another, or you risk deforming the glasses and changing their regularity or their curvature.

7.4.2 Precautions for Smoothing

During smoothing, you must guard against scratches and sticking. Scratches are caused by an abrasive grain that is larger than that used during the current wet. For example, a grain of #80 carbo that falls in the jar of 5-micron aluminum oxide will

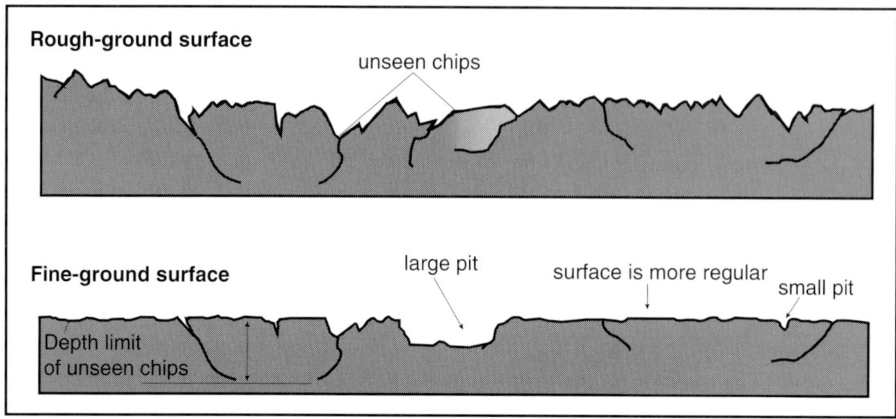

Fig. 7.6 Highly magnified cross-section of a mirror surface. During rough grinding, the carborundum fractures the glass, creating a multitude of small, rough-sided chips in its surface. The pits thus created are sometimes several tenths of a millimeter deep and may last through the first several hours of smoothing.

cause a very large scratch, at least 0.1 mm deep in the glass. Scratches are almost always formed at the beginning of a wet, rather than during one, because the errant larger grains cannot move in between the mirror and the tool after the wet has begun. When you find a scratch on the smoothed mirror, examination with a loupe (or an inverted eyepiece) will help you determine its depth. In the best case (a scratch smaller than 5⁄100 mm) the scratch may disappear after a few more wets using the current grade of abrasive. If it persists you must return to a coarser grit. If it is particularly deep you may have to move backwards by several abrasive grades to remove it in a reasonable amount of time.

Scratches, which cause considerable time-losses for the mirror-maker, are most often the result of negligence in cleanliness. To eliminate them, you must carefully clean up between each change of abrasive. It is not too extreme for the mirror maker to take a shower, clean under the finger nails and change to freshly laundered clothes between each abrasive grade. Still, it is good to know that a good mirror with a scratch or two provides images of the same quality as does a mirror without scratches (except in some very rare solar applications); scratches are therefore mostly a cosmetic problem.

The two disks may stick together suddenly during a wet and refuse to budge. Such sticking generally happens toward the end of a wet when the abrasives have been ground down to a very small size and then quickly dry out. They can also stick if you neglect to separate the disks before turning your attention to another task or your work is interrupted—if the phone rings, separate the disks before answering! See the box on the next page on hot separate stuck disks.

7.4.3 When to Change Abrasive Grains during Smoothing

Smoothing begins with 500 carbo. After a few wets, the glass will be covered with

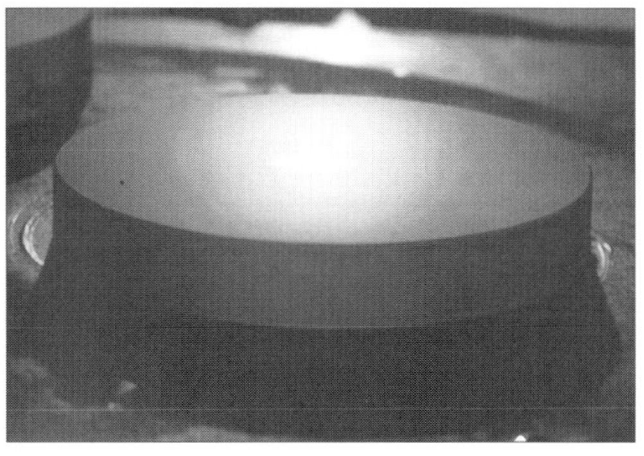

Fig. 7.7 To make sure that smoothing is proceeding correctly, observe the reflection of a light filament from a lamp at a low angle on the surface of the mirror and the tool. A defect in smoothing (an under-worked center or edge) appears darker than the rest of the surface. This condition should be tested before changing grains. The photograph shows a surface well-smoothed with W5. At this stage, the coincidence of the faces of the mirror and the tool should be better than a few microns.

a multitude of pits, the largest remaining from rough grinding, and the smallest from #500 carbo. When you have finished several wets, the small pits will start to be in the majority.

Before moving on to a finer abrasive, all of the defects caused by the previous abrasive must have disappeared. At the start of smoothing, you must be patient in order to remove the large pits left by rough grinding—these can remain for several hours. When you finish with #500 carbo, no pit caused during rough grinding should remain on the mirror surface.

7.4.4 Duration of Smoothing

To smooth the 130-mm mirror, use the following example as a guideline. It is effective and economical in terms of time and abrasives (preferred abrasives are in bold type).

- **W240**: Work 2 hours (25 wets) before examining the surface with a loupe. If large pits, from the #80 carbo, persist, continue working for a half hour. Also verify the radius of curvature with a spherometer. If it has changed, see Section 7.4.5 on page 100.
- **#400 carbo** or **W1 emery**: Work one hour (about 10 wets)
- **W3** or **W4**: Work 30 to 45 minutes (about 6 wets)
- **W5** or **W6**: Work for one hour (6 wets) and carefully examine the surface with a loupe. There should be no pits visible to the naked eye and no scratches or holes visible in the eyepiece. If pits remain, continue smoothing for an

How to Separate Two Disks that are Stuck Together

To unstick the mirror from the tool, you could first try to push the side of the disk with the palm of your hand. If nothing moves, you must refrain from using force, because you risk damaging the mirror with a scratch or chip. The quickest solution is to submerge the tool in a tub of hot water (50° C), holding the mirror above the water. The expansion of the glass should permit the water to flow between the surfaces, and they should separate by themselves after a few minutes.

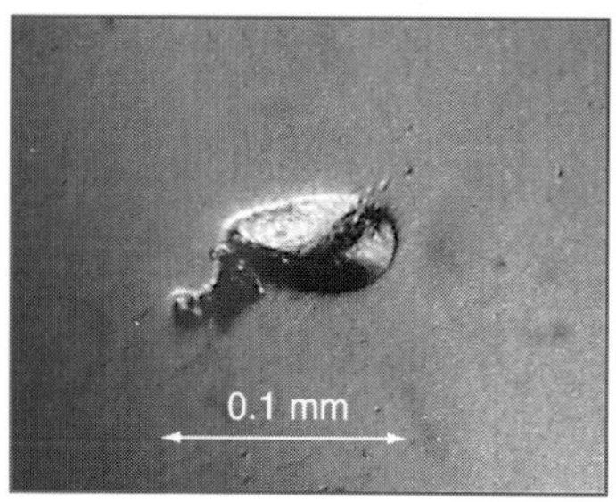

Fig. 7.8 Pit left from smoothing seen on a polished surface. The interior borders of this defect are rounded, which allows us to assume that the glass has been pushed from the surface to the interior of the hole during polishing.

hour or two with W6.

During smoothing you should check the sagitta from time to time, to make sure that your work is not changing the radius of curvature from that of the original plan.

7.4.5 How to Change the Radius of Curvature during Smoothing

If the radius of curvature measured with the spherometer is too short (the sagitta is too deep), you should work more often with the tool on top (mirror on the bottom) and increase the amplitude of your strokes (passing progressively from ⅓ to ⅔ diameter, Figure 7.3d on page 94). If the radius is too long (sagitta too shallow), work with the mirror on top and make the same strokes. This technique will cause the formation of an air bubble between the mirror and the tool, because the work creates a different radius of curvature on the two pieces. To bring both surfaces into contact, change back to normal smoothing strokes (⅓ diameter, W-shaped) every four or five wets.

7.4.6 The Last Wet before Polishing

The surface is ready for polishing if there are no more visible pits when the mirror's surface is inspected through a loupe. The last wet before you move on to polishing should last longer than normal. By stretching out this final wet you can progressively reduce the abrasive to produce a super fine surface. During this process no new abrasives are added; only add water from time to time to prevent sticking. A wet with W5 can thus be pushed to a half-hour duration.

Finish smoothing with W5, W6, or #600 and #1200 carbo. These abrasives are equivalent in grain size, but carbo is harsher than emery and creates slightly larger pits in the glass.

When smoothing is finished, thoroughly clean both the tool and mirror, rinsing them in running water and wiping the face, back, and sides with a clean sponge, a nylon brush (*not* a steel brush) or paper towels; change sponges and clean the mirror case (repaint the interior) to completely rid yourself of polluting abrasives. During polishing, if any of the abrasives used in grinding and smoothing show up anywhere in the work area they will scratch the mirror.

7.5 Polishing

Polishing is the most interesting step in making a mirror. The best substance for high quality optical polishing is pitch, a black resin secreted by Nordic conifers.

Pitch: Use and Testing

After unwrapping the pitch, melt it slowly in an old saucepan over very low heat. As it melts, stir your pitch with a wooden spoon; don't use utensils made of metal (they conduct heat) or plastic (they may melt).

For melting the pitch, an electric hot plate is preferable to an open gas flame, and one that is outdoors is best. **Pitch and its vapors are flammable.** Do not use maximum heat, because the pitch will boil and lose too much of its volatile components which are needed to promote flow rate once formed into a polishing lap. Overheated pitch will make a dead lap. Pitch should never be heated to the point where it smokes.

NEVER ALLOW AN OPEN FLAME NEAR THE PAN WHILE MELTING PITCH!

NEVER ADD WATER TO MELTED PITCH!

The flow rate of the room temperature pitch is what makes it possible to figure a telescope mirror. You can quantify this attribute by measuring the depth of the imprint made by an 8-mm stainless steel ball attached to a 200-g weight. After 30 seconds, the mark left by the ball should be between one and two millimeters deep (at 22°C or 72°F). If the imprint is shallower, the pitch is too hard; if it is deeper, the pitch is too soft. Pitch that is too hard or too soft will make polishing and figuring difficult to impossible. To harden pitch, warm it over medium heat for an hour or two, without quite letting it boil. Just a hint of smoke should escape from the pitch, which is evidence that it is losing some of its solvents. Occasionally take a sample with a spoon, cool it under running water for about 10 minutes, and test its hardness with your thumbnail or the stainless steel ball.

To soften pitch that is too hard, add a little turpentine to the melted pitch in the saucepan. A teaspoon full is a good starting point for a kilogram of pitch. Stir for ten minutes before testing a sample. If still too hard, incrementally add 4 or 5 drops before testing. A recipe for pitch that's the right consistency, under normal conditions, for polishing spherical mirrors is found on page 103.

Fig. 7.9 You can recycle used pitch from a polishing tool already charged with polishing compound (cerium oxide, rouge, etc.) by melting it in a saucepan. Decant the mixture without stirring after an hour or two, then filter it through a very fine sieve or a stocking.

Pitch melts around 60°C and has a degree of plasticity at room temperature. It is malleable and takes the form of the glass mirror quite precisely. In contact with a pitch lap charged with polishing compound, the first few molecular layers of the mirror's surface behaves like a liquid; there is not only removal, but also displacement of the glass. Experiments have shown that the depth of glass that must be removed to make pits or scratches disappear is less than the depth of the defects. The glass is thus "pushed" by the polisher during work, and a thin layer of silicon (called the Beilby layer) is created which covers the entire surface of the polished glass. This fragile layer begins forming after several dozen minutes of work and continues to evolve during every polishing wet or retouching.

During polishing, your work takes on a new dimension because the result depends not only on the quantity and type of polishing compound used, but also on the speed of your strokes, the nature of the pitch, atmospheric conditions, and most of all, on your experience and the way you work.

Before you begin polishing, clean your work area well. Any trace of abrasive—destined to scratch the mirror surface—should be eliminated.

The 130-mm mirror is spherical. A sphere is easier to achieve than a paraboloidal form. Normal polishing strokes tend to produce a sphere, while a parabola, with its variable radius of curvature, requires a variety of special techniques.

7.5.1 Making the Pitch Lap

The tool used to grind the mirror will serve as a base for the polishing pitch lap. Pitch is poured over the face of the tool, then made convex by pressing it against the mirror. Next, the lap is faceted by cutting channels in the surface of the pitch in order to permit the polishing agent to freely circulate between the mirror and lap. The fabrication procedure is as follows:

Preparing the Pitch

1. Slowly heat the pitch just enough to melt it.
2. While waiting for the pitch to melt, soak the mirror in cold water and warm the tool by putting it in a tub of very hot water for a few minutes.[1] When the pitch is ready, take the mirror out of the water and clean and dry its surface. Make a dam with masking tape around the edge of the tool, see Figure 7.10A.
3. Pour the pitch onto the tool, proceeding from the center to the edge. You can gently tilt the tool so that the pitch flows evenly thick over the whole mirror. A depth of 5 to 10 mm is sufficient for a good polisher. See Figure 7.10B and C.
4. Before the pitch completely hardens, remove the masking tape and press the concave mirror, covered with cerium oxide and water, on top of it.
5. After pressing firmly, remove the pitch that has flowed over the sides of

[1] Do this only with Pyrex® or other low-expansion glass.

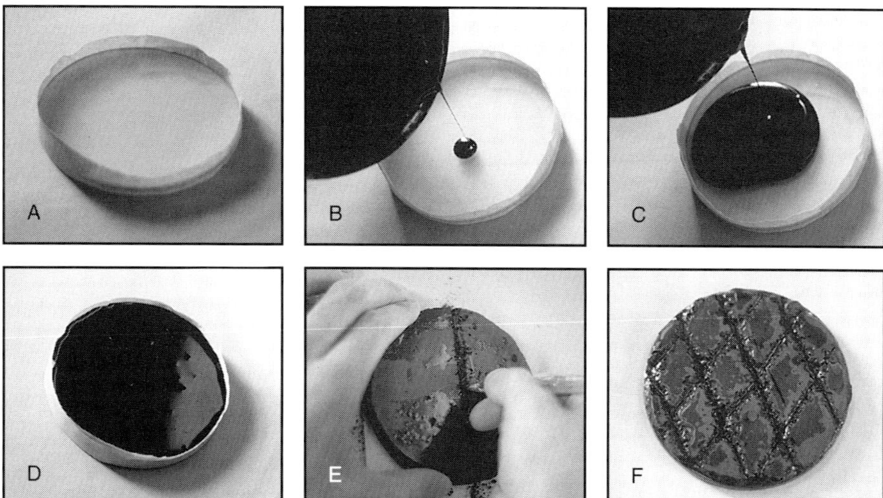

Fig. 7.10 Making the pitch lap. A. The tool is encircled with a band of masking tape. **B.** and **C.** Pouring the pitch onto the tool. **D.** Waiting for the pitch to harden before removing the masking tape and pressing it with the mirror. **E.** The lap is formed, and channels are cut into the pitch. **F.** The finished lap. The roughness of the pitch will disappear quickly after work is begun.

the tool with a single-edged razor blade or an old knife. Repeat the pressing as many times as necessary; the pitch tool should touch the mirror at every point of its surface.

6. After the tool is formed, dunk it in a tub of cold water to fix its form. Cut channels in the shape of squares using a knife or a razor blade. The pattern should be off-center, see Figure 7.10E and F. **Do not place a pitch**

A Tempered Pitch Polisher for Flat or Spherical Mirrors (for Beginners)

This is the mixture used by Jean Texereau at the optical workshop of the Paris observatory. It requires the following ingredients:

- 1 kg of pitch, hard and used, already covered with polishing powder
- 2 kg of rosin (to add to the hot pitch)
- 345 cc of castor oil

The polisher obtained from this mixture is rather hard; at 22°C, the imprint of an 8-mm diameter steel ball (with a 200 g weight, pressed for 30 seconds) is only 1.1 mm deep. To make it more fluid, add some castor oil (a total volume of 414 cc will give you an imprint 2.1 mm deep).

This tempered polisher deforms less than pure pitch of similar hardness. This allows you to make large strokes with the tool on top without creating a turned-down edge.

Fig. 7.11 Before placing the mirror on the tool, apply some polishing compound with a brush. Here, a lot of cerium oxide is used because it is the first polishing step. In the glass container, the polishing compound tends to be diluted near the surface and concentrated at the bottom.

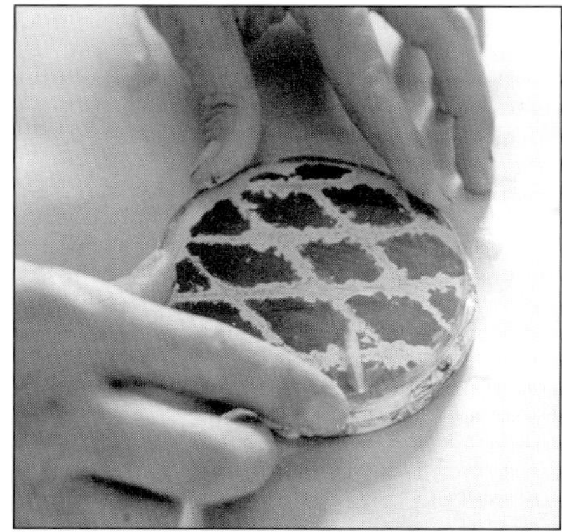

Fig. 7.12 Polishing strokes. As in smoothing, make W-shaped strokes with a total amplitude of ⅓ the mirror's diameter. When the mirror is on top, you can check the lubrication of the polisher visually. Work with your fingers on the sides of the top disk, pushing as little as possible.

facet at the exact center of the lap; if you do, you will polish rings into the face of the mirror. These channels will allow the water-polisher mixture to circulate freely over the whole tool and the pitch to flow and conform to the mirror's surface. As you polish your mirror, these channels will gradually flow and will need to be re-cut before they flow together.

Note: Before cutting the channels in the polisher, quickly pass the surface of the warm pitch over the flame of a gas burner for a few seconds; this eliminates any bubbles that have marred the surface. This way, you get a more homogenous polisher, which allows more regular work.

7.5.2 Beginning Polishing

Before beginning, reheat the surface of the pitch lap by dunking it for a few minutes in a tub of warm water, between 35° and 40° C. During this time, mix in a small glass container (e.g., a baby food jar) one part polisher (cerium oxide or zirconium oxide) to four parts water. Stir this with a brush.

You will begin work with the mirror on top. This allows you to observe the slow flowing pitch and the movement of the polishing compound through the mirror's back (if the back of the mirror is rough, wet it with water to make it more transparent).

Using a brush, spread a small amount of thick polishing compound (from the bottom of the jar) over the surface of the pitch, and then carefully place the mirror

Pitch Flow

Pitch flow is a convection phenomenon similar to that which occurs when a liquid is heated in a pan. After a prolonged polishing spell, the pitch warms up and begins to flow outward into the groves, exposing new pitch at the center of the squares. The center of these squares, where this "new" pitch appears, is not yet charged with polishing compound and appears darker. At this point you need to recharge the lap with more polishing compound. Pitch's ability to flow and to conform to the mirror's constantly changing surface produces an "optical grade" polish. In order to maintain pitch flow, you must avoid interrupting a polishing session for more than five minutes, because the pitch must not cool down too much.

on top (see Figure 7.12). Begin polishing with strokes similar to those used in smoothing (W-strokes, ⅓ of the mirror's diameter—see Figure 7.5 on page 97). As during grinding, working with the mirror on top concentrates the polishing action at the center of the mirror which will then polish out first. Placing the tool on top will concentrate the polishing action towards the mirror's outside edge. Inverting the mirror and tool about every 30 minutes will therefore tend to polish the surface evenly.

After polishing for about 30 minutes, check to make sure that the entire surface of the tool is still in contact with the mirror; the pitch lap should have a uniform grayish tint, which is a sign that the mirror is in full contact and that the polishing compound has fully charged the lap (if this is not the case, warm the lap and press it with the mirror until all the pitch facets contact the mirror).

During the course of polishing, make sure the tool never runs dry. If the mirror becomes dry, polishing becomes difficult, and the surface will become optically rough. It is useless to spread too much polishing compound on the lap; polishing does not speed up, and the polishing compound is wasted. Moreover, if polishing compound is allowed to build up at the edge of the mirror, it can cause a turned-down edge, which will be very difficult to correct and is damaging to image resolution. After the tool becomes fully charged with polishing compound, you need only occasionally wet the polisher with a few drops of polishing solution taken from the top of the jar to polish continuously for several hours. Optimal polishing sessions last about 30 to 40 minutes.

When the mirror is removed from the lap after a polishing session, rinse the polisher and mirror under clean, running water. Cover the polisher (face up) with

Trimming the Polishing Lap

During polishing, the pitch warms up and spreads out. Because of this, the outside edge of the polisher increases by a few millimeters every hour. In a similar fashion, the channels start to close. When this happens the outside edge and the pitch facets must be trimmed back to their original position or zones will appear in the mirror.

A Few Essential Rules for Polishing

- During your work, regularly check and maintain the width of the channels. To assure that the abrasives are circulating correctly over the polisher, trim the pitch facets well before they flow together.
- At the start of a wet, moisten the surfaces of the mirror and the tool with water, and then place the disks one on top of the other. Wait a minute or two, then off-set the mirror and spread the polishing compound on the lap; then begin work.

  This method tends to prevent large scratches, which are difficult to eliminate. When the mirror is placed on a tool that is covered with polishing compound, any contamination is likely to embed in the pitch where it can repeatedly scratch the mirror.
- Beginning work with a tool that's too cold (below 60°F, for example) or too dry can cause sleeks (very small but visible scratches). To avoid this, the surface of the polisher should be heated for a few seconds before each polishing wet (use a tub of hot water, a hair dryer, etc.).
- Clean the mirror before each polishing session, as dust particles from the mirror can become embedded in the pitch and then scratch the surface of the mirror. The polisher should never be stored without covering it with clean plastic film (like Saran® wrap) to keep it dust free.
- To store a pitch lap between polishing sessions during hot days, immerse it in a basin of cool water (up to a maximum of 48 hours at a time). This assures that the pitch won't melt and flow under its own weight. This method also helps to preserve soft pitch.

a large pan or plastic wrap and dry the mirror with paper towels or a clean cloth.

7.5.3 Polishing Out a "Gray" Surface

Opticians refer to a fine ground surface ready for polishing as being "gray." This is a milky veil caused by countless very small pits that remain even after the finest abrasives have been used to grind and smooth a surface prior to polishing. The first hours of polishing are exclusively dedicated to elimination of these very small pits. For this, you use a fast-acting polishing compound like cerium oxide. Cerium oxide, under heavy pressure, sometimes creates a squeaking sound near the end of smoothing. Don't worry—the noise is not a symptom of scratches, but it should not be allowed to persist—add more water to the lap and decrease pressure.

During the first five or six hours of polishing, you remove the grayness from the surface in order to make it completely reflective. While you can test the surface of the mirror with the Foucault tester after polishing for just only an hour or two, **you cannot stop polishing at this point even if the mirror is a perfect**

sphere. You have not completely polished out the mirror even though it looks shiny. If you stop at this point and send your mirror off to be aluminized, you will be disappointed when your mirror returns with a dull surface caused by the remaining very small pits now revealed by the highly reflective aluminum coating.

Polishing times depend on the type of glass, polishing compound, and the condition of the fine ground surface with which you begin. A Zerodur mirror takes twice as long to polish as a Pyrex or Duran 50 mirror. Plate glass polishes most quickly (compared to Pyrex, about 30% faster). A mirror smoothed with 25-micron aluminum oxide is more quickly smoothed than a mirror worked with #600 carbo (carbo fractures the glass while aluminum oxide tends to shear it). Cerium oxide smooths out the fine ground surface twice as fast as rouge. Zirconium oxide is close to cerium oxide in its polishing speed but gives a somewhat better surface without the mess of rouge.

8 Testing and Finishing the Primary Mirror

8.1 Making a Foucault Tester

8.1.1 Principle

The Foucault tester shown in Figure 8.1 has both the light source and knife edge joined together. This assembly moves in two directions: the first is along the optical axis which moves the tester toward and away from the mirror; the second is across the optical axis.

The head of this apparatus is situated about 15 cm above the base, has a lamp to illuminate the mirror, an adjustable slit, and a knife-edge to analyze the reflected light beam.

8.1.2 Lighting

The light source in this tester is provided by a 4.5-V to 12-V bulb with the filament oriented parallel to the opening of the slit. The choice of bulb is left to the reader; nonetheless, here are a few pointers:

- Bulbs from pocket flashlights are not very powerful, but they have the advantage that they do not get too hot. They are practical and economical for

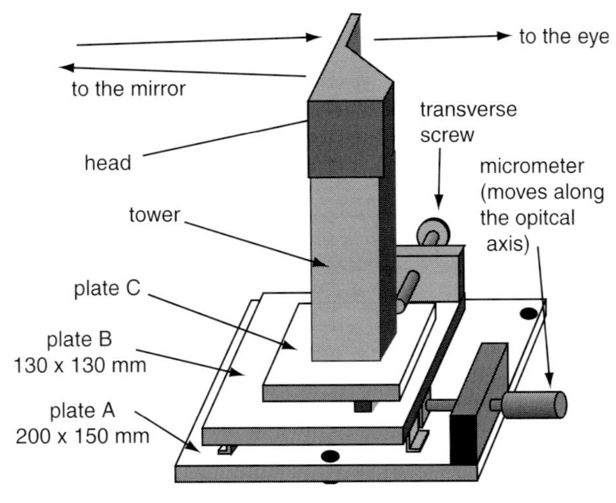

Fig. 8.1 Overview of the Foucault tester. The base (plate *A*) rests on three pressure points spaced 120° apart. Plate *B* moves over plate *A*, parallel to the optical axis. Plate *C* moves over plate *B*, perpendicular to the optical axis. The tower is an aluminum tube about 15 cm tall. Light from the head is reflected back from the mirror where it can be analyzed with the knife-edge. Parts *A*, *B* and *C* can be made from wood, aluminum or PVC. Note: This is a moving-source tester. Measurements must be doubled before using them in the formulae of Chapter 5 starting on page 49.

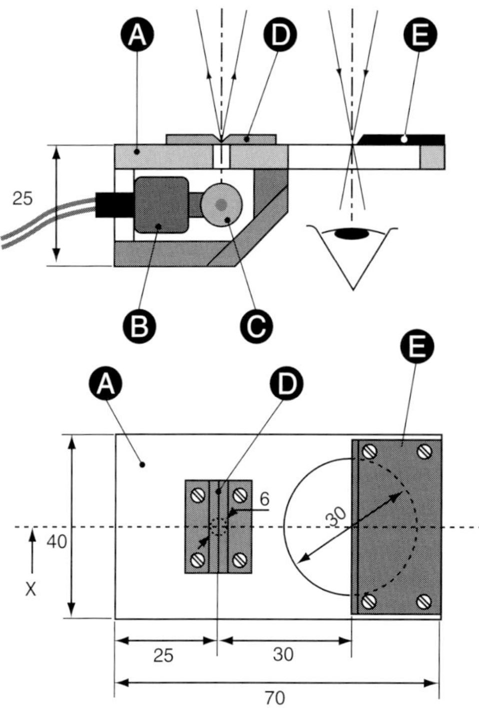

Fig. 8.2 Foucault Tester Head Top, a cross-section of the Foucault tester. Center, the slit and knife-edge as seen from the mirror. All dimensions are in millimeters.

Legend:

A: Aluminum sheet metal plate
B: Cap
C: Light bulb
D: Slit
E: Knife-edge

The eye observes the mirror from behind the knife through a 30-mm diameter circular hole. The two lips of the slit are placed in front of a hole light source 6 mm in diameter.

Notes:

- The filament of the lamp must be parallel to the slit.
- Do not paint the inside of the head (it will get too hot).
- To test mirrors with short focal lengths, you can place the knife-edge 20 mm away from the slit.

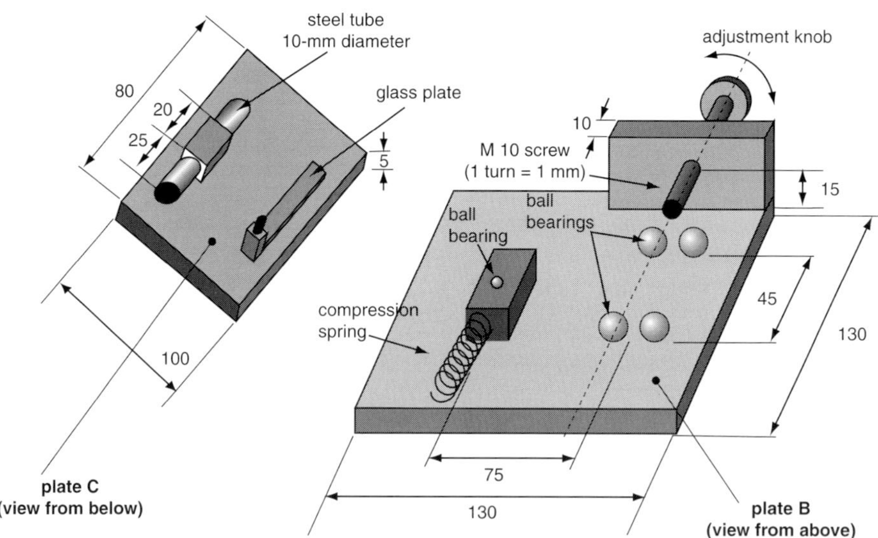

Fig. 8.3 Parts layout on plates *B* and *C*.

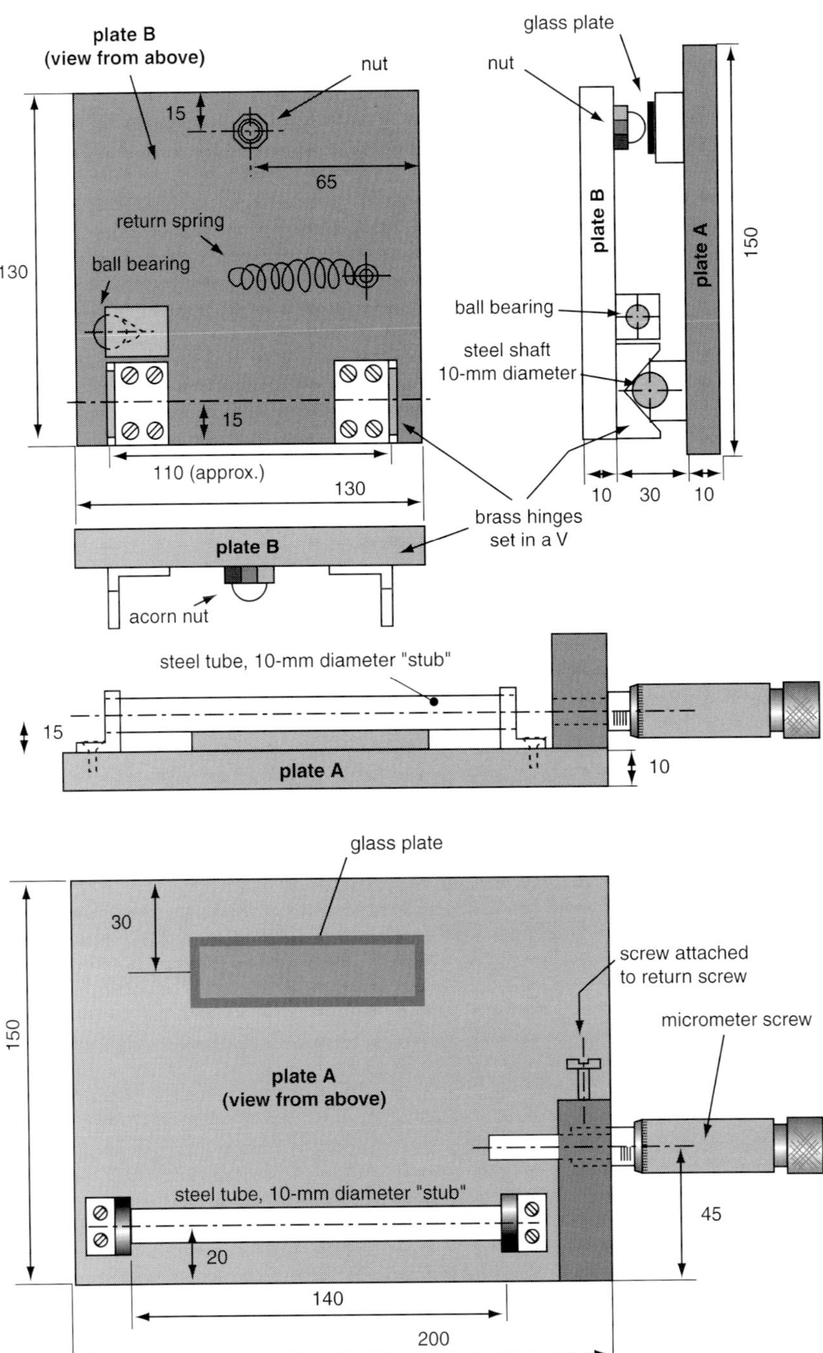

Fig. 8.4 Schematic of the movable stages (plates *A* and *B*).

Legend for Figure 8.4 (previous page)

Plate A is the base of the Foucault tester. It measures 150 mm by 200 mm; it is equipped with a micrometer screw, graduated in hundredths of a millimeter, and a ground stainless-steel rod (10-mm diameter) to guide the moving plate. The ensemble of plates *B* and *C* move over plate *A*. Brass hinges (V-shaped) and an acorn nut are attached to the bottom of plate *B* (130 mm x 130 mm). The return spring holds the ball (of plate *B*) to the end of the micrometer screw (on plate *B*). The ball provides for a point contact and can be left out if the end of the micrometer screw is round.

Note: For greater stability and better contact between the tube and the hinges, shim plates *B* and *C* with small pieces of lead.

use in the Foucault tester.

- Halogen bulbs, much more powerful, require a filter to block ultraviolet light (place the filter between the bulb and the slit). These bulbs get *very* hot and will quickly overheat. Without a fan, you can use a 20-watt bulb. Above that wattage, you can install a lamp—with a fan—in a box independent from the test apparatus and "pipe" the light to the slit with a fiber optics bundle to keep the heat and vibration away from the tester.
- Light-emitting diodes (LEDs) provide sufficient light, are very small, use little energy, and don't get hot.
- Do not use a laser source because it is both dangerous to your eyes and creates "speckles," little grains in the light pattern that are caused by interference that make knife-edge shadow interpretation difficult to impossible.

8.1.3 Dimensions of the Light Slit

A slit suitable for detailed zone testing measures 5 to 6 mm high by about 10 microns wide. The width and parallelism of the slit are maintained by the fixing screws (make sure that your light filament is at least 1 mm wider than the diameter of the slit). Since the slit assembly is non-adjustable once the slit jaws are screwed down, you must also make a second head with a larger slit opening (0.5 to 1 mm wide) to test the roughness of the mirror.

Important Note: *If you use a halogen lamp, the Foucault tester must be equipped with a heat-absorbing glass filter to block the infrared and ultraviolet light created by this bulb. This is sold by Edmund Scientific and others.*

8.1.4 Making the Knife-Edge and the Slit.

The knife-edge and the slit are made from 2 mm thick brass stock which can usually be found at a hobby store, especially those who cater to model railroad enthu-

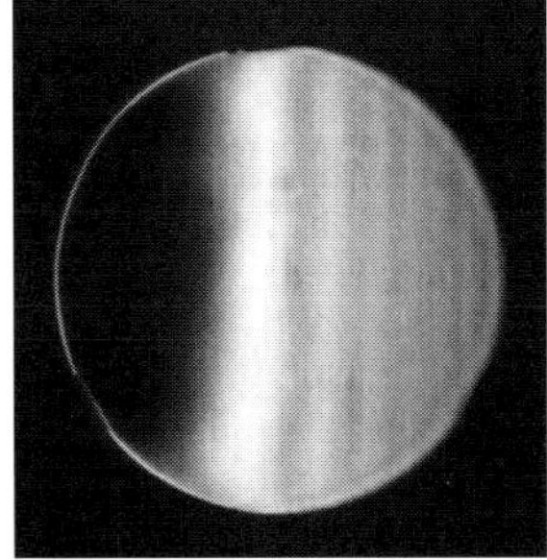

Fig. 8.5 Left: Spherical mirror photographed in the Foucault test. The knife-edge, placed behind the center of curvature of the mirror, is not exactly parallel to the slit. No interference fringes are observable on the mirror. **Right:** Same mirror with the slit and the knife-edge parallel. Vertical fringes are easily visible. These fringes are caused by the diffraction of the light on the edge of the knife. If the slit is very thin and the mirror is well-polished, the fringes appear very colorful.

siasts. The 45° beveled edges are formed on a sheet of window glass using a fine abrasive like #500 carbo and then lapped with 12 or 5 micron aluminum oxide. A perfectly straight edge, with no rough edges, is necessary.

8.1.5 Parallelism of the Knife-Edge and the Slit

The parallelism of the knife-edge and the slit is confirmed and refined when you test it against a mirror. When you place your eye about 20 cm behind the knife-edge, you should see near the knife-edge a focused bright image of the slit returned from the mirror. If you have trouble finding this image, hold up a sheet of thin, transparent tracing paper behind the tester and move it around until the image shows up on the paper (this is easier to do if the room is dark). If this is the first time you have done this, it may be necessary to adjust the mirror's position (up, down, right and/or left) to get the returning beam near the knife-edge. Once seen, it is easy to adjust the knife-edge against this image. When you cut a properly aligned knife-edge into the returned beam of light, you should see the simultaneous disappearance of both the top and the bottom of the slit image.

With a thin slit, the alignment can be further refined by observing the contrast between the interference fringes that appear on the mirror (see Figure 8.5).

Adjusting parallelism between the knife-edge and the slit is facilitated by temporarily taping the knife-edge to the assembly and inspecting the returned slit image. Once the knife-edge is parallel to this image, it may be permanently fixed with screws.

8.2 Testing the Mirror during Polishing

The 130-mm mirror described in this book is spherical. The test of its form is confirmed visually; there is no need for detailed Foucault tester measurements. The

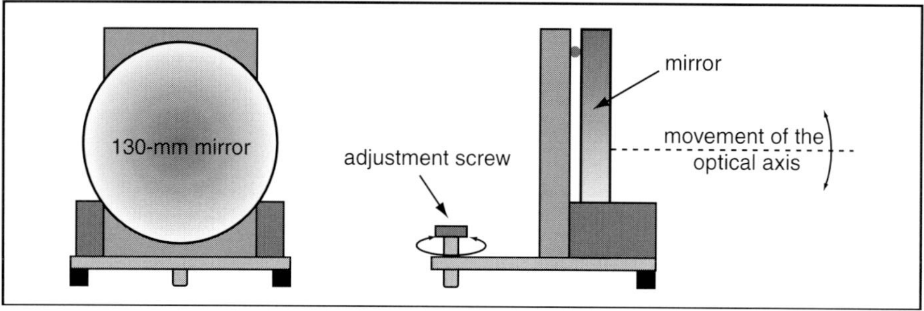

Fig. 8.6 Simplified wood support to keep the 130-mm mirror vertical during the Foucault test.

goal is to obtain a null—a simultaneous graying of the mirror's entire surface.

You can do the first test of the mirror as soon as the surface begins to reflect light. In general, two or three hours of polishing are sufficient to make a first test. At this stage, the mirror will still be gray and unusable, its surface covered by small pits. The surface, observed under a bright light, presents a milky appearance.

8.2.1 Setting Up

The testing area should be dimly lit and draft free because air currents make the images dance and testing difficult. Place the mirror on its test support (see Figure 8.6) and wait until the mirror reaches ambient temperature. During this time, you can set up the Foucault tester as described below.

8.2.2 Positioning the Foucault Tester

1. Place the tester at the same height as the mirror, which is set vertically on its own adjustable support, at a distance equal to its radius of curvature (measured with a spherometer at the end of smoothing).
2. Turn on the Foucault tester's lamp. Start with the removable head that has the wide slit to project the greatest amount of light onto the mirror.
3. Turn off the lights in the test area and find the reflection of the light slit in the mirror by holding a sheet of paper in front of the Foucault tester (do not cover the light source). When you find the reflected slit, adjust the orientation of the mirror support (with the adjusting screw at the mirror's optical axis) until the reflection of the slit is projected onto the knife-edge (see Figure 8.7). If the reflection is fuzzy, move the Foucault tester closer or farther away until it is sharp. The image of the slit on the knife-edge is sharply focused when the distance between the slit and the mirror is exactly equal to the distance between the mirror and the knife. This distance is the radius of curvature of the mirror.
4. Start testing by turning the transverse displacement screw so that the re-

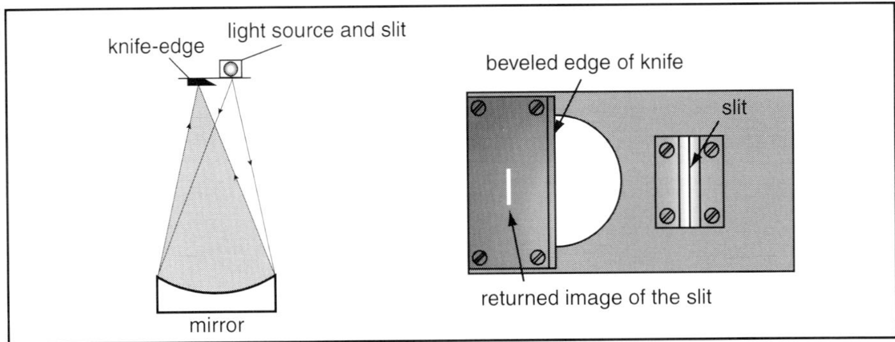

Fig. 8.7 How the image of the slit is reflected off the mirror and then focused on the surface of the knife-edge. You are at the center of the mirror's radius of curvature when the returned image is exactly the same size as the slit. To observe the returned image from behind the knife-edge, turn the transverse displacement screw until the image moves across the back of the knife-edge and then appears just outside and parallel to the sharp edge of the knife-edge.

flection of the slit is formed close to knife-edge. By both cutting into the returning beam and moving the knife edge toward and away from the mirror, there should be a point where the illuminated mirror instantly goes gray (nulls). This null is the exact center of curvature for a sphere.

The location of your eye behind the Foucault tester is important. Your pupil should receive all the light that passes by the plane of the knife-edge, which you see as a uniform illumination of the mirror. If your eye is too far from the knife-edge, the mirror will not be uniformly illuminated.

As soon as you get a null, measure the mirror's radius of curvature with a tape measure, carefully placing one end in direct contact with the center of the mirror and the other over the knife edge. You should be able to measure the radius of curvature to within a millimeter.

8.2.3 What You Should See

When the apparatus is correctly set up, you can examine the surface of the mirror in detail. In our case, the light source is situated to the left and the knife-edge to the right of the observer's eye. When you move the knife-edge to the left, the de-

Making Fine Adjustments by Pressing on the Foucault Tester

If you press lightly with your finger on the bottom part of the left side of the Foucault test apparatus, the knife-edge moves, by a fraction of a millimeter into the light beam reflected by the mirror. This causes the zone of the mirror that you are testing to uniformly darken. On the other hand, when you lift lightly on the left side of the tester, that zone becomes brighter. This allows you to move ever so slightly the knife-edge without having to turn the transverse screw, which is sometimes too much.

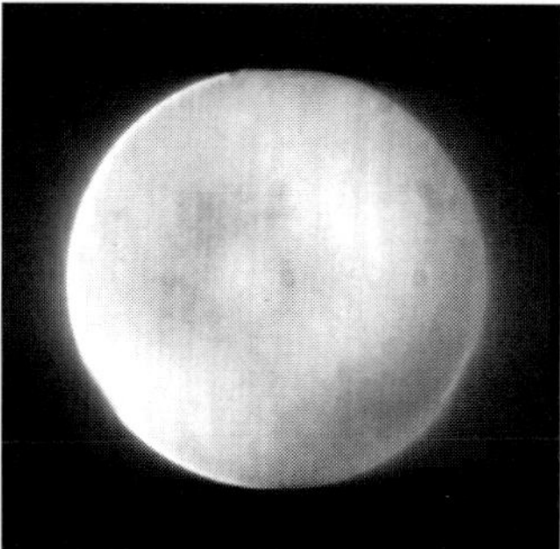

Fig. 8.8 Photograph of a130-mm spherical mirror, at the point where the knife edge nulls the mirror. This is not an absolutely perfect null. A few defects still persist, but their amplitude is less than 5⁄100 micron at the surface of the mirror (or λ⁄5 wave on the glass). The mirror has a slight hole at the center and has a turned-down edge. The darkness at the lower right is the result of a poor alignment of the knife-edge and the slit. This mirror is perfectly usable.

fects of the mirror will come into view and seem to be lit from the left.

In the next section we will review the major mirror conditions that you might encounter during polishing. Remember that the condition you want to end up with is a sphere; that is, when the knife-edge is exactly at the center of curvature, the appearance of the mirror is the "null" (no. 1 on page 115)

Each of the described mirror conditions (except no. 1) corresponds to a particular form of the mirror surface that must be retouched with a full-size tool in order to get a null.

8.3 Retouching during Polishing

The times given for these retouching steps are only approximations and will differ from one person to another. Temperature, hand pressure, and the hardness of the pitch used are all variables that will affect these times. During retouching, always make complete turns around the polishing post.

1. **Null, Form 1 on page 117.** The mirror is perfectly spherical and it's not necessary to retouch it. *If the surface still has gray areas, you must extend polishing until they have all disappeared;* gray areas are signs of incomplete polishing and have considerable negative effects on image quality; continue working, making the same strokes as you did at the start of polishing. Test the mirror hourly.
2. **The mirror has a raised edge, Form 2 on page 117**. After soaking the lap in lukewarm water (about 35°C) for a few seconds, work for 5 to 10 minutes with the mirror on top, using off-center W strokes with a ¼-diameter amplitude. Press *lightly* with the hand on the overhanging edge. See top left drawing in Figure 8.9 on page 118.
3. **The edge of the mirror is turned down, Form 3 on page 117.** First, size the lap to the exact diameter of the mirror and cut very clean channels in the pitch. Then, work with the tool "hot"; that is, with pitch that is beginning to flow (it will have the characteristic odor of warm pitch and conform rapidly to the mirror's surface). With the tool on top, make regular ¼-diameter W-strokes. You must work for an hour so the pitch can stabi-

Principal Forms Seen in the Foucault Test

The figures on this page show the different possible aspects, and the corresponding profiles, of the 130-mm mirror behind the Foucault tester.

Form 1: null

Form 2: raised edge

Form 3: turned-down edge

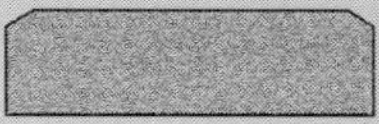

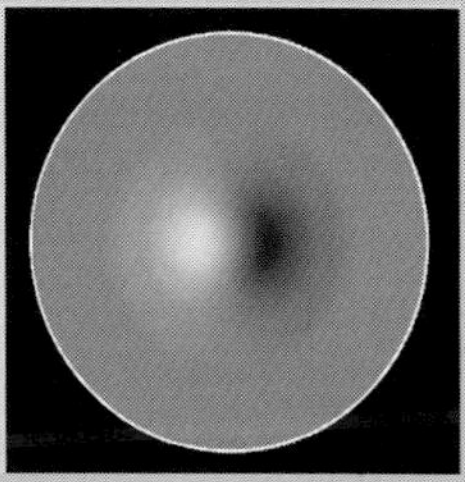

Form 4: raised center

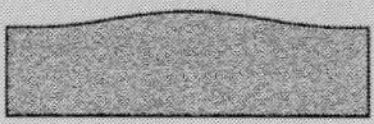

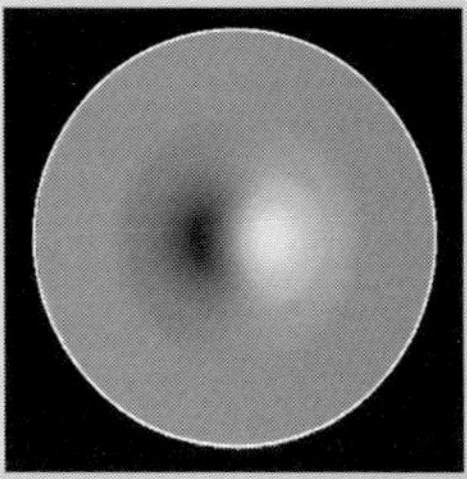

Form 5: central depression

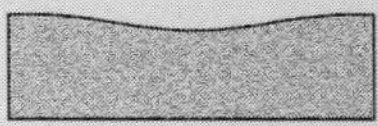

Form 6: central hole and turned-down edge

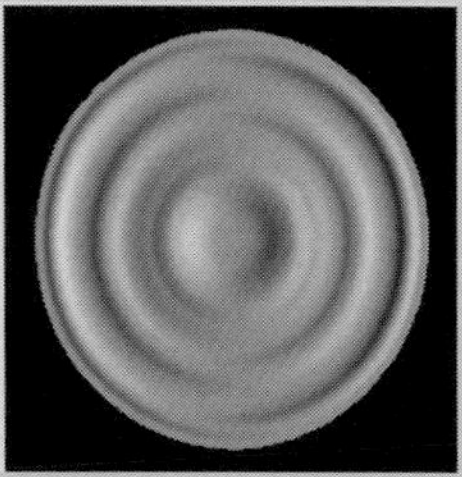

Form 7: zones

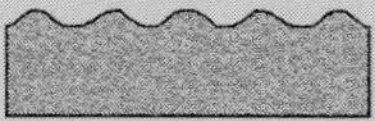

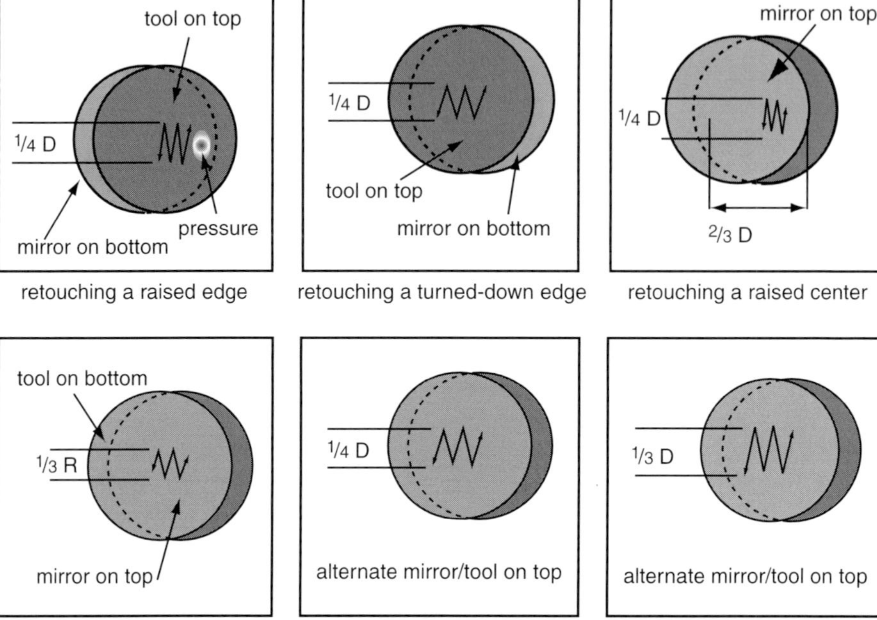

Fig. 8.9 Correcting strokes for six common mirror defects.

lize and flow properly. You'll add about 1 mm of longitudinal aberration (as observed in the Foucault test) per hour at the edge of the mirror. See top center drawing in Figure 8.9.

4. **The mirror has a raised center, Form 4 on page 117.** Starting with a cold or lukewarm lap, work with the mirror on top, the center of the mirror at about ⅔ the diameter of the tool, and make small, ¼-diameter W-strokes. Don't work very long—it only takes about 5 minutes per millimeter of longitudinal aberration. See top right drawing in Figure 8.9.
5. **The mirror has a central hole, Form 5 on page 117.** Work with warmed pitch (the lap should be soaked in hot water and then pressed for a few minutes), making ⅙-diameter W's. Do 2 wets with the mirror on top for each single wet with the tool on top. Work for an hour to polish out 1 mm of longitudinal aberration. See bottom right drawing in Figure 8.9.
6. **The mirror has a central hole and a turned-down edge, Form 6 on page 117.** The mirror is deformed; it is no longer spherical but is perhaps elliptical, paraboloidal, or hyperboloidal. This will happen if your strokes were too large or the pitch was too soft. To correct this defect, begin by sizing the tool to the exact diameter of the mirror and make central ¼-diameter W strokes, alternating the tool and mirror on top. This meth-

Dealing with a Severe Turned-Down Edge

I'm polishing my primary mirror, and it has a severe turned-down edge. A friend advised me to work with the mirror on top. After several hours of work, the problem is still there. What should I do?

By working only with the mirror on top, you have reduced the radius of curvature of your mirror over its central and middle zones. You haven't worked on the edge of the mirror, and it has kept its original radius of curvature. By keeping the mirror on top, you are maintaining the defect. To make it disappear, begin by shaping the tool to the same diameter (bevel the outside edge of the lap 3 or 4 mm) as the mirror and trimming the pitch squares; then, raise the temperature of your workroom (5° to 10°). Polish alternately with the mirror on top and the tool on top, making W-strokes of ⅙ the mirror's diameter in amplitude.

od works the 70% zone of the mirror and gradually returns it to spherical form. See bottom center drawing Figure 8.9.

If the total difference measured with the Foucault tester (with the source and knife-edge moving together) is about 0.8 mm, the mirror is paraboloidal and can be used as is, assuming it doesn't have any dull spots.

7. **The mirror has zones or grooves, Form 7 on page 117.** Alternate mirror and tool on top, making ⅓-diameter W-strokes with a warmed tool and lots of polishing compound. Avoid working with too thin a pitch lap; pitch works best if it is about 4 to 5 mm thick. Grooves can be caused by repetitive strokes of the mirror over a defective area of the polisher; if this is the case, reheat the surface of the pitch and re-press the tool to eliminate the defect. See bottom left drawing Figure 8.9.

As you proceed to polish out your mirror, test it from time to time and adapt your strokes to deal with the conditions uncovered by the test. Normally, it takes about six hours to fully polish a 130-mm mirror.

8.4 When is the Mirror Finished?

This 130 mm *f*/9.2 spherical mirror is finished when the outside edge is fully polished and the mirror shows no grayness or pits. A null should be found covering the surface in a Foucault test, but a null is easier to demand than to accomplish. Since few mirrors perfectly null, try for a condition where the null for the outside of the mirror (see Figure 5.13 on page 57, image at C_2) is farther from the mirror than the null at the center (C_1). If the null like that at C_1 is farther than the null C_2 from the mirror, the figure is drifting into a correction region opposite from that of the parabola, called the oblate spheroid. In addition to the direction of the shift, the difference from this shallow spherical mirror, $C_2 - C_1$, should be less than about one millimeter for a moving-source tester, and twice that for a fixed-source. Check that the edge is not rolled and surface does not display zones and primary ripple. You should wait at least an hour before testing, because the correction of a hot mirror has been known to drift.

8.5 Sending the Mirrors to be Aluminized

Polished glass reflects about 4% of the light it receives. That's not enough if you want to observe dim stars in a telescope. A fine layer of reflective metal film deposited on the optical surfaces—primary and secondary—will increase their combined powers of reflection to about 80%.

The application of a reflective coating to the mirror's surface is the last step in mirror making—this is not within the capabilities of the average amateur and therefore is usually best done by a company that specializes in it.

8.5.1 Reflective Coating Basics

Depending on how the telescope will be used, a very thin film of aluminum, silver, or gold is applied to the mirror's front surface. These coatings are applied by evaporation in a vacuum, except in the case of silver, where the deposit can be chemically precipitated from a solution of silver nitrate. The thickness of the layer is on the order of 1⁄20 of a micron.

The evaporation is done in a vacuum of lower than 10^{-6} mm Hg, in a bell jar; with the optical surface facing down, the mirror (substrate) is placed just above crucibles filled with evaporate (powdered or as thin strips of metal). The crucibles are heated to over 1000° C, depending on the nature of the substance to be evaporated; the metal thus boils and evaporates, and evenly deposits itself on the mirror (and everything else within the bell jar).

Sometimes a sub-layer of chrome is deposited on the mirror, and the silver or aluminum is layered on top of it. This sub-layer does not increase the reflective power of the mirror, but it helps the reflective coating to adhere to the glass. It is optional for aluminizing, but it is necessary for a silver or gold coating, as these materials do not adhere well to glass.

After the reflective coating, sometimes a protective layer of silicon or magnesium fluoride is applied to the mirror. This layer is optically neutral—it does not affect the reflective curve of the mirror. Its role is to provide some protection against environmental chemical attack and very light physical abuse.

8.5.2 Choosing a Coating

Most mirrors are aluminized. This treatment reflects at most 89% of incident light. It resists both moderate physical (touching and cleaning) and chemical (pollution, salt air) wear relatively well.

Silver is more fragile than aluminum. It oxidizes rapidly (forming purple or brown spots) when exposed to moisture. On the other hand, it does have a greater

Latex Gloves and Acetone

Is it safe to use Latex gloves when cleaning a mirror with acetone?

Uncoated mirrors can be cleaned with a cloth dipped in a little acetone. To protect your hands (and to keep the mirror surface free of oil from your hands), you should wear Latex gloves. The small amount of acetone used for this operation will not dissolve the Latex provided you do not soak the cloth with acetone which then "wets" your gloves. Do this with plenty of ventilation.

reflective power (95%), an important advantage for deep-sky observing.

Finally, a gold coating is reserved for infrared applications. The reflective power of gold is as high as 99% at wavelengths longer than 0.6 μm.

In Appendix B beginning on page 289 you will find addresses for companies that provide this service.

8.5.3 Arranging the Details for Your Reflective Coating

An order should always be preceded by an estimate. These are generally free and can be obtained through a phone call or by letter. You must provide the following details:

- The type of glass (Duran 50, Pyrex, BK7, etc.).
- The exterior diameter of the mirror and its thickness (you can send a sketch along with the order).
- The type of coating: aluminum or silver, sub-layer of chrome or not, protective coating (not required for aluminum), etc.
- The clear aperture of the optic. If you want to coat the mirror all the way to the edge, then this number is simply the diameter of the mirror; if the outside edge of the mirror will be diaphragmed, then the clear aperture is smaller than the diameter of the mirror.
- A statement of surface quality (optional). For normal astronomical uses a military specification of "80–50" is justified for a mirror that is very well polished and has no large scratches or grayness.

How Do Coating Companies Clean Optics?

I have heard that mirrors are cleaned with polishing agents before they're aluminized. Is this true? Will my mirror return from its treatment with scratches?

Some professional treatment companies do clean the optical pieces with a solution containing a polishing agent (cerium or zirconium oxide). This procedure has no effect on the quality of a mirror's surface. They rub the surface far too lightly to have any effect.

- You gain nothing by requesting a reflectivity evaluation curve after coating from the overcoater.

The price is determined by the size of the mirror. Aluminizing costs (in 2002) about $60 for a 130-mm mirror, $80 for a 200-mm mirror, and more than $120 for a 300-mm mirror (these prices include the secondary mirror but not return shipping).

8.5.4 Cleaning the Mirror Before Shipment to the Aluminizer

Before sending the mirror off to be aluminized, clean all surfaces well enough to eliminate all traces of pitch, grease and polishing compound.

Professionals use a series of ultrasonic baths and solvents. Amateurs can effectively clean the mirror with acetic acid (to eliminate oxidized spots on the glass) and pure ethyl alcohol or acetone to remove traces of grease (work with latex gloves in a well ventilated area; these chemicals are a fire hazard).

Begin by cleaning the chamfer and the sides of the mirror by rubbing them with a thick cloth towel. If bits of polishing powder remain, scratch them off with a razor blade. In a vacuum, impurities lodged in the chamfer can escape and pollute the metallic surface of the mirror. The metal coating thus becomes more fragile (it degrades faster) and diffuses light. This is another reason to finely grind the chamfer (see Figure 4.11 on page 46).

To clean the mirror's optical surface, rub it lightly moving outward from the center to the edge with a very clean cloth soaked in acetone or alcohol. To locate oxidized spots, breathe on the mirror; the spots will appear as dull points in the haze that forms. To eliminate them, use a drop of diluted reagent grade acetic acid on another cloth, rub the spot, and wipe it off quickly with the acetone-soaked cloth (to avoid leaving traces of acetic acid). The clean mirror should then be handled only with latex gloves and placed directly in its box, optical face toward the bottom.

Problems With Coatings

I sent my mirror to a professional aluminizing service. When I got the aluminized mirror back, it had gray spots and pinpricks on its optical surface. What should I do?

First, when a mirror returns from treatment with flaws in the surface, *never* polish the optical surface in order to clean it. Polishing can cause hundreds of scratches and can even remove the coating.

When the mirror comes back, it can have any of several defects:

- **Oxidation spots.** These appear as small circles when you breathe hot-air fog onto the mirror. These spots result from oxidation of the glass when it is in open air too long, or even from water spots that were not cleaned off before coating. To eliminate these problems before the mirror is sent off, clean the surface with a cloth dipped in strong acetic acid, and then remove any residue with acetone.

• **Cleaning spots.** These are traces left by the drying cloth right before metallization. These spots are the responsibility of the company that did the treatment.

• **Pinpricks, gray spots, and scratches.** These defects, often overlooked on an uncoated polished surface, reappear when the mirror is treated. In most cases, the problems occur because the mirror was not sufficiently polished. Gray spots appear as dull areas, usually near the center of the periphery of the mirror. If the nature of the defect is not apparent, observe it with a strong loupe under a bright light. If it's a gray spot, you should easily see the pinpricks.

Most of the time, defects noticed when the mirror is returned from treatment are not the fault of the aluminizer. Fortunately, most problems are cosmetic and usually have no effect on the image quality.

In any case, before sending your mirror back to the aluminizer and blaming them for the problem, make sure that you have performed all of the surface-quality tests before you send the optic the first time. Among these methods, we recommend the Foucault test (see Section 5.4 and subsequent sub-sections beginning on page 52), the phase-contrast method (see page 81), and the black-background (black box test) method (see page 79) which is employed throughout the optical industry because it detects defects only a few microns in size on uncoated mirrors.

When doing the black-background test, take special care to use a unfrosted lamp which will allow you to see very small pinpricks and scratches which will show up best very near the reflection of the filament as long as the mirror is also reflecting the black background.

The black-background test is more efficient than the Foucault test or the phase-contrast test. In effect, these tests do not show all the scratches and pinpricks on a mirror at the same time. They show defects whose visibility varies with the orientation of the knife-edge and the slit source. To analyze the whole surface of a mirror, you would have to turn the slit and the knife-edge simultaneously and make several observations over a full half-turn.

Only after you have performed all of these tests and are certain that you have cleaned your mirror correctly should you send your mirror back. It will then be much harder, if there is any problem, for the coating company to argue that these tests were not done. On the other hand if you do these tests and correct any problems that they reveal, you almost certainly will receive back a perfect coated mirror without any blemishes that you did not know were there beforehand.

8.5.5 Packaging the Mirror for Shipping

The mirror must not be allowed to move about in its shipping box. To prevent movement the space around the mirror may be filled with foam rubber or tightly crumpled newspaper. If the mirror is being sent by mail or a parcel service, put the box in the center of a larger carton filled with foam rubber, bubble wrap, foam peanuts, etc. If in doubt, contact the shipper to determine what kind of packaging is required for them to insure the safe delivery of your mirror. Note that if you insure your mirror and it is not properly packed, you will not be paid if you experience damage.

8.5.6 Checking the Reflective Coating

Reflective coating usually takes from one to two weeks, depending on the company. You must always check the mirrors upon their return. There are rarely gray spots (possibly because of poor cleaning) or finger marks under the coating. If the defect is a problem, and if *you are certain that it's not your fault*, it is always possible to negotiate a free removal and re-coating as compensation.

> *Note: Be careful of scratches! The coating operation does not create scratches on the glass and does not make them bigger either. It simply makes defects appear that were invisible before treatment. Highly reflective coatings make gray spots strikingly obvious... was the mirror polished for a sufficient time?*

9 Making the Secondary Mirror for a 130-mm *f*/9.2 Newtonian

A complete plan for making a Newtonian secondary diagonal mirror is detailed below. However, *this step is not essential in the construction of the 130-mm telescope*. Although purchasing a finished mirror may seem to go against the spirit of this project, beginners are advised to buy their secondary mirror. A high-quality diagonal mirror with a 25-mm minor axis costs from \$15 to \$50. You can order one, with perhaps a certificate to guarantee its precision, from makers and sellers of optical components.

9.1 Bill of Materials and Organization of the Work

A Newtonian secondary mirror is inclined at a 45° angle to the optical axis. It can be rectangular, octagonal, or elliptical. For the 130-mm telescope the specifications call for a clear aperture of 25 by 36 mm; in reality, you're going to make it slightly larger (about 28 by 40 mm) to compensate for the chamfer and the turned-down edge that will appear on the piece after polishing. Making a rectangular diagonal mirror is within the reach of any amateur, requiring only the purchase of a glass cutter. The mirror can also be cut out using a 40-mm buffing wheel (1 mm thick), that you can buy in modeling shops and mount in a Dremmel® tool. Cutting an octagonal mirror is somewhat more difficult and requires some experience in glass cutting. Making an elliptical mirror, the most complex and difficult shape, is described in Section 15.2 on page 177.

The following materials will be necessary:

- A 5-mm thick piece of plate glass, 100 by 100 mm square, to make the diagonal blank
- Two glass disks, 100 mm in diameter, to be used as smoothing tools
- An aluminum disk, 100 mm in diameter and 10 mm thick, to be used as a blocking support (if you can't get this, use a glass disk) for the diagonal blank
- Optician's blocking pitch (see the recipe on page 128)
- Smoothing abrasives and polishing compound, as used on the primary mirror

The work is organized in the following manner:

1. Draw the desired outline of the mirror on the plate glass with a permanent marker.
2. With a glass cutter and using the outlines as a guide, cut out the mirror, then bevel the edges with a grinding stone.
3. Using optician's blocking pitch, pitch the diagonal blank to the middle of the aluminum disk. Also pitch pieces of scrap glass (same thickness) around the diagonal blank to approximate a circle (this assembly is called a *block* and the surrounding scrap glass *surrounds*. See Figure 9.5 on page 128.).
4. Smooth and polish the block using the two 100-mm diameter glass tools according to the three disk method described in Section 9.4.1 on page 129.
5. Test the flatness of the fully polished blocked diagonal mirror.
6. Disassemble the block and again test the diagonal for flatness.
7. If necessary, retouch and retest the diagonal.

9.2 Cutting the Diagonal Mirror Shapes

9.2.1 Cutting out a Rectangular Mirror

Smooth one of the faces of the square glass plate with #220 carbo, by grinding it for a few minutes on another piece of plate glass. When the face is completely frosted, draw the pattern shown in Figure 9.1 with a permanent marker. You can make six 33 x 40-mm rectangles (shown in gray). The rectangle that has the fewest chips after cutting it out will become the secondary mirror.

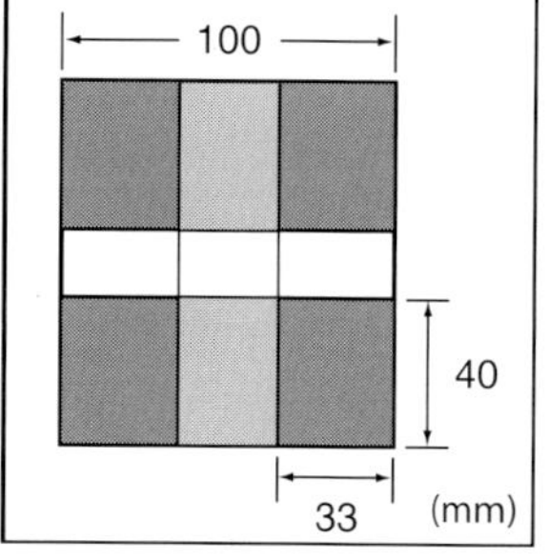

Fig. 9.1 Layout for cutting six diagonal mirror blanks.

Using a ruler as a guide, cut along the lines with the glass cutter, pressing hard and without going all the way to the edges. Place the glass plate on a table, aligning the cut lines with an edge. Cover the plate with a cloth and, holding the larger part of the plate on the table with your hand, strike the overhanging part with the side of your other hand or a block of wood. The glass should break along the scribed line. Repeat this operation as many times as necessary to get a good piece. Next, smooth all of the sharp edges and corners with a fine grinding stone (180 or 240 grain). Then grind a 0.5-mm chamfer around the outer edges. Finally, re-measure the dimensions and retouch, with a carbo stone if necessary, to arrive at the mirror's specified length (40 mm) and width (28 mm).

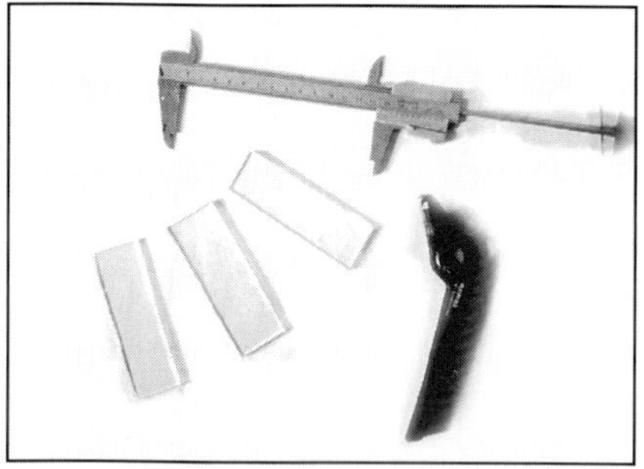

Fig. 9.2 The 100 x 100-mm glass plate cut into three identical pieces.

Fig. 9.3 Left: Cutting the glass blank with the diamond-tipped glass cutter. **Right:** The best rectangular piece is chosen to become the secondary mirror.

9.2.2 Cutting Out an Octagonal Mirror

To make an octagonal mirror, remove the corners from a rectangular blank. Draw in the corners to be removed. Next, make a 45° chamfer on the sides of the rectangle to avoid chipping when making the "corners" of the octagon. See Figure 9.4.

With a #60 or #80 carbo stone, grind the corners of the mirror down to the drawn reference lines. It will take 15 to 20 minutes per corner to obtain a clean, regular octagon. At the end of your work, make a 45°, 0.5-mm bevel around the entire piece.

9.3 Blocking the Diagonal Mirror Blank

Heat about 100 g of optician's blocking pitch (per the recipe on page 128), making a viscous paste, that is warm enough to comfortably mold with your hands, but not sticky; if it is sticky, this is a sign that it is too hot.

Warm the pieces of cut plate glass (the mirror and the surrounding scraps which are called "surrounds") and the aluminum disk to the same temperature as the blocking pitch. For this, place the pieces, optical face down, on a hot plate cov-

Fig. 9.4 Left: Making an octagon from a rectangle with a carbo stone. This operation is repeated 4 times. **Right:** The octagonal mirror is cut out; it is now time to work on the surface.

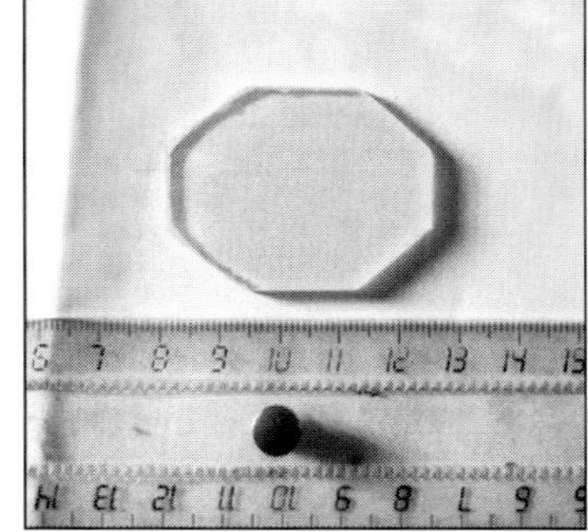

Making Optician's Blocking Pitch

Optician's blocking pitch is a 9-to-1 mixture of black sealing wax and pitch. Heat the wax first to 60–70° C in a large saucepan, and then add the pitch before stirring.

Optician's blocking pitch is used to make cloth polishers (for rapid polishing) and to attach several pieces of glass on the same support in order to work them together (this operation is called *blocking*).

The blocking pitch adheres to hot metal and to polished and unpolished glass. It releases when tapped by a mallet (done very carefully) or after several minutes in a freezer. It dissolves in trichloroethylene or acetone. The principle property of the blocking pitch is its plasticity; it can mold itself slowly to the pieces that it is working, deforming them as little as possible. By varying the proportions of the wax and the pitch slightly, or by adjusting the cooking time, you can modify the plasticity of the blocking pitch. It becomes harder if it cooks longer; adding a little soft pitch makes it softer.

ered with several layers of aluminum foil. Next, form in your hands small balls of blocking pitch 10 to 20 mm in diameter and place them, without pressing, on the backs of each piece of glass. Next, take these pitched pieces of glass immediately off the hot plate and let them slowly cool. When the blocking pitch hardens, arrange each piece with its optical surface face down on one of the glass disks. Carefully inspect the contacting glass surfaces and remove any dust, abrasive grains, and stray blocking pitch. The diagonal mirror blank is placed at the center of the disk (see Figure 9.5).

To finish the block, place the warmed (60–70° C) aluminum disk in contact with the exposed pitch balls. The blocking pitch should slowly flatten *under the weight of the disk alone*. Make sure the aluminum disk is centered over the glass pieces. This operation is called *blocking*. You must wait until the block is completely cooled (1 to 2 hours) before moving it.

Carefully inspect the blocked pieces of glass. If any of the blocking pitch somehow has crept onto the face of one or more pieces of glass, it should be immediately removed. If this condition is severe, that is, if there was a great deal of pitch and the glass plate is tilted, the block should be rebuilt. Tilting can be detect-

Fig. 9.5 A block made to work an octagonal mirror (at center) surrounded by similar scraps of plate-glass arranged to make a more or less circular disk. The black color comes from the blocking pitch.

Blocking Pitch—How Thick?

What thickness of optician's blocking pitch should I use to pitch the glass pieces to the block?

Optical glass and surrounds are pitched to the block with small balls of optician's blocking pitch. These balls measure about 10 to 20 mm in diameter. The thickness of these balls is a function of the size and number of the balls. The optimum thickness is between 5 and 10 mm.

Remember that when it is cooled, optician's blocking pitch should be hard, dull and coarse. (The recipe on page 128 will produce such a blocking pitch.) If it is shiny when it dries, it's too rich in pitch, and it may not be sufficiently rigid.

ed early when, after about 10 minutes of smoothing, portions of one or more glasses on the block show no wear.

It is equally important not to get any blocking pitch on the sides of the blocked glass. This can become a problem during smoothing or polishing and frustrate those processes. In the case where this happens, soak the block in a strong solvent like trichloroethylene or acetone for a few minutes and then clean the affected areas with a clean cloth (**Be Careful**—these solvents are very toxic and flammable; make sure you take the appropriate precautions when using them).

9.4 Smoothing the Diagonal

When you make a diagonal mirror, you start with polished plate glass which is fairly flat to begin with so there are normally no rough- or fine-grinding steps. To begin smoothing, check to see that all of the glass pieces on the block are at the same height by doing a few smoothing wets. Begin work with 500 aluminum oxide, and, using one of the glass tools, do several quick wets. This will show the high (abraded) and low (abrasion free) areas on the block. If the glass is not abraded evenly across the entire block, continue smoothing with a coarser abrasive, like 220 or 320 aluminum oxide; but don't use carborundum—the glass fractures caused by carbo are very deep and hard to remove.

9.4.1 The Three-Disk Method

To produce a flat mirror by hand, the time-proven method is to systematically grind three disks of the same diameter against each other—the block and two 100-mm glass tools. Working with the block and a single tool only, you will create complimentary spherical surfaces with curvatures that are weak—too curved for this application. The block and the two glass disks should be numbered 1, 2 and 3 on their backs or sides with a felt-tipped marker—the block is number 1. To make a flat, you must work in set order without favoring the block or the two glass tools.

An Alternate Blocking Method

An alternate method for blocking the pieces is to use several smaller drops of optician's blocking pitch, from 2 to 3 mm in diameter, instead of the 10- to 20-mm drops. Here's how to do it:

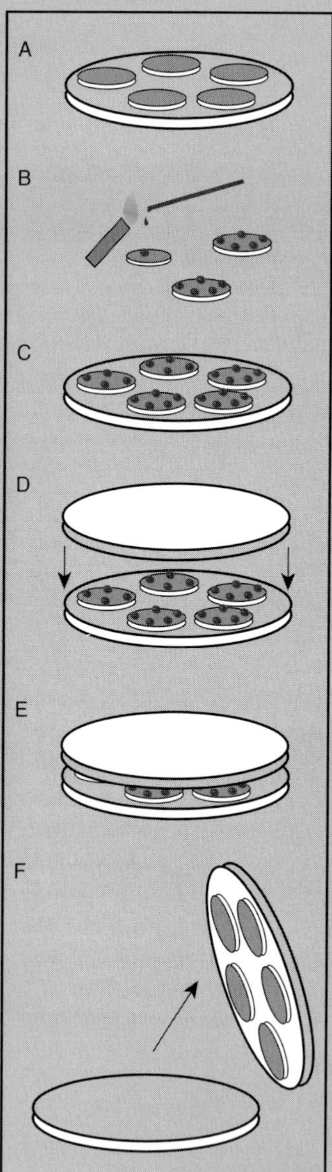

Fig. 9.6 *Blocking using* small drops of optician's blocking pitch.

- In a sauce pan, heat some blocking pitch just enough to soften it somewhat, but *do not melt it into a liquid.* Test the temperature by touching the pitch with a wooden block. The warm pitch should be viscous but not stick to the wood.
- Remove a little bit of blocking pitch with your hands (the temperature must be high enough to make the pitch malleable but low enough that you do not burn your hands; (it should not stick to your fingers which is a sign that it is too hot) and roll it into a cylinder about 1 cm thick and 15 to 20 cm long. Let this cylinder cool down and harden.
- Place the glasses to be blocked, optical face down, on a reference flat that's clean, degreased with a solvent, and dust-free. This flat support should be smoothly ground if the glass pieces are, and polished flat to ¼λ if your glass is polished (Figure 9.6, step A).
- Using a butane lighter or a gas burner, melt the end of the cooled pitch cylinder. When the pitch liquefies, place 5 or 6 drops, evenly spaced, on the backs of the optical pieces (Figure 9.6, *steps B and C*).
- Using an electric hot plate, heat the aluminum plate to about 60°C. Test the temperature by rubbing a piece of blocking pitch on the plate; if the pitch leaves a trace on the plate, like a crayon, the temperature is right. If it leaves no mark, the plate is too cold; if the pitch melts and sticks to the plate, the plate is too hot.
- Carefully place the heated aluminum plate, centered over the ensemble of pitched pieces (step *D*). *Do not press on the plate.* Let the weight of the plate do the pressing (step *E*). If you press on the plate, you will introduce mechanical stresses that you'll rediscover when you remove the pieces. If the aluminum plate is not centered over the optical pieces, you can push it laterally to true it up. Let the assembly cool down completely before removing the block (step *F*). Do not try to cool it down faster by dunking it in water, the glass will crack.

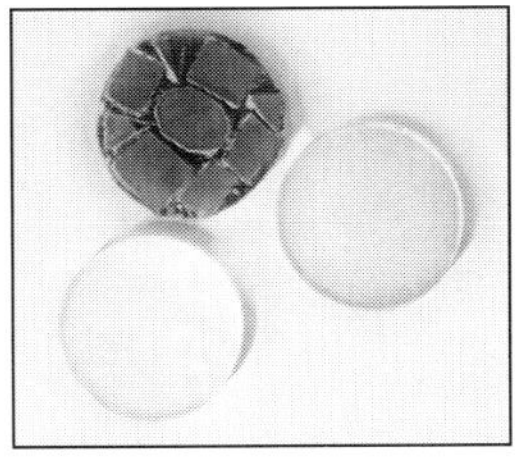
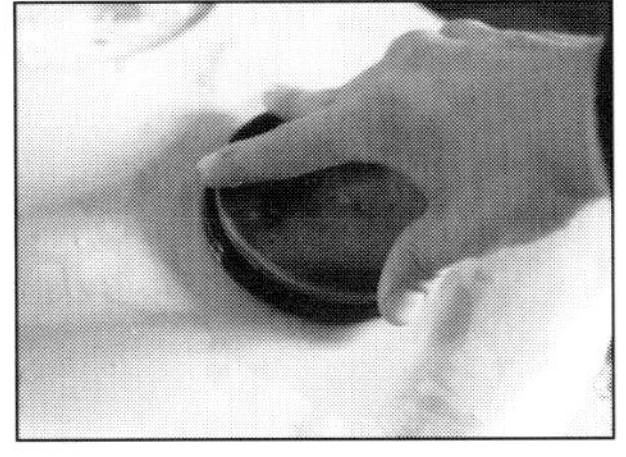
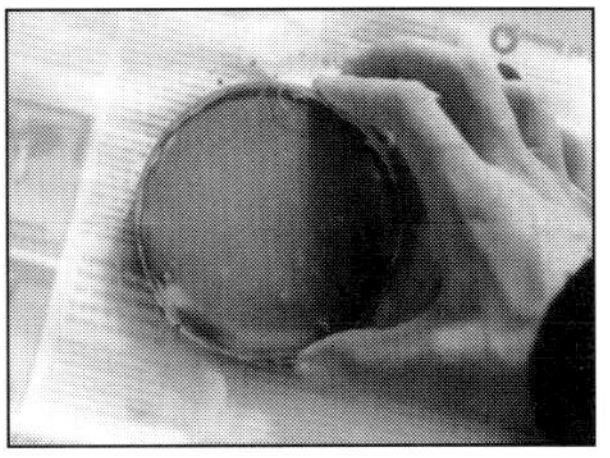

Fig. 9.7 Left: The block and the two smoothing tools. **Center:** Pressing the polisher with one of the glass tools. **Right:** After pressing, the pitch contacts the glass disk over its entire surface.

An example of such a work plan is as follows.

Wet #	**1**	**2**	**3**	**4**	**5**	**6**	**7**	**8**	**9**	**10**	**11**	**12**
Disk on top	1	3	2	2	1	3	1	3	2	3	1	2
Disk on bottom	2	1	3	1	3	2	2	1	3	2	3	1

As you did when making the primary mirror, grind with small W-strokes of ¼- to ⅓-diameter amplitude. The lateral movement of the strokes should not exceed ⅙ of the diameter. As you work, lightly hold the pieces by their sides, putting your fingers flat on the back. Turn the top disk in your hands from time to time.

9.4.2 Testing the Flat Mirror During Smoothing

As you work on the secondary, you should regularly test the surface's uniformity with an inverted eyepiece or photographic lens (see Figure 5.5 on page 51) and the flatness of the faces with a spherometer or machinist straight edge.

The surface should be uniform before you move to a finer grade of abrasive. Because you are using a much finer grade of abrasive to start with and the glass itself was polished and nearly flat, the surface should have fewer pits than the primary did at this stage.

When you test the flatness of the faces, use a spherometer equipped with a precise micrometer screw with which you can measure differences in sagitta of a hundredth of a millimeter. An experienced user can eventually estimate to a ½ hundredth on a screw with gradations of 1/100. If you do not have a spherometer, a precision-ground machinist square can be used.

If, during a wet, an air bubble appears between disks 2 and 3, it means that there is a difference in the curvatures between the two disks. Put the spherometer on each of the disks and turn the screw just until it is in contact with the glass (this will take practice, the contact must be very light or the spherometer will either deflect or lift off the glass). Note the value measured on each of the three pieces; the highest corresponds to the most convex piece and vice versa. The defect in curvature is corrected by working with the most convex disk on top of the most concave for 4 or 5 wets.

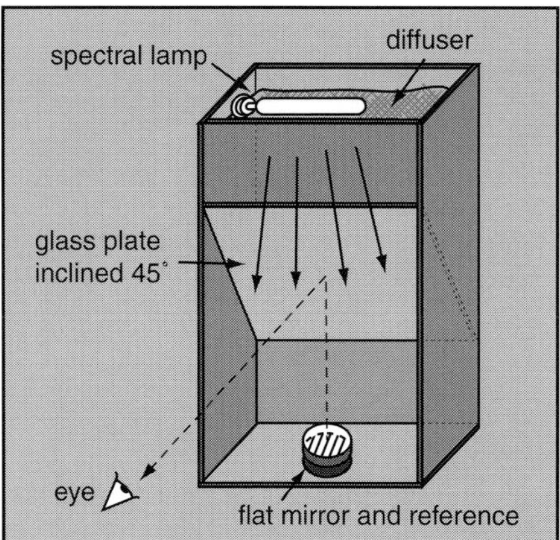

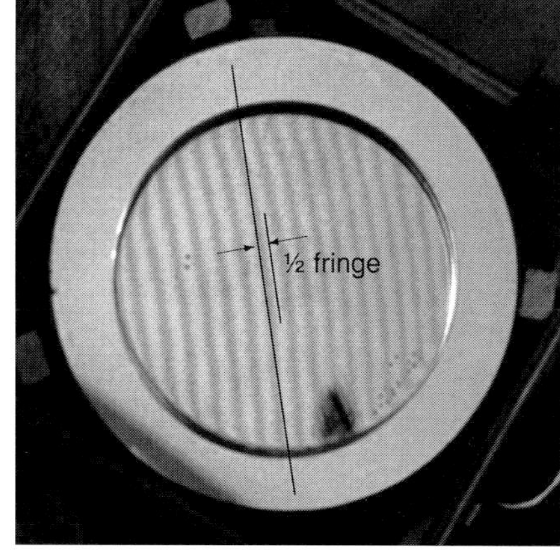

Fig. 9.8 Left: An easy to construct interference tester. The mirror is set on the reference flat under a glass plate inclined 45° which acts a beam splitter. The lamp is placed behind a diffuser above the glass plate, and the resulting fringe pattern between the test and reference optics is observed via reflection off the inclined glass plate. This device allows the observer to see the fringe pattern straight on without distortion. **Right:** Test of a 100-mm mirror on a 150-mm reference flat, using a sodium lamp. The defect observed measures ½ fringe, or $\lambda/4$ on the glass.

9.4.3 Testing for Flatness at the End of Smoothing

With a flat reference mirror, you can test the flatness of the block during the final stages of fine grinding by giving the surface a flash polish on a pitch lap charged with cerium oxide. A flash polish is reflective enough to obtain faint interference fringes against a polished reference flat.

If the defect in curvature is larger than 10 to 12 fringes, it is best to continue abrasive smoothing, where you can correct the curvature in three or four wets. Use 5-micron aluminum oxide and work with the most concave disk on top of the least concave disk. It is easy to confuse the appearance of concave and convex fringes. Here is how to tell the difference—when you press at the center of the rings, the circular fringes of a *convex* mirror get *larger*, whereas those of a *concave* mirror get *smaller*.

During the first few hours of polishing, the curvature of the mirror may vary by several fringes; this variation is caused by the Twymann effect. Grinding sets up stress in the glass which "push" the molecular structure of the glass outward, making the glass convex. When the ground surface is polished, this stress is gradually relieved and the mirror becomes slightly concave.

9.5 Polishing the Flat Mirror

Making a flat polishing tool is similar to making the polisher for the primary mirror. Pour polishing pitch on one of the two glass tools. Make sure to warm this tool before pouring the pitch to prevent thermal shock. The remaining glass tool should also be warmed and then pressed against the lap to give it a flat surface.

Note: It is not necessary to offset the pitch facets when working a flat—in fact, you do not even need facets—all that is needed is a way for the pitch to flow. For example, you can use concentric circular channels laid out with a compass and "opened up" with a razor blade.

The strokes employed in polishing the flat mirror are similar to those used in polishing the spherical primary mirror—W-strokes of ⅓ diameter amplitude and ⅛ diameter lateral movement. Press as little as possible on the superior disk, because you run the risk of deforming the mirror in an irregular manner. In general, when polishing small mirrors, the hands should manipulate the top disk only from its sides.

9.6 Testing and Retouching the Flat Mirror

9.6.1 Testing

Testing is done using a reference flat. For our 130-mm telescope diagonal, you can be satisfied with a reference flat that is precise to *at least* ⅛ of a fringe on a 30-mm diameter, which is superior to the minimal precision obtained at the end of smoothing (⅓ of a fringe on the minor axis). Testing with the interference-fringe method can begin at the start of polishing. This permits you to become familiar with the glass's behavior under differing strokes (concave or convex) and to adjust your polishing accordingly.

Before testing the mirror, you must thoroughly clean the optical surfaces that will come in contact with each other, using a soft cloth and a bit of moisture from your breath. Position the block on a stable horizontal support. Carefully place the reference flat on the blocked secondary without allowing the pieces to slide against one another. To prevent scratching, place 3 pieces of cigarette paper, which is about 25 μm thick, between the disks at 120° intervals. The glass pieces should not be allowed to strike each other, and you should never hear the crunch of even one abrasive grain as the flat is lowered on the block. When the reference flat has been placed on top of the diagonal mirror, interference fringes will appear (under a monochromatic light: a neon tube, mercury- or sodium-vapor lamp, etc.). Wait half an hour before touching either piece so they can come to thermal equilibrium—a "cold" reference placed on a "hot" mirror will give false results.

9.6.2 Retouching during Polishing

Interference fringes between a test plate and a known standard allows you to determine the form of your diagonal mirror. Only the glass at the center of the block, the secondary mirror, matters. Press lightly on the reference flat, or raise it slightly, to make 5 or 6 fringes appear across the flat mirror. Draw their shapes, to scale, on a piece of paper. The fringes will show an enlarged profile of the flat mirror. If the fringes are curved, press on the reference flat: if the fringes shrink, the mirror is concave; if they expand, it is convex.

It is easy to correct a convex defect: slightly lengthen the amplitude of your polishing stroke with the mirror on top. Work slowly and test often.

Correcting a concave defect requires a different technique. Fringe testing will tell you where the raised and lowered zones are on the glass. Cut out a circular or annular paper mask the size of the defect; this mask will be applied to the tool at the same position as the low zones of the mirror. Heat the polisher and press the paper mask, using one of the glass tools, until you have a depression about 1/10 mm deep.

To retouch a minor defect, use a razor blade to scrape the lap lightly where the interference fringe test shows a low zone on the block. Return to polishing for about ten minutes, making short strokes (1/4 radius total amplitude) and turning the top piece frequently in your hands. To retouch a regular (not zonal) defect, cut the edges of the paper mask in a sawtooth pattern. This eliminates the creation of a "stair-step" on the mirror, as the transition between the high and low parts of the polisher is smoothed out.

Stop polishing when the largest defect is smaller than 1/3 of a fringe on the minor axis and 1/6 of a fringe on the major axis.

9.7 Removing the Mirror from the Block

Removing the diagonal mirror from the block is a delicate procedure, but it's not dangerous for the mirror. Be careful of the solvents that are used to dissolve and clean up the optician's blocking pitch; **these products are both toxic and flammable and must be handled with great care.**

The mirror can be removed from the block by one of two methods. You can submerge the block into an acetone or trichloroethylene bath and wait a day for all the blocking pitch to dissolve, or you can place the block in a freezer at a temperature of –18°C for at least an hour. When the block is removed from the freezer, the pieces will come apart easily when pinched with the fingers.

Most of the time, the mirror deforms after it is removed; the curvature may vary by several tenths of a fringe, depending on the quality of the blocking pitch and the care with which the pieces were mounted. In our case, the fabrication tolerances are rather large, and a mirror that conformed to them before being removed will most likely conform afterwards.

9.8 Retouching of the Flat Mirror after Removal

After you have removed the diagonal from the block you may find that it has deformed and is outside the limits of tolerance; you will need to attempt retouching.

The flat mirror is very thin, and it is not worth the risk of trying to retouch by hand because it will almost certainly deform. There are two possible solutions, and both unfortunately are somewhat problematic.

The first method is to lay the mirror, without constraints, on a flat support and to polish it directly with a finger. For a small mirror, the tip of the index finger makes an ideal polisher. Use a little bit of zirconium oxide or rouge. Work only the zones that are too "high" (as revealed by the interference fringes), pressing only lightly with the finger. How long you should polish depends on the pressure of the finger, but you must take care not to press hard; you risk cutting a zone which can happen quite quickly. It is best to work a little and then test to see how the defect responds to your efforts.

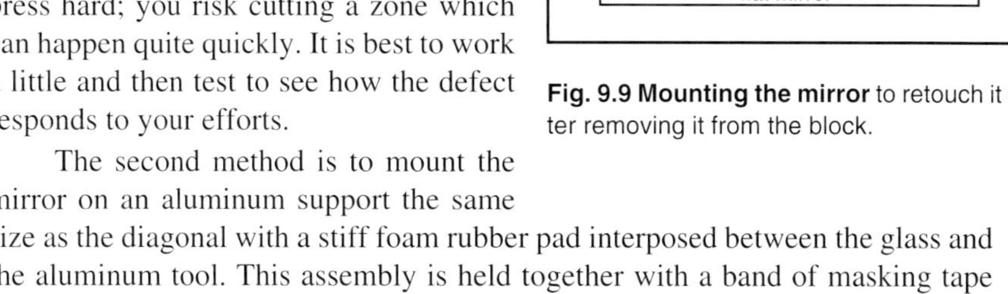

Fig. 9.9 Mounting the mirror to retouch it after removing it from the block.

The second method is to mount the mirror on an aluminum support the same size as the diagonal with a stiff foam rubber pad interposed between the glass and the aluminum tool. This assembly is held together with a band of masking tape wrapped around the edge of the ensemble (see Figure 9.9). Make a flat pitch polisher 50 mm in diameter. To retouch the piece, hold it by the metal backing plate. The major disadvantage in this operation is that it almost always causes a large turned-down edge. You should attempt this only as a last resort.

When the secondary mirror is finished, send it, along with the primary mirror, to the aluminizer. See Section 8.5 starting on page 120 for details on how to go about this.

10 Making the Tube and Optical Component Supports

The tube and optical component supports described here are intended to be primarily indicative and illustrative of a functional telescope. They can be modified according to your taste, tools and woodworking abilities. All measurements are in millimeters.

10.1 The Octagonal Tube

Our proposed tube is octagonal; that is, it's built using eight boards. This type of tube is very rigid and thus well-suited for a long focal length telescope.

The primary mirror is 130 mm in diameter and its focal length is 1200 mm. The telescope tube will be 1280 mm long. There is a 30-mm space between the mirror and the interior walls, so that air can freely circulate within the tube. The interior diameter of the tube therefore measures 190 mm.

10.1.1 Making the Octagon Tube

- Cut out eight 1280 x 84-mm boards from 5 mm thick plywood, bevel the two long sides of each at a 45° angle with a table saw (see Figure 10.3 on page 138). Make sure that all eight boards end up with the same identical width and length.
- Cut out the three baffles (e.g., baffles *E*, *J*, and *N*). Note the differing central hole size, E = 140, J = 150, and N = 160 mm. See Figure 10.1. Piece *E* measures 15 mm thick. It supports the primary mirror holder with three metal

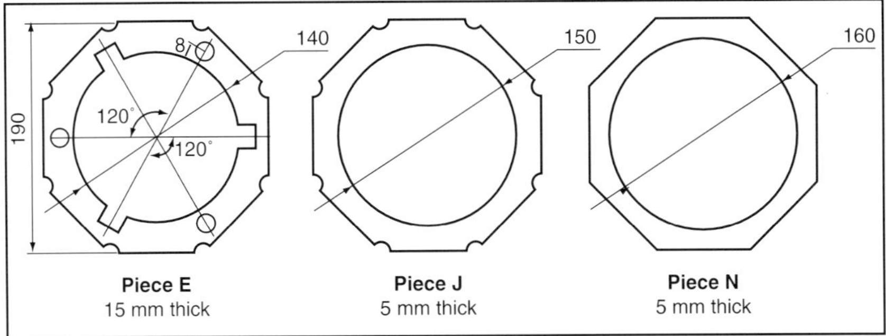

Fig. 10.1 View of the faces of baffles *E*, *J*, and *N*.

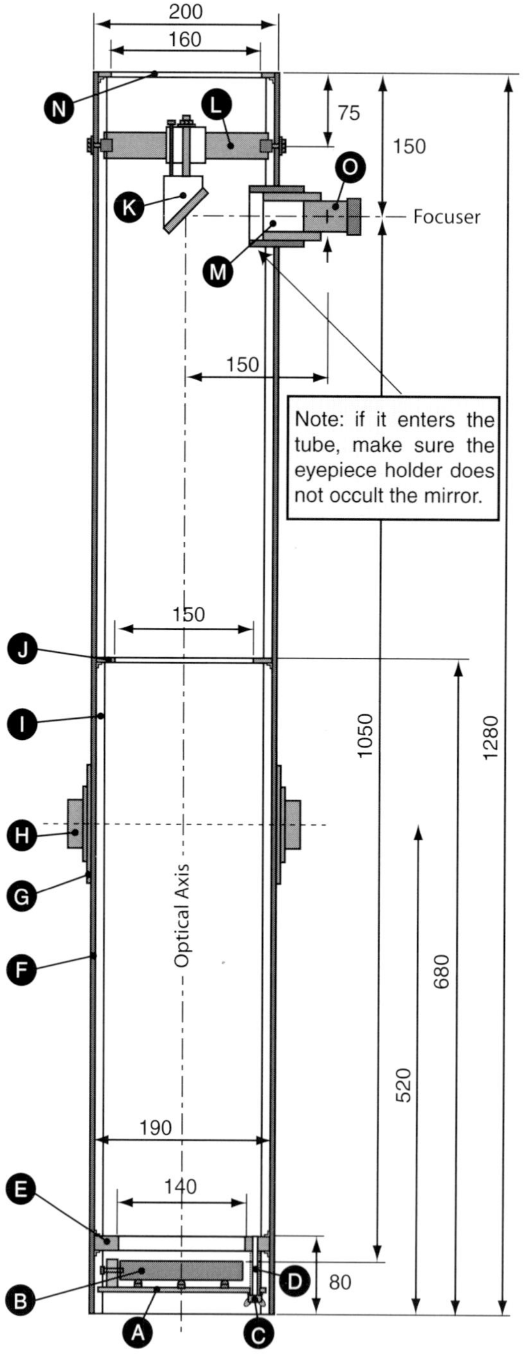

Note: if it enters the tube, make sure the eyepiece holder does not occult the mirror.

Fig. 10.2 (Left) The Telescope Tube

All dimensions in millimeters.

Scale is about 1:7.

Legend:

A. Primary mirror cell.

B. Primary mirror.

C. Wing nut (8 mm diam.).

D. Adjustment screw for primary mirror cell.

E. Support for primary cell and baffle, 140 mm diam., 15 mm thick.

F. Side of tube (8 panels, 84 mm x 1280 mm, 5 mm thick).

G. Side bearing reinforcement pad (5 mm thick).

H. Side bearing (vertical axis bearing) 70 mm diam.

I. Girder (half-round, 15 mm diam. wood trim, 1275 mm long). Note that the inside surface (side opposite the curve) is shaped with a 135° angle to butt tightly against the tube wall.

J. Baffle, 150 mm diam., 5 mm thick.

K. Secondary mirror mounted in its support.

L. Spider vein (5 to 10 mm thick).

M. Eyepiece holder.

N. Baffle, 160 mm diam., 5 mm thick.

O. Eyepiece.

Fig. 10.3 (Below) Cross-section of the tube. Note the overlapping of the panels. Part *I* is one of the eight girders glued between two panels.

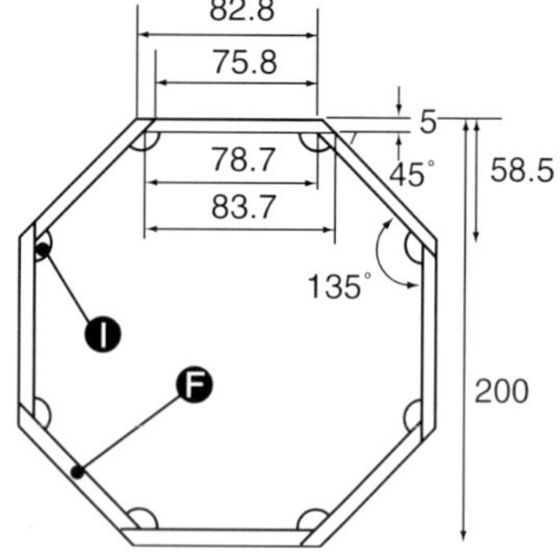

How to Assemble the Tube Parts

Use wood glue to attach the various elements to one another. A well-glued mount is more rigid and more stable in varying weather conditions than a nailed or screwed mounting. Before gluing two pieces, clean and sand their surfaces with fine sand paper. Coat the surfaces with glue and press them together using 5- to 10-kg weights or clamps. You must always wait for the glue to dry completely before removing the weight or clamps and moving the pieces. The assembly of the various parts of the tube should be done with care and patience.

bolts, 6 mm in diameter and 70 mm long (see *D* in Figure 10.2). Pieces *E*, *J*, and *N* determine the clear aperture of the telescope and allow the primary mirror to be fully illuminated. These diaphragms baffle extraneous light and increase the tube's rigidity.

- Attach pieces *E* and *J* to one of the eight octagonal boards. Next, assemble the panels with the half-round girders (see *I* in Figures 10.2 and 10.3) which have their underside shaped to form a 135° angle. These girders are shorter than the panels by 5 mm. This difference corresponds to the thickness of piece *N* that caps the top of the tube. The half-round girders should therefore be positioned flush with the bottom of the tube as you build your assembly.

Where the side bearings will be attached to the tube, reinforce the tube by gluing a plywood reinforcing block, 5 mm thick and about 120 mm wide.

10.2 The Primary Mirror Cell

The mirror cell is a circular plywood plate between 8 to 10 mm thick (see *A* in Figures 10.2 and 10.4 on page 140). Its diameter is about 180 mm. This piece supports the primary mirror through three pressure points in the back of the mirror (piece *Q*) and three bolts fixed in the lateral blocks (piece *R*). These bolts permit the centering of the mirror in the telescope tube. Leave 0.5 mm of play around the mirror, because the mirror absolutely must not be mechanically constrained (it is completely normal to hear the mirror move in its mounting during observations). You can change the orientation of the holder and the mirror with the three regulating bolts (*D*). The security pads (*P*) prevent forward movement of the mirror if the telescope points below the horizon. These pads are made of aluminum or of plastic from milk cartons 2 to 3 mm thick. In normal usage conditions, they should not touch the surface of the mirror; therefore, allow 0.5 mm of space that you will fill with small pads of cork. This way, the mirror will not easily move out of alignment.

10.3 The Spider and the Secondary Mirror Support

The spider has four veins (*L*) which centrally position the secondary mirror (*K*)

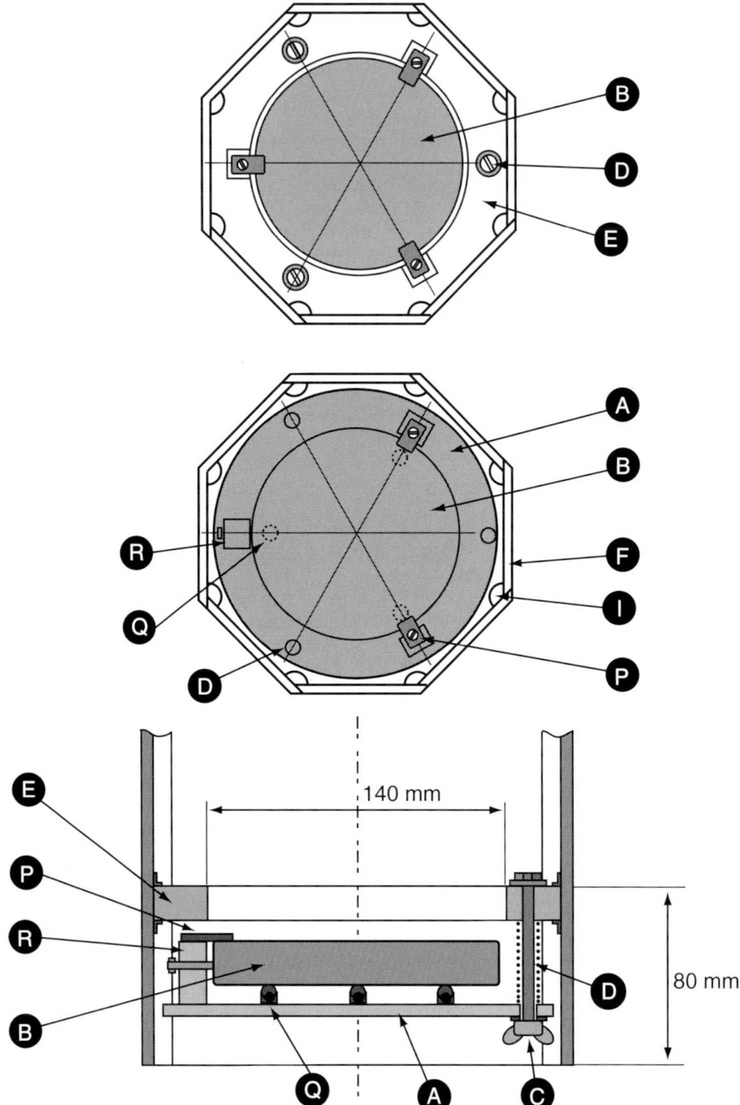

Fig. 10.4 Top: The view looking down the tube from the top showing the mirror (*B*) and the diaphragm (*E*) **Middle**: Same view with the diaphragm (*E*) removed. Beneath the mirror are three support points for the mirror (*Q*) that can be brass bolt heads, small pieces of Teflon or Nylon. Below these three support points is a circular disc of plywood 180 mm diameter and 10 mm thick (*A*) which is the base of the cell. The mirror is restrained laterally by three Nylon bolts spaced at 120° apart, which pass through a hole (bored for a smooth but tight fit with the bolt) in lateral holding blocks (*R*) to a steel Tee nut. Above the mirror, three plastic security pads (*P*) prevent the mirror from falling forward. **Bottom:** Cross-section of the mirror cell. Note the position of the cell's control bolt (*D*), only one of the three are shown here. The nut (*C*) controls the mirror cell tilt to facilitate alignment while tension is maintained by a compression spring around the bolt (*D*).

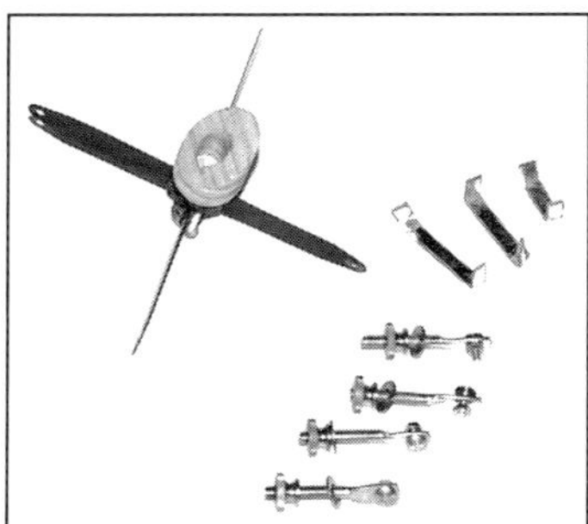

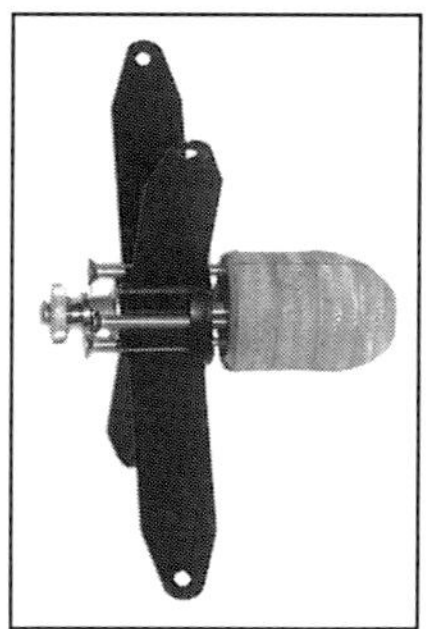

Fig. 10.5 Left: The spider, along with its four fixing bolts (6 mm diameter), and the three retaining clips hold the flat diagonal mirror. **Center:** The spider assembled and ready be installed in the tube. **Right:** Side view of the mirror showing the adjusting bolts.

within the tube. To make the legs, you need two sheets of brass or soft steel about 170 mm long and 20 mm wide by ½ mm thick. Bend these two plates as shown in Figure 9.6 (center) to conform to the central body of the spider (piece *U*, made using a square of aluminum tube) to form the four branches, and attach them to it with screws or small rivets (*Y*). The legs are then attached to the sides of the telescope tube with a flathead screw.

The diagonal mirror support is made from a wooden dowel drilled with a central hole (slightly less than 6 mm) and cut at an angle of 45° on one end (*S*).

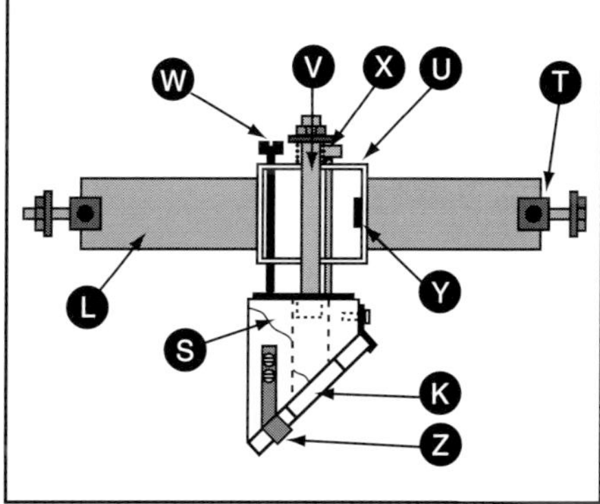

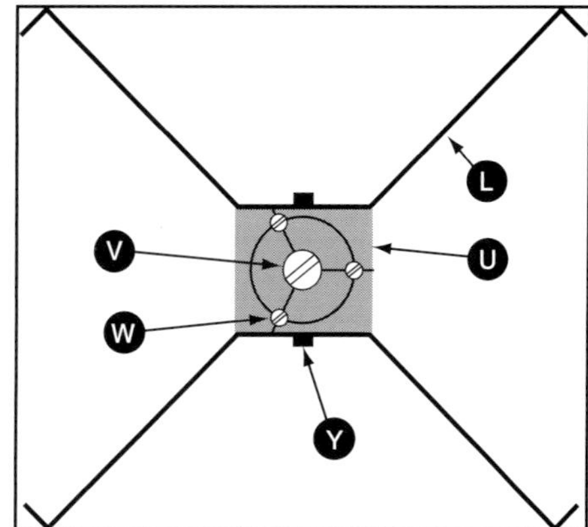

Fig. 10.6 Top: Side view of the spider. **Center:** View of the spider from the front of the telescope tube. **Bottom:** Wood support destined to receive the secondary mirror. Note the metal disk, which the adjustment screws push against.

Legend: K: Secondary mirror; **L:** spider arms (2 pieces); **S:** support for the flat mirror; **T:** attachment screws for the spider arms; **U:** body of the spider; **V:** central attachment screw; **W:** attachment screws for the flat mirror (3 screws—each 3 mm in diameter by 60 mm long at 120°); **X:** compressed spring; **Y:** attachment screw or rivet for the spider branch on the central body; **Z:** holding arms for the flat mirror (3 aluminum pieces, at 120°).

Three metal clips (*Z*) spaced 120° apart hold the mirror against this wooden cylinder with a little slack—1⁄10 or 2⁄10 of a millimeter. The cylinder (*S*) is held on the spider by a 6-mm diameter bolt that screws into the top of the piece (*U*). A spring holds the mirror support (*S*) against the three screws (*W*) used to align the secondary mirror.

Fig. 10.7 Commercial eyepiece holder.

10.4 The Focusing Mount

The focusing mount holds the eyepiece that is used to examine the image formed by the primary mirror. You can buy this part ready made or make it yourself if you have a lathe and other metal working equipment.

The barrel holder size you choose for the eyepiece holder depends on the eyepieces you want to buy or that you already have. Eyepieces with barrel diameters of 24.5 mm are less expensive, but those with 37.75-mm barrel diameters are of better quality and come in a wider selection of types and brands. The eyepiece holder is mounted, after the octagonal tube is finished, by cutting a hole in the side of the tube 150 mm from the front end, provided that the primary focal length is exactly 1200 mm. Otherwise, this position will have to be recomputed. You can place it on the circumference of the tube according to your personal taste.

Fig. 10.8 Example of a side bearing.

10.5 Side Bearings

Cut two side bearings of equal length (at least 25 mm) from 150-mm diameter metal or PVC pipe, using a hacksaw. Attach these rings to the tube reinforcers (part *G* in Figure 10.2 on page 138), making sure to center them exactly opposite each other. The side bearings are situated on the center of gravity of the tube, which should be about 520 mm from the bottom of the tube. These bearings allow the telescope to pivot up and down (altitude axis) while the entire telescope rotates right or left (azimuth axis) upon its base support.

11 Making the Mounting

The 130-mm telescope rides on a simple alt-azimuth mounting. It is made of two elements: a fork and a platform. This telescope mount has two motions—up and down (altitude) and right and left (azimuth).

11.1 Plans for the Mount

The mounting is composed of a fork and a base. The fork is built using four plywood or particle board panels, each 15 mm thick, and assembled according to the plan illustrated in Figures 11.2 and 11.3 on page 144. The arms of the fork are 260 mm apart on the outside, making them 230 mm apart on the inside. Before cutting and assembling the fork arms, make sure that the tube will fit in this space (this is especially important if the side bearings are thick). A 10-mm diameter hole is drilled in the center of the fork's bottom panel. The vertical panel, 370 x 250 mm (panel A in the diagram) adds substantial rigidity to the mounting.

The mount's base is a 500-mm diameter disk of 20-mm thick plywood. A 10-mm diameter hole is drilled in the center of this disk; the fork and the base are at-

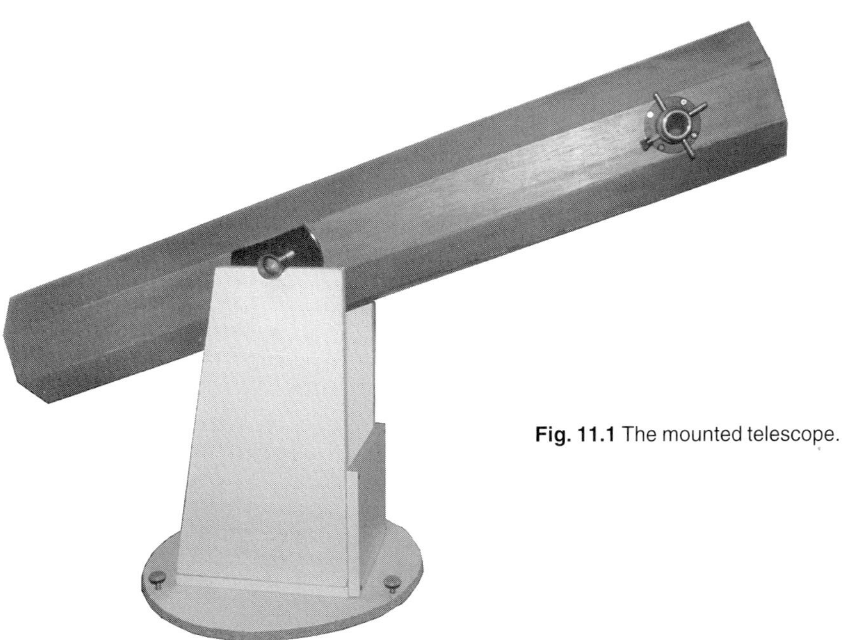

Fig. 11.1 The mounted telescope.

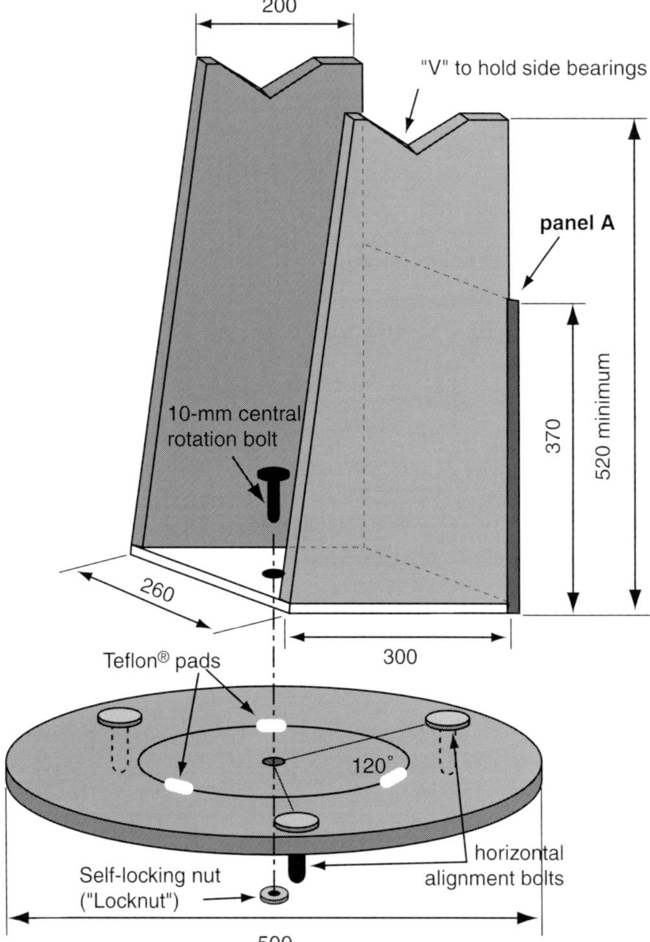

Fig. 11.2 Schematic of the altazimuth mount for the 130-mm telescope. In this figure, the mounting and the fork are separated to show the different pieces of the ensemble.

Fig. 11.3 The three adjustment screws, set at 120°, for the leveling of the base, are 200 mm from the central rotation bolt. Thus placed, the screws will not hit the fork when it rotates on the base. The screws are held onto the base by nuts attached below the base.

Painting the Telescope

Baffles inside the telescope tube help to eliminate stray light but baffles alone are not enough to sufficiently darken the interior of the tube. The interior of the tube must be painted flat black to obtain maximum image contrast. In general, two or three coats are applied to the insides of the eight octagon panels, except where they will be glued, before they are assembled. It is a waste of effort to paint the spider legs or body

The exterior of the tube and mount must be protected from humidity and dew. Varnish or paint all of the exterior parts of the telescope (your choice of color; white is preferable, since it will reflect the heat from your body). A minimum of two coats is necessary for effective protection.

tached by a bolt and nut that pass through and join these holes. The base also has three adjustable feet made of 10- to 12-mm diameter threaded bolts topped by large adjustment knobs. The bottom of the fork is covered with Formica®, and the top of the base is equipped with three equally spaced Teflon® pads, 20 mm x 30 mm and 5 mm thick which are positioned on a circle 200 mm from the base rotation bolt.

11.2 Mounting the Telescope

With the base on the ground (and with three wood blocks under its feet), level it up by adjusting the feet. Place the fork on top of the base, and attach it to the base with the 10-mm screw. Finally, install the tube on the fork, letting the side bearings rest in the V's cut in the top of the side boards. You are now ready to fine balance the tube and collimate the optics.

Part 3: A 250-mm *f*/4.3 Newtonian

Plans for a 250-mm *f*/4.3 Newtonian reflector on a motorized equatorial mount.

This Newtonian telescope is powerful, stable, and easy to use. Its equatorial mount allows the user to track and explore the night sky from the planets to the deep sky, both visually and photographically. It is transportable—the whole instrument weighs 40 kg, is easily disassembled, and fits in the trunk of a car.

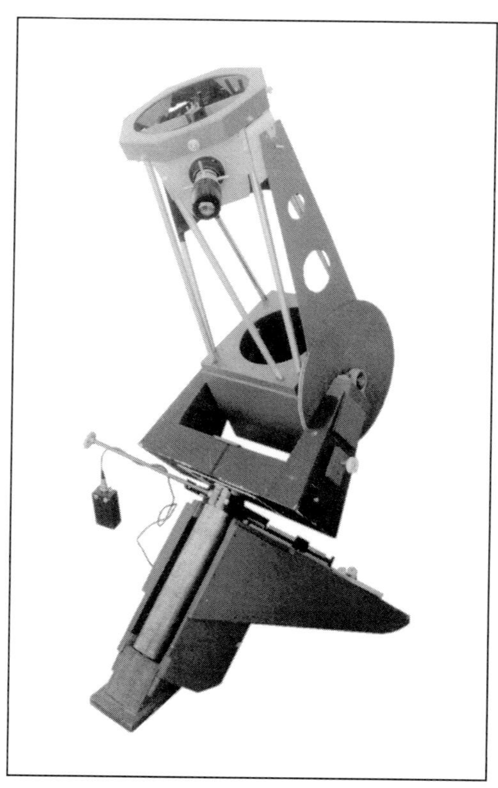

12 The 250-mm *f*/4.3 Newtonian Mirror

12.1 Optical Characteristics

The primary mirror has a 250-mm outside diameter and an optical diameter of 248 mm. Its focal length is 1,092 mm (*f*/4.3). The polished face is concave, its radius of curvature is 2,180 mm, and its sagitta is 3.68 mm. Its form is paraboloidal and, compared to a spherical mirror of the same radius of curvature, the difference in sagitta at the edge is 3 μm.

The secondary mirror is flat with a minor axis measuring 70 mm. It may be either octagonal or elliptical. The secondary mirror is placed 250 mm inside the focus, which exits the tube at about 95 mm. The central obstruction is 28% of the diameter of the primary mirror (0.28 ≅ 70/248); the fully-illuminated field of view at focus measures just under 17 mm in diameter, about 53 minutes of arc (twice the diameter of the full Moon).

12.2 Theoretical Performance

The radius of the diffraction spot measures 3 μm at the focus, which is equivalent to a separating power of 0.6 seconds of arc. The primary mirror gathers 1,225

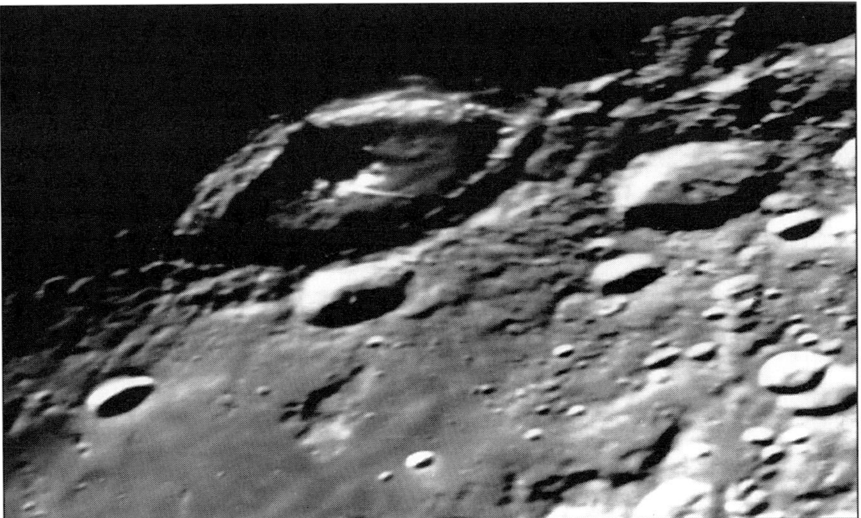

Fig. 12.1 Petavius's Crater, photographed through the eyepiece of a 250-mm telescope.

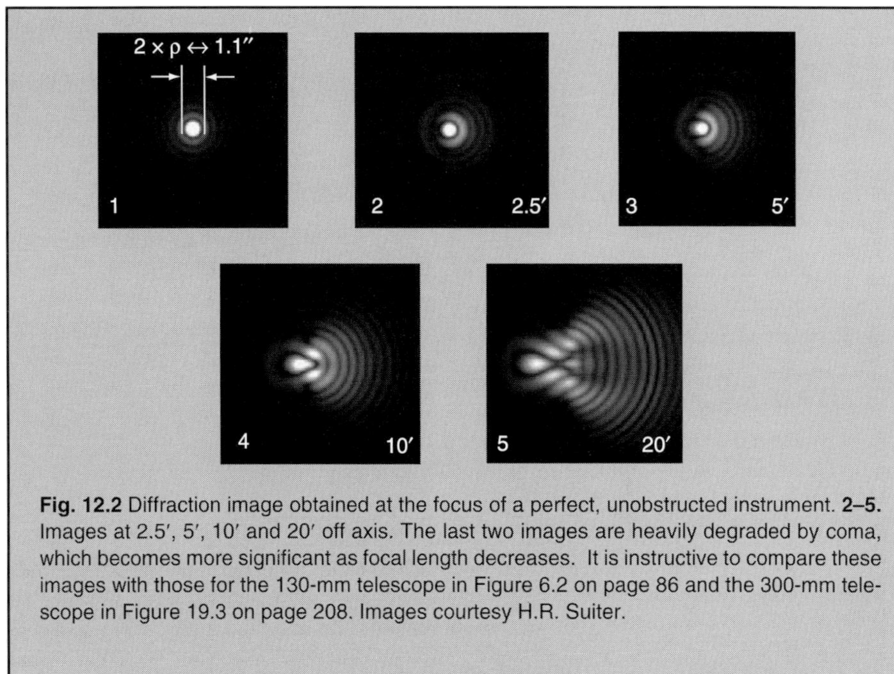

Fig. 12.2 Diffraction image obtained at the focus of a perfect, unobstructed instrument. **2–5.** Images at 2.5′, 5′, 10′ and 20′ off axis. The last two images are heavily degraded by coma, which becomes more significant as focal length decreases. It is instructive to compare these images with those for the 130-mm telescope in Figure 6.2 on page 86 and the 300-mm telescope in Figure 19.3 on page 208. Images courtesy H.R. Suiter.

times as much light as the human eye. Under dark skies it reveals stars to magnitude 14.3, and diffuse objects (galaxies, nebulae) to about magnitude 13.8.

12.3 Magnification

This instrument has an effective magnification range between 35x and about 450x. Equipupilar magnification is 35x with a 30-mm focal-length eyepiece. Maximum magnification can vary between 450x to 500x (with a 5- or 4-mm Barlow or a 2.5-mm focal-length eyepiece), depending on the quality of optical alignment and atmospheric turbulence.

12.4 Mirror-Making Tolerances

12.4.1 Tolerancing the Primary Mirror

The primary mirror is a paraboloid which must be figured to a precision of at least ¼λ to achieve its theoretical resolving power. At the end of figuring, you can eventually tolerate a defect larger than ¼λ at the center of the mirror if it is within a 70-mm diameter circle (which will be covered by the secondary mirror); however, you must be vigilant to avoid at all costs a turned-down edge. A turned-down edge will throw unfocused light back into the diffraction rings at the expense of the central disk.

12.4.2 Tolerancing the Secondary Mirror

The focal length of this instrument is shorter than that of the 130-mm telescope, and the secondary mirror is larger.

Theoretical calculations (of the astigmatism caused by a residual curvature of the secondary mirror) impose exceptionally strict tolerances for making the flat mirror. In practice, we can tolerate a curvature defect no larger than ¼ of a fringe, or ⅛ λ on the minor axis. This tolerance translates to less than $\lambda/5.5$ on the refocused wavefront—anything less precise and images will have astigmatism and be unusable at high magnification. Of course, such residual curvature is less troublesome at low magnifications and in photography at the focus.

13 Making the 250-mm *f*/4.3 Primary Mirror

13.1 Rough Grinding, Fine Grinding, and Smoothing

The experience gained from making the 130-mm mirror as described Chapter 7 beginning on page 89 is a good preparation for this mirror. But this mirror is a more challenging project because it has a considerably larger diameter and a much shorter focal ratio. The sagitta is 3.54 mm.

13.1.1 Rough Grinding

Rough grinding proceeds as it did for the 130-mm mirror. The strokes and their amplitude (in relationship to the diameter) do not change. Be careful when handling the glass disks: you must carry each disk with both hands and never hold the mirror and tool together, because you risk striking them together and causing a large chip.

Occasionally rough grinding begins with a tool that has already been used on another mirror (in astronomy clubs, the same tool might make several mirrors). If this is the case, when you begin rough grinding, work with the mirror on top with centered strokes of ⅔ diameter, pressing hard, until the mirror attains the same curvature as the tool. Then switch to using normal grinding strokes (see Section 7.2 on page 90).

Operation	Type of Abrasive	Quantity of Abrasive	Time	Other Material
Rough Grinding	# 80 carbo	1.5 kg	4 hours	
Fine Grinding	#120 carbo #180 carbo	500 g 400 g	3 hours	
Smoothing	W1, W3, W6 #400–#600 carbo	150 g 150 g	12 hours	
Polishing	Cerium Zirconium	100 g 100 g	15 hours	pitch, 1 kg

13.1.2 Fine Grinding and Smoothing

For fine grinding and smoothing, use the same strokes as described in Section 7.3 beginning on page 92. The most critical times during smoothing is at the beginning and end of a wet. When you place the mirror on the tool at the beginning of the wet, hold it firmly by its sides, using both hands, keeping the mirror's face parallel to the tool's face (see Figure 13.1, left on page 154). Never rest the edge of

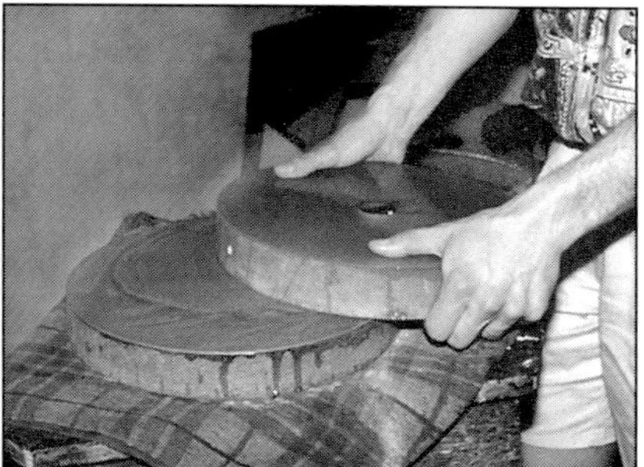

Fig. 13.1 Two critical stages in grinding. **Left:** Placing the mirror on the tool at the beginning of a wet. **Right:** Removing the mirror from the tool at the end of a wet. Photos: © Arnaud Gaillon.

the mirror on the tool; all the weight of the mirror would then be pressing down on a very small area of glass, and you run a great risk of chipping.

At the end of a wet, the spent abrasive and glass have formed a very fine sticky layer, this acts as if there is no air between the mirror and tool and the two disks are held together by suction. To separate the mirror from the tool, do not pull up. Slide the top disk smoothly to the side of the bottom disk without any downward pressure (see Figure 13.1 Right). This technique should be practiced from the very beginning of rough grinding because as you move to finer abrasives the disks become progressively more difficult to slide apart.

A Glass Tool Substitute

You may not always have a glass tool with which to grind and smooth your mirror. It is sometimes necessary to make a tool of wood or plaster, covered with tiles. The tiles can be either square or hexagonal, and can be made of ceramic or enamel.

The ceramic tiles wear away faster than glass.Therefore, to have enough tool to finish the job, you must make two ceramic tools and alternate them through the end of smoothing.

Grinding and smoothing operations are done using the same abrasives and methods as a glass tool.

During smoothing, the tiled structure of the tool allows for a better distribution of the abrasive on the mirror. The work is thus often faster and more uniform than it is with an ungrooved glass tool.

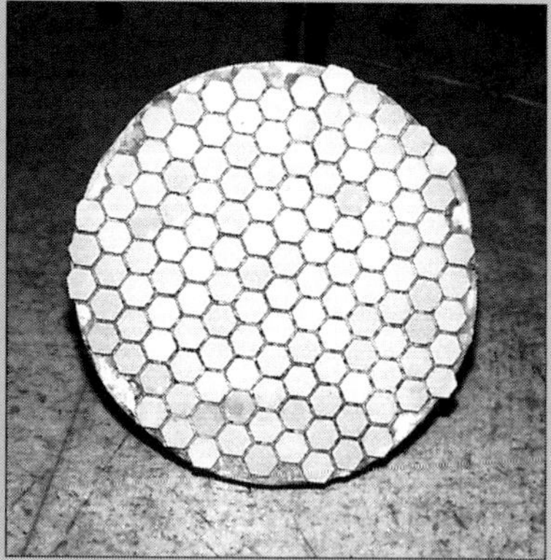

Fig. 13.2 250-mm diameter tool made with hexagonal ceramic tiles. The body of the tool is a plaster mixture.

Grinding and Smoothing Procedure in Pictures for a 300-mm Mirror

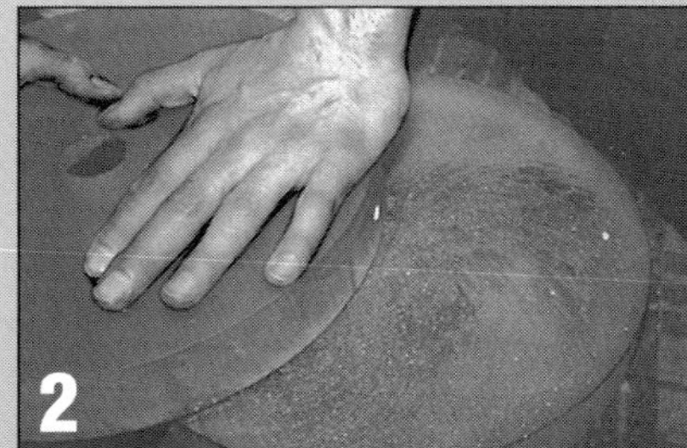

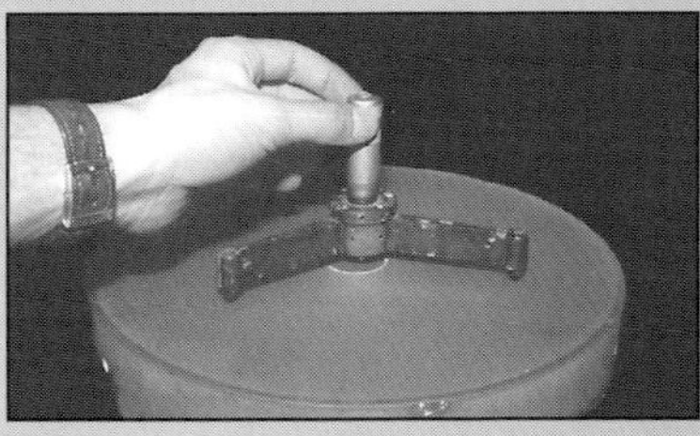

Check the mirror's radius of curvature regularly throughout grinding and smoothing.

1. Spreading #80 carbo at the beginning of rough grinding.
2. Tangental stroke with the center of the mirror on the edge of the tool.
3. Center-over-center strokes to correct the defects caused by rough grinding.
4. Increased hand pressure with W180 to increase the efficiency of the abrasive.
5. Beginning with W1, hold the top disk with your hands on its sides.
6. An air bubble between the mirror and the tool. To correct make very regular strokes, 1/3 diameter amplitude and 1/8 diameter lateral motion defect.
Photos: © Arnaud Gaillon and Jean-Marc Lecleire.

What Happened When the Author Made His 250-mm Mirror

Optical diameter of the mirror: 255.5 mm
Thickness of the mirror: 50 mm
Focal length desired: 1,280 mm, about *f*/5.

Rough grinding began with a used tool that had a radius of curvature of 2,500 mm (the tool had already been used to make an *f*/5 mirror).

Procedure:

Chamfer for the mirror and tool: 10 minutes. Formed chips on the border of the mirror (0.2 mm deep) because the stone was too course for the chamfer.

Back of the mirror: 10 minutes with #80 carbo, mirror on top, using two faces of the tool, and with strokes ⅓ to ½ diameter on amplitude.

Rough grinding: 2 hours and 30 minutes with #80 carbo. End of rough grinding, beginning of fine grinding: 1 hour and 15 minutes with #180 carbo using small-amplitude W-shaped strokes. Focal length obtained at end of fine grinding: 1,295 mm.

Fine Grinding and elimination of chips on the edge of the mirror: 2 hours and 15 minutes with #180 carbo, 8-part W-shaped strokes, with the mirror on top and on the bottom. Focal length obtained: 1260 mm.

Smoothing: with W 240 for 2 hours and 15 minutes; the surface was correct in the center but several pits remained at the edges. Focal length obtained: 1,625 mm. Still using W 240, 1 hour more work. The surface is smooth and the largest pits are the same size as the largest abrasive grade.

Changed to W 1 for one hour. Focal length obtained: 1,263 mm.
#400 carbo for 2 hours, tool on top only. Focal length: 1,265 mm.
#400 carbo for 30 minutes, mirror on top.
#600 carbo for 45 minutes, tool on top only. Focal length: 1,275 mm.
Last retouch with #600 carbo for 45 minutes. Focal length: 1,264 mm.

Polishing for 7 hours with cerium oxide to eliminate gray spots (had the surface been smoothed with W 6 polishing would have been shortened by about one hour).

Finish polishing and parabolization took 23 sessions. Approximate duration: 5 hours and 30 minutes. Polishing compounds used: cerium oxide at the beginning, then rouge or zirconium oxide.

Final result:

Focal length obtained: 1,260 mm
Mirror finished to $\frac{1}{16}\lambda$, surface very smooth and diffuses very little light.

Time spent making the mirror: About 27 hours (without counting preparation time):

14 hours for grinding and smoothing;
13 hours for polishing and figuring.

13.2 Polishing

13.2.1 Making the Pitch Lap Polisher

When fine grinding is finished, the pitch lap will be built up on the curved surface of the tool. Thoroughly scrub the entire tool under running water to remove any remaining abrasives. The lap for the 250-mm mirror is made of 20-by-20-mm squares of 10 mm thick pitch. Begin by tracing the pitch square layout on the face of the tool (with the central facet off-set from the lap's center), then make the tiles using a silicon rubber mold. Finally, attach the pitch squares to the tool and press the assembled lap with the mirror to get the mirror in contact with the entire lap.

13.2.1.1 Drawing the Outline of the Facets on the Tool

Position the cleaned tool on a level surface, with the curved face up. With a flexible ruler, at least a foot long, and a hard-lead pencil (no. 2 or harder) mark the center of the tool. Using this centered mark as reference, draw the outline of an off-centered facet. Using this facet as a reference make a grid, spacing the lines alternately at 7 and 20 millimeters. When you finish, the surface of the tool will be entirely covered by the layout grid to guide you when you apply the pitch facets. See Figure 13.3.

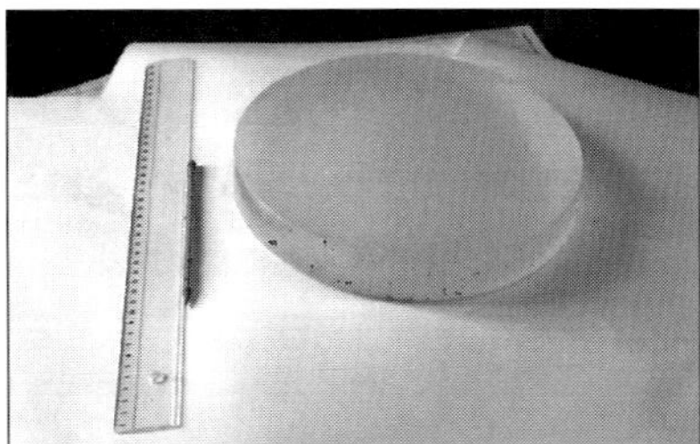

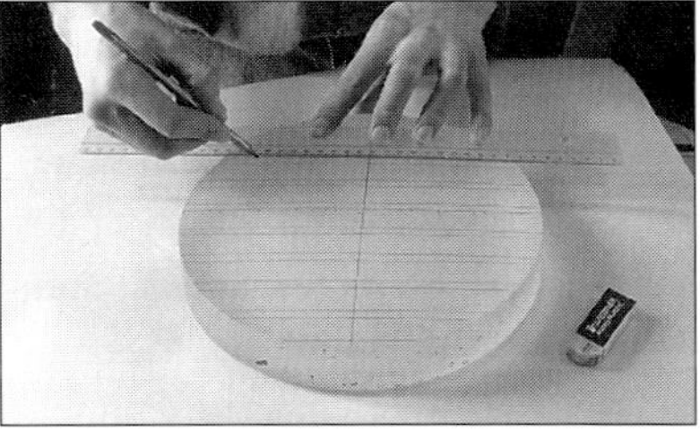

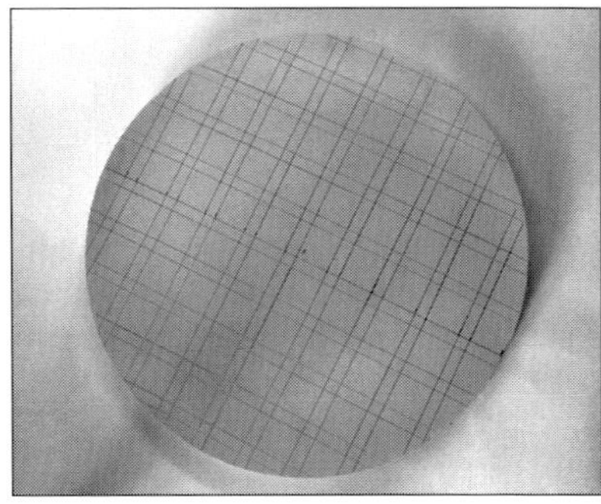

Fig. 13.3 Top: The tool has been cleaned. Next to it are assembled a ruler and pencil. **Center:** Trace the grid on the surface of the tool. **Bottom:** The tool is ready for the pitch tiles. The dot marks the center of the tool; the off-center placement of the squares prevents polishing ring zones into the mirror.

13.2.1.2 Preparing the Mold for the Pitch Squares

The easiest way to make pitch squares is to use a silicone rubber mold which can be made from a wood or plastic master. Silicone rubber is a supple, elastic material that is resistant to heat and impermeable by water or pitch.

13.2.1.3 Preparing the Pitch

As described in Section 7.5.1 on page 102, heat the pitch in a pot. But since you must parabolize this mirror, use a slightly softer pitch. Softer pitch will flow more easily and will produce a better aspheric surface than pitch that's slightly too hard. Softer pitch will

Fig. 13.4 1. Preparing the pitch in a pot. **2.** Pouring the pitch into silicon rubber mold. **3.** Letting the pitch cool. **4.** Removing the tiles from the mold by hand.

also tend not to create zones as the pitch facets thin, which commonly occurs towards the end of polishing.

13.2.1.4 Pouring the Pitch, Making the Facets

Carefully pour the warmed pitch into the mold. Do not allow the pitch to overflow the mold. If you do, there will be much more of a mess. Cooled pitch breaks easily; however, it warms up when it comes into contact with your hands which will soon be covered with hard-to-remove sticky pitch. So it is good to be neat to begin with.

13.2.1.5 Attaching the Pitch Facets

Attaching the pitch squares to the tool requires patience and care. You will need a candle, matches, and a tub of hot water to dunk the glass tool (if this tool is made of wood or plaster, it's best not to submerge it; warm it with a hair dryer). You'll also need large sheets of wax paper or newspaper to protect your worktable.

Here's the procedure:

- To warm the glass tool submerge it in a tub of hot water for five to ten minutes.
- Thoroughly dry the tool (pitch will not stick to wet glass).
- Place the pitch facets on the wax or newspaper, close to the tool; light the candle and bring it close to your work area.
- Pick up the first pitch facet with your fingers and bring it over the candle flame, without placing it directly in the flame (be sure to warm the face that will be

attached to the glass!). Be careful not to burn yourself.

- When the back of the pitch facet has liquefied, move it away from the flame and place it immediately on the tool (do not wait for the pitch to cool). Place the facets inside one of the outlines on the tool's grid and press it down with your thumb. Do not press too hard; if the pitch is soft, your thumb will make an imprint on it.
- Follow the same steps for the remaining facets, including the partial facets that lie around the edge of the lap.

13.2.1.6 Pressing the Polishing Tool

- Prepare a large basin filled with warm water (30° to 40° C, depending on the hardness of the pitch).
- Place sponges or wood blocks on the bottom of the basin and place the tool on top of them.
- Sprinkle a little polishing agent (cerium oxide) into the water; use just enough to make it somewhat cloudy. This way the pitch facets tend not to stick to the mirror.
- When the pitch softens, place the concave face of the mirror on top of the tool (still submerged). Press and turn the mirror all about the tool, especially at the beginning of the operation, to make sure the pitch doesn't have time to stick to the mirror.
- If the pitch facets flow together, remove the tool from the basin, run it under cold water, and trim it.
- Once the lap is in general contact with the mirror, remove the mirror and the lap from the water. Put a little bit of cerium oxide on the tool and place the mirror on top of it. With both hands, continue pressing and turning the mirror until it is in full contact with the lap.
- When all of the pitch facets touch the surface of the mirror (observable by the color of the squares which will become darker when fully in contact with the mirror), separate the mirror from the lap and dunk the lap in a tub of cold water to harden the pitch.
- Finally, trim the pitch facets as described below.

Fig. 13.5 Top: Heating the square with the candle flame. **Center:** Pressing with the thumb. **Bottom:** The polisher is almost finished.

Fig. 13.6 Top: Pressing the tool with the mirror. **Bottom:** All of the squares do not yet touch the mirror.

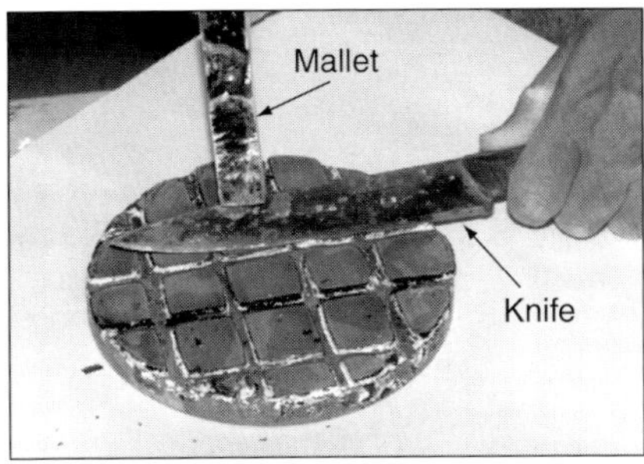

Fig. 13.7 Trimming the tool.

13.2.1.7 Trimming the Tool

You can use a chisel to trim the squares one by one, or you can use a large kitchen knife and a small rubber mallet (see Figure 13.7). The knife will allow you to trim several tiles at once: this method is much faster and more regular than trimming with a chisel. Pitch trimmings and dust from the tool should be collected and returned to the pitch pot where they can be melted and used again for another tool. It makes no difference if this pitch has been charged with polishing agents.

13.2.1.8 Pressing before Polishing

Hold the pitch lap under cold running water long enough to thoroughly rinse all of the squares. Next spread some cerium oxide on the lap and then dunk the mirror in hot water for several minutes. Remove the warmed mirror and lay it on the tool, without pressing down on it.

Another method is to warm the surface of the pitch squares before beginning work, either by dunking the tool for a few minutes in warm water or by blowing hot air on it with a hair dryer. This softens the top layer of the polisher before you lay the wet mirror on it.

Note: You cannot count on warming up the pitch during polishing to compensate for pitch that's too hard.

13.2.2 Beginning Polishing

As you polish you must work towards a spherical surface. Do not begin parabolizing before all grayness is completely polished away. Use W-shaped strokes, ⅓-diameter in amplitude and ⅛-diameter in lateral displacement. Invert the mirror and tool about every 30 minutes. For a 250-mm mirror, you will spend about 6 to 7 hours polishing with cerium oxide before eliminating all of the grayness. During work, check your progress by examining the surface under a bright light. If the surface polishes more quickly at the center, work with the tool on top; if the edge progresses more quickly, work with the mirror on top. It is not productive at this stage to exhaustively Foucault test the mirror; wait until the mirror is well pol-

Making a Plaster-Backed Polishing Lap

If you do not have a glass tool, or if you decide to re-polish an old mirror, you can make a plaster backed polishing lap

Note: you cannot polish a curved mirror with a tool that is flat even if you curve the pitch. Pitch has to flow equally well over the tool and this can happen only when the pitch is of equal thickness. The plaster polisher therefore has to have the same diameter and the same curvature as the mirror.

To make a plaster polisher:

- Cut a large band of rigid plastic or thin cardboard with which to encircle the mirror. Make the band twice as high as the mirror is thick.
- Wrap the mirror in a large plastic bag and attach the band to the edge of the mirror with tape. Make sure there are no folds in the plastic covering the surface of the mirror. Verify that the mold is watertight around the edge.
- Oil the plastic surface and the interior of the plastic or cardboard band.
- Prepare the plaster and pour it into the mold.
- Wait for one day before removing the plaster from the mold. When the plaster is dry, it should sound hollow when you tap on it.
- Cover the entire plaster tool with several coats of paint, to "waterproof" it. Use a water-based paint that contains no solvents (otherwise, the paint may dissolve the pitch).
- Trace a grid outline for the pitch squares with a fine marker on the tool's curved surface. Make sure to offset the center facet.
- As described for the glass tool, heat the pitch squares and apply them to the grid outline. See Figure 13.5 on page 159.

Notes:

- Check the surface of the tool regularly; small holes may appear in the paint, allowing hard plaster particles to fall through and scratch the mirror. When trimming the pitch squares, you will damage the paint. It's therefore necessary to reapply a layer of paint between the squares.
- If you have a lathe, you might be able to make a convex plywood tool with the same radius of curvature as the mirror. Skilled woodworkers may also be able to use a belt sander to rough in a sufficiently accurate curve.

Notes on Pitch

- Pitch adheres better to a painted tool than a glass one.
- To remove traces of pitch from your hands, use turpentine, mineral spirits, or paint thinner; *use no other solvents. Some combine with pitch to make a toxic substance.* Do not rinse your hands with trichloroethylene, which is also toxic.
- A somewhat thicker polisher will flow better than one that is thinner, and thus facilitate parabolization. A very thin lap will not flow and as a result will create zones.

Fig. 13.8 Above: Polishing the 250-mm mirror. When the water/polishing compound is very thin on the surface of the pitch squares, you will occasionally see colored interference fringes if the lap is lit by a fluorescent tube. This is a sign that you are forming a good surface. **Right**: Spreading a bit of wetted polishing compound on the mirror with a brush, without separating the disks.

ished. You should only check for a sphere and adjust your technique to maintain it. If polishing is going well, you will feel a resistance proportional to the amount of contact between the mirror and lap. Do not let the lap run dry to make the mirror squeal at the beginning of polishing; the pressure required to do this can pull the pitch squares off of the tool.

13.3 Parabolizing the Mirror

Parabolizing begins when the mirror is fully polished. Carefully check the outside edge, where wear tends to be minimal. Beginning with a more-or-less spherical mirror, we will progressively work towards a paraboloidal curve.

At the beginning of parabolization, you will change the form of the polishing stroke and increase its amplitude, which will have the effect of rapidly deforming the surface of the pitch polisher. The mirror will then deform through local wear. Working with the mirror on top will mostly wear the mirror's central zones. With the tool on top, most of the action will be concentrated on the peripheral zones if the lap has been given time to flow.

Do not seek to parabolize the mirror all at once; you must proceed in steps. Work for 15 to 20 minutes at a time, so that you can follow the evolution of the parabola with the Foucault tester. During polishing press evenly with your hands flat on the back of the mirror or on its side. Turn the top disk frequently in your hands, and remember to turn the bottom disk about every 15 minutes.

It is possible to make an infinite number of parabolas from a sphere. Let's look at three examples:

13.3.1 Parabolizing from the Center

Polishing is carried out with the mirror on top with large strokes, 4/5 diameter maximum (never bring the center of the mirror all the way to the edge of the tool). The edges of the polisher will be depressed, and only the center of the tool will touch the mirror (this change in the lap is invisible to the eye and imperceptible to your hands because it is a very slight deformation). The repeated passage of the mirror over the center of the tool causes wear inversely proportional to its distance from the center of the mirror. In other words, the center of the mirror is worn and the edges are almost never touched (see Figure 13.9). You can use three types of strokes to create this effect (see Figure 13.10):

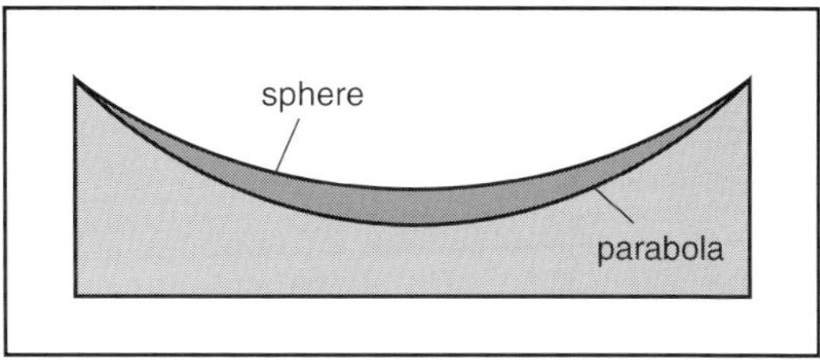

Fig. 13.9 Parabolization at the center. In this drawing, the deformations are exaggerated to show the material differences between the sphere and the parabola. With this polishing regime, work is maximal at the center and minimal at the edge. This method is not the fastest to recover from surface defects (grayness, scratches, milky spots) at the edge of the mirror.

- **Stroke A** tends to produce a uniform paraboloid. The wear, while concentrated at the center of the mirror, gradually moves out and diminishes toward the edge. This stroke is well adapted for all kinds of mirrors and is used frequently to smooth the mirror's overall surface after it's been retouched (for zones or a ridge, etc.).
- **Stroke B** narrows as the center of the mirror approaches the edge of the tool. Wear is strongest at the center of the mirror. This stroke is useful for retouching a central hill caused, for example, by too-short strokes. Be careful not to create a central depression.
- **Stroke C** concentrates most of the action over the center of the mirror and produces a reduction in the radius of curvature of the interior zones of the mirror (< zone 0.7) with little or no edge action. With this stroke, the difference between the curvatures of the interior and exterior zones gives the impression that the periphery of

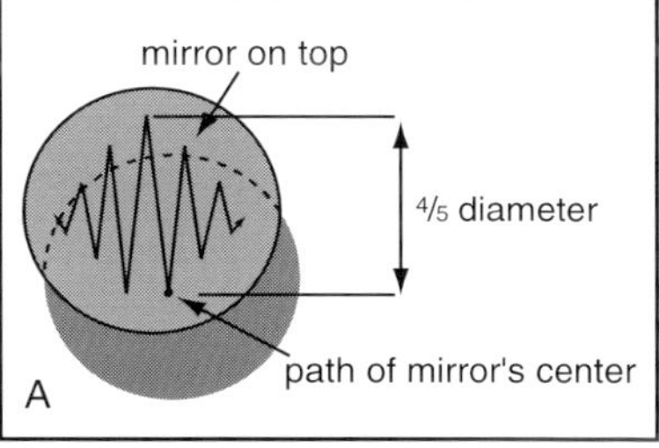

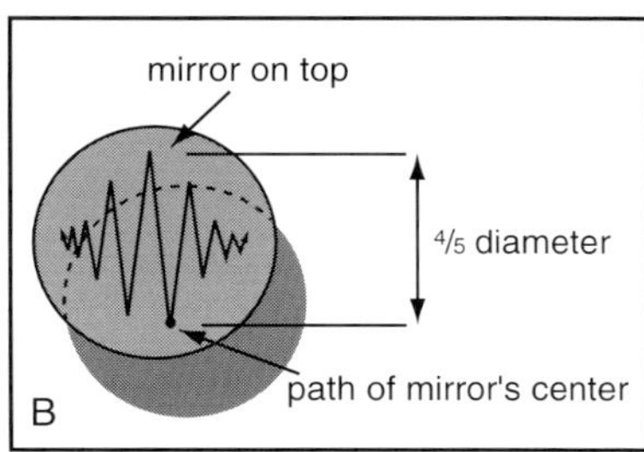

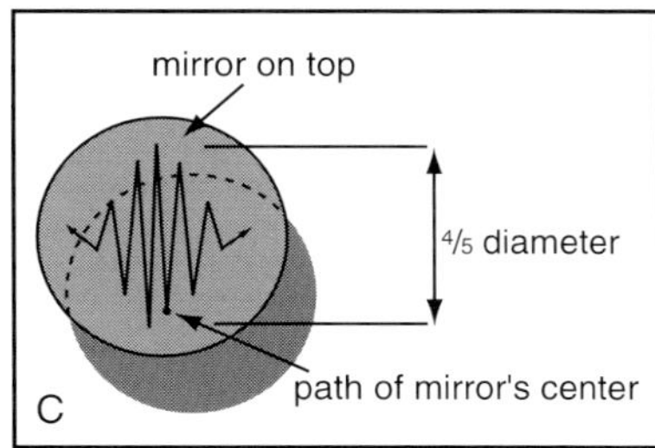

Fig. 13.10 Parabolizing strokes.

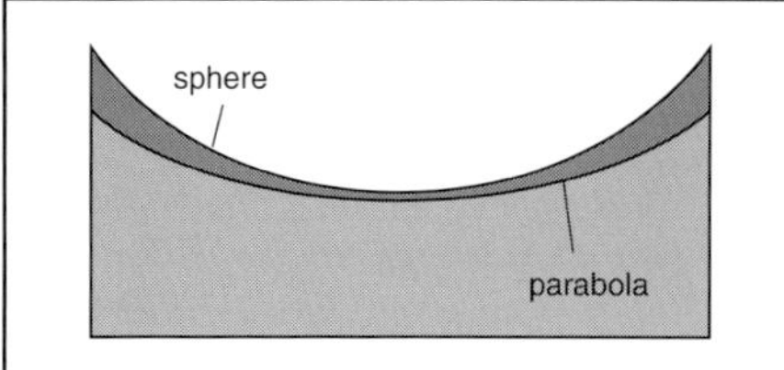

Fig. 13.11 Parabolizing at the edge. Spherical and paraboloidal forms compared. The center of the mirror is not touched; only the periphery is worked.

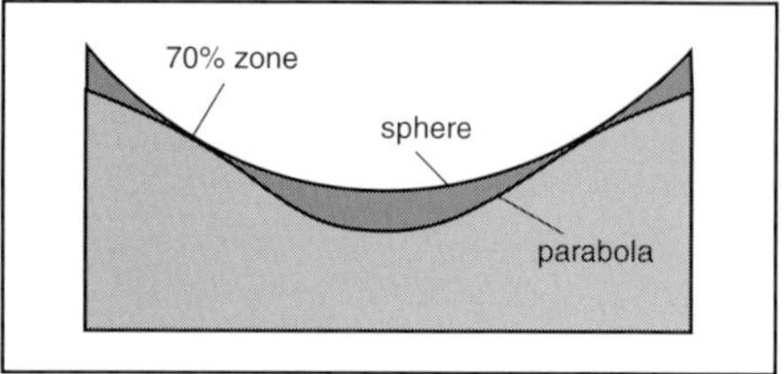

Fig. 13.12 Parabolizing at the center and the edge. Work is spread over the central and peripheral zones of the mirror. This method removes only half as much glass as the other two methods do. The 70% zone is called the *zone of least deformation.*

the mirror has been worn away. This stroke can be used to recover from a flat edge, but it does not allow improvement of the surface quality at the edge of the mirror. This stroke, if it is prolonged, can cause a turned-down edge.

13.3.2 Parabolizing at the Edge

You can make the parabola by working with the tool on top. The tool is therefore worn at the center and touches the mirror only at its exterior crown. Work on the mirror is localized between zones 0.7 and the edge of the mirror, and the center retains its initial radius of curvature (see Figure 13.11).

The strokes are W-shaped and their amplitude varies between ⅓ and ½ of the diameter. From time to time, make straight strokes to regularize the form of the mirror. The major inconvenience of this method is the risk of creating a turned-down edge. It is not recommended for beginners.

13.3.3 Parabolizing at the Center and the Edge

In this case, zone 0.7 retains its initial radius of curvature. This "economical" method minimizes the quantity of glass that must be removed—an important advantage for a large mirror (see Figure 13.12). This paraboloidal form is created by alternating methods 1 and 2 above.

13.4 General Remarks on Pitch Polishing

The removal of glass during polishing depends on localized pressure and the speed of the lap. The amount of glass removed is proportional to the strength of the pressure. Optimal pressure is about 20 g/cm^2. It is increased considerably (200 g/cm^2) during local retouching with a small tool. Fast strokes, lubricated well with polishing compound, increase the space between the polisher and the tool. This permits greater deformation of the mirror, at the risk of causing a central depression or a turned-down edge and a lower surface quality.

Some Advice on Polishing

How do I avoid a turned-down edge?

- The diameter of the polishing lap must be slightly less than the optical diameter of the mirror. The pitch squares at the edge of the lap should be trimmed back by 2 or 3 mm.
- When you work with the tool on top, avoid making strokes greater than ½ diameter. You can also consciously reduce pressure at the end of a stroke as you approach the edge of the mirror.
- Do not use an especially large amount of polishing compound.
- When the pitch is soft (on hot days or when it's heated on purpose), work 90% of the time with the mirror on top.

How do I avoid surface ripple?

- A pitch lap that is too thick increases the likelihood of ripple. The ideal is to work with pitch squares 3 to 4 mm thick (unless the pitch won't deform enough to continue parabolization). If the squares are thinner than 1 mm or it no longer flows in a reasonable time, remake the tool.
- Vary the size and spacing of the pitch facets to the mirror size: the squares should be about ⅒ of the mirror diameter on a side (20 mm for a 200-mm or 250-mm mirror, 50 mm for a 400-mm or 500-mm mirror).
- Do not stroke too fast: you should work smoothly without making sharp motions. In effect, the tool will run up against the hills on the mirror and you will feel the mirror jerk in your hands. Suppression of ripple is accomplished through smooth strokes, and it may take several hours.
- Gently round off the points of your W-strokes.

How do I avoid scratches?

- Clean the tool under running water before and after work.
- At the beginning of a wet, dampen the tool with a little pure water before placing the mirror on top of it; never put abrasive on the tool before putting the mirror on it, because the hard particles will encrust themselves in the pitch. Apply the polishing compound to the lap with a brush while the mirror is moved to the side; never separate the disks.
- Avoid working in a drafty area—wind transports dust.
- Scratches usually form at the beginning of a wet. If you must make a short-duration retouching, use rouge—not cerium oxide.

How can I reduce microripple?

- Microripple is always present on the surface of a mirror. It's impossible to eliminate, but you can at least reduce its amplitude. To do that, use a thin mixture of rouge and water. Work smoothly without jerking and never let the surface dry out. Rinse the tool and the mirror immediately under running water when polishing is complete.

- With the mirror on top, the tool, charged with abrasive and water, sometimes forms a foam that circulates between the pitch tiles. This foam helps the polisher stay uniformly and constantly covered with polishing compound.

What are the best conditions for polishing?
The ideal temperature for a workshop is between 18° and 22°C, depending on the composition and hardness of the pitch. The pressure to exert varies as a function of the type of retouching you're doing: pressing hard at the beginning of polishing eliminates grayness; pressing only moderately near the end helps to avoid ripple. For a 200- to 250-mm mirror, the pressure from the weight of the tool or mirror alone (about 10 to 20 g/cm^2) will give a good surface. For mirrors with diameters greater than 300 mm, it is sometimes wise to make a varnished plywood polisher, which is lighter and less cumbersome than a glass tool. A heavy tool is sometimes a handicap when working with the tool on top (as when correcting a turned-down edge, etc.).

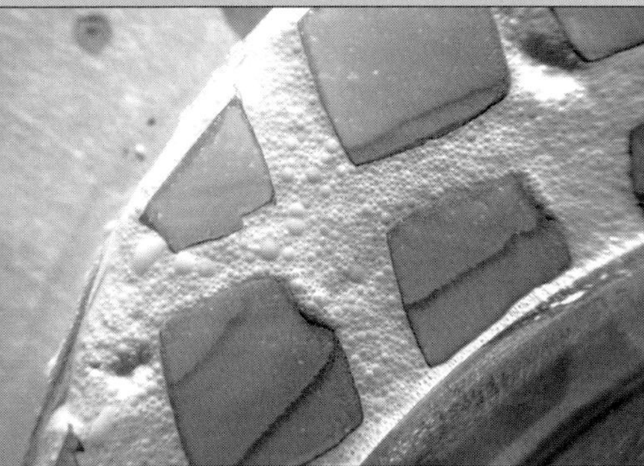

Fig. 13.13 When the polishing compound foams between the pitch tiles, the tool is perfectly lubricated and produces a better mirror surface.

In general, about ⅔ of polishing time is done with fast acting polishing compounds like cerium oxide or zirconium oxide. These compounds bond to the pitch and give the lap a uniform pink or gray tint. When parabolization nears completion, you should use a finer powder, such as rouge. After a few dozen minutes of polishing, the mirror surface will noticeably begin to show a polish at the center, less subject to pits, scratches and ripple. The grayness caused by final grinding will take from two to three hours to completely disappear from the surface of the mirror. The outer periphery of the mirror is usually polished out last.

For good polishing action, the pitch should always show traces of polishing compound. If the lap becomes black, recharge it with polishing compound. If the lap has to be remade during polishing, always resume work for at least 15 minutes with cerium oxide to charge the pitch, and then switch to rouge or zirconium oxide.

Temperature and its Effect on Various Mirror Materials

The time that it takes a mirror to cool is a function of its thickness, its volume and its composition. A 200-mm Pyrex mirror that's 35 mm thick takes a half hour to an hour to reach equilibrium. A 250-mm mirror that's 45 mm thick takes at least an hour. A 600-mm mirror, 50 mm thick, takes about three hours. The cooling time for a plate-glass mirror (Saint-Gobain glass) is two to three times that of a Pyrex mirror. A 300-mm disk of Saint-Gobain glass must rest for a night in its testing stand before any measurement can be taken. The handling of a glass mirror requires the use of insulated gloves to prevent your hands from heating the glass. Zerodur glass, for all practical purposes, is unaffected to ambient temperature variations.

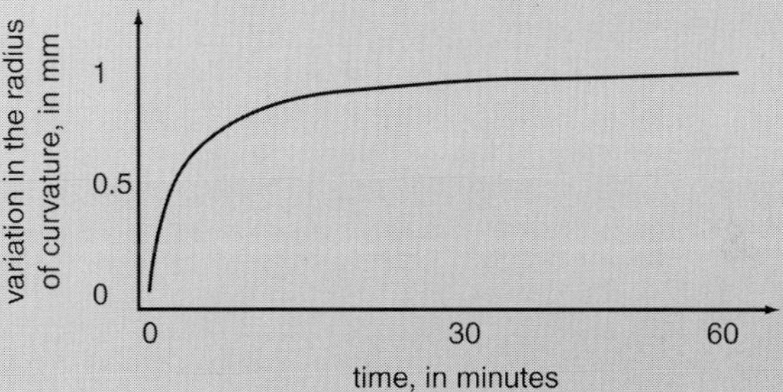

Fig. 13.14 The curve shows the variation of the radius of curvature of a Duran glass mirror, 250 mm in diameter and 45 mm thick after the end of a polishing session and removal from the lap. The ambient temperature was 20°C.

14 Testing and Finishing the 250-mm *f*/4.3 Newtonian Mirror

Before reading this chapter we recommend that you re-read Section 5.4 beginning on page 52 which deals with Foucault testing.

14.1 Testing with the Foucault Apparatus

Set the mirror in its test support and align it with the Foucault tester. Next, make a general evaluation of the mirror. Is it spherical? Does it null? If not, where are the zones, holes and hills? Precisely measure the radius of curvature *R*. Prepare a Couder screen (see Section 5.4.6 on page 61). The test report in Table 5.1 on page 68 is for a mirror of this aperture and focal length.

14.2 Measurement Procedures

Before conducting a detailed Foucault test, you must wait for the mirror temperature to stabilize. Make two sets of measurements for each Couder mask zone. Make one set of measurements working from the center out, and the other from the outside edge in. For each zone, calculate the average of at least two measurements and then reduce your data so that you can produce a graph of longitudinal aberration to provide a pictorial representation of the mirror's profile. You can then calculate the value of the transverse aberration as well as the peak-to-valley difference on the wave front. If the mirror is not tolerances, seeSection 12.4.1 on page 150, then it will be necessary to retouch it.

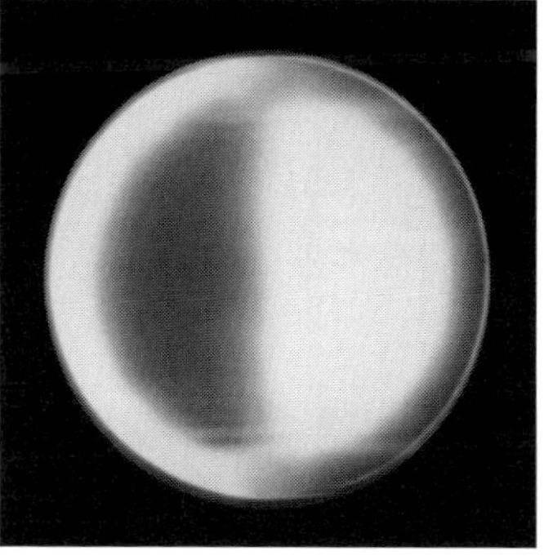

Fig. 14.1 250-mm *f*/4.3 parabolic mirror after polishing. The surface is very smooth and there is no ripple.

Note: the longitudinal aberration graph corresponds only to half of the mirror (see Figure 14.2 on page 170). By symmetry, it is very easy to imagine the other half of the mirror and to highlight zones that need to be retouched.

14.3 Retouching the Parabola

Retouching corrects defects in the mirror's figure. Most defects can be corrected with a well-trimmed, full-diameter lap that makes good contact with the mirror.

If the pitch facets are glazed with polishing compound, you should lightly scrape the surface with a single-edged razor blade, re-charge the lap with polishing compound and cold press the mirror on the lap until the lap is in full contact with the mirror.

The times recommended here for certain corrective actions are only guidelines and can vary from half to double from one person to another, temperature, pitch hardness, etc. Analysis of mirror tests, correcting strokes, pitch thickness, working time and other variables should be carefully recorded in a notebook so they may be available to help you plan future corrective actions.

On the following pages we will examine a number of common defects that you are likely to encounter as you proceed to finish your mirror. In each instance we provide a graphic representation of the mirror profile, a verbal description of the defect, and a corrective stroke pictorial. The times given are approximate; work time depends on the pressure exerted on the mirror or the tool and on the speed and shape of the strokes.

Up to this point mirror fabrication has been more or less a mechanical operation where general instructions provide predictable results. But when it comes to finishing a mirror, especially if it must deviate from a sphere and the surface finish must be of high quality, you must be much more active in the analysis of the situ-

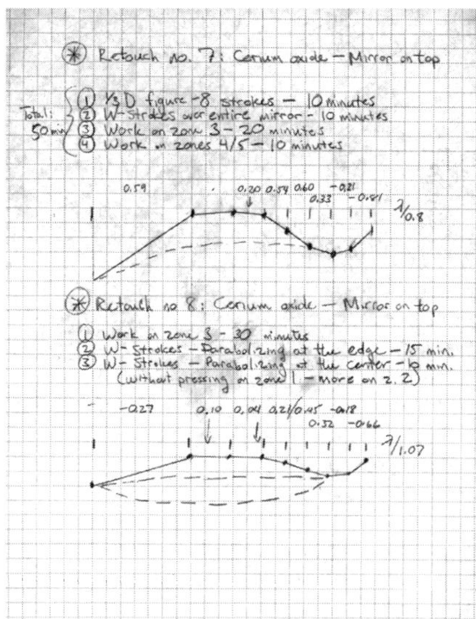

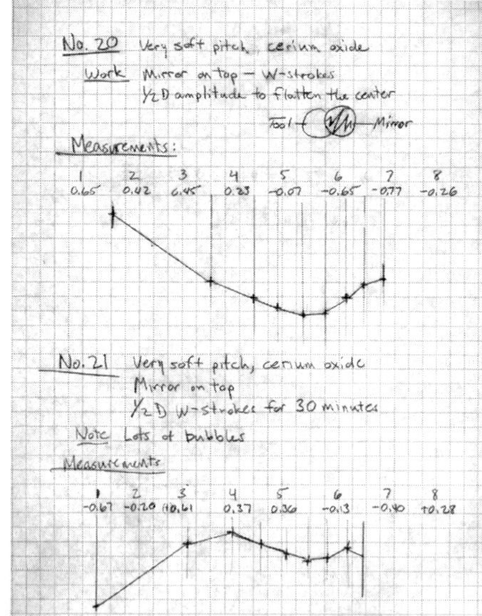

Fig. 14.2 Pages from a retouching notebook for a 250-mm mirror with 8 zones. The longitudinal aberration graphs enable you to follow the evolution of the profile of the mirror. Reduced measurements are notated above the curves. The arrows show the zones that need to be retouched.

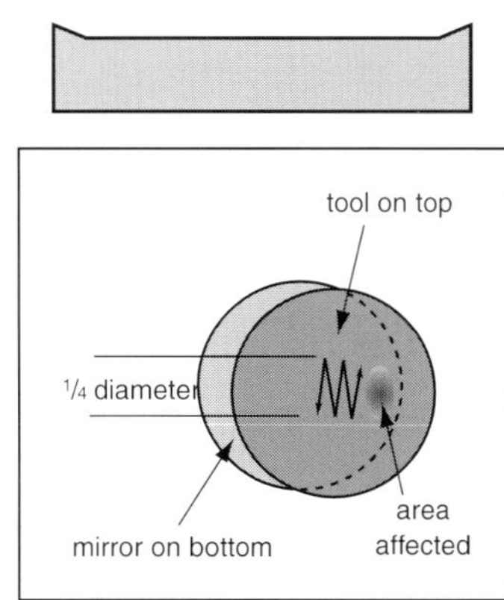

Defect no. 1: Raised Edge or Outside Zone

Description: The radius of curvature of the exterior zone is too short. Probable cause: too short strokes, the polishing tool is smaller than the mirror or is too worn at its edge.
Recommended treatment: If the raised area is only at the edge, work with the tool on top using 1/4-diameter W-strokes. Press **lightly** with your hand on the side of the tool that crosses the edge of the mirror.
Duration: Work for a short period to avoid creating a turned-down edge; 5 to 10 minutes, making a number of complete turns around the workpost.
Note: If the raised zone extends over the entire outside zone of the mirror, use the stroke shown in Figure 13.10A with the mirror on top, for several minutes, without pressing at all at the end of the stroke. The lap should be well lubricated.

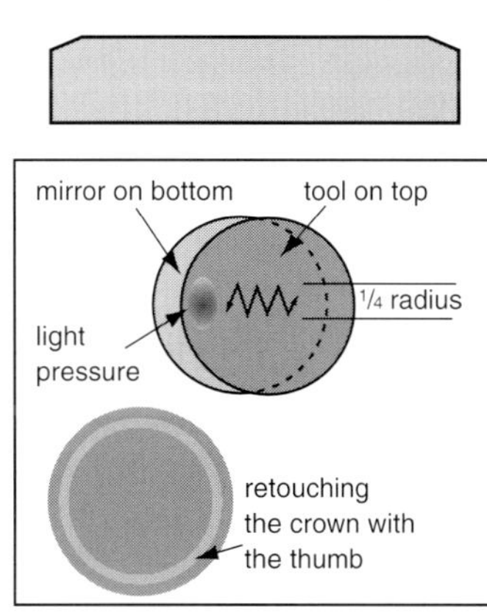

Defect no. 2: Turned-Down Edge

Description: The radius of curvature of the exterior zone is too long. Probable causes: the tool is larger than the mirror, the pitch is too soft, or the strokes were too long.
Recommended treatment: Trim the tool to the *exact* diameter of the mirror. Work with the tool on top, making 1/4- to 1/3-radius strokes. Change the pitch if it is too soft. Press lightly on the edge of the tool that is nearer the edge of the mirror.
Duration: Work for 20 to 30 minutes at least for a weak defect; double or triple the time for a stronger one.
Note: If the defect extends over the whole exterior zone of the mirror, consider retouching with your thumb at the height of the zone for 20 to 40 minutes, pressing as little as possible so as not to cause a depression in the glass and making sure to change the movement of your thumb frequently (several times each time you circle the post).

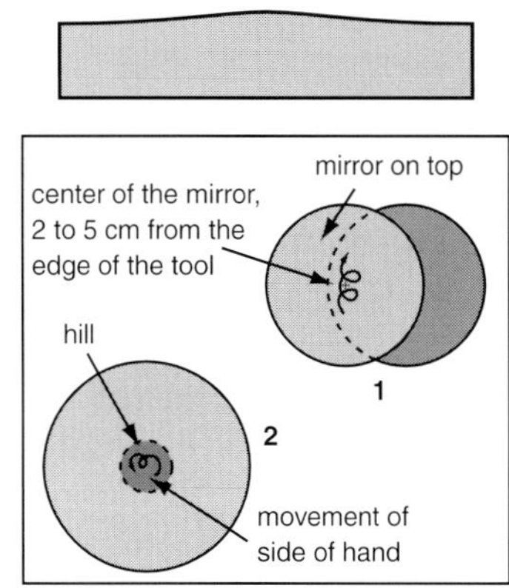

Defect no. 3: Central Hill (Boss)

Description: The radius of curvature of the central zone is too long. In the Foucault test the center of the mirror appears flat or domed, depending on the amplitude of the defect.
Recommended treatment: If the hill is large (*1*), work with the mirror on top, with the center of the mirror 2/10 of its diameter from the edge of the tool (5 cm for a 250-mm mirror). Make large circles with the tool, trying to center the circles on the edge of the hill. If the hill is small (*2*), work with the side of your hand, pressing very lightly. If the hill extends over two zones, make loops centered on the exterior zone.
Duration: Work for a few minutes between each test.
Note: Check the state of the pitch squares at the center of the tool. If they are worn (or missing), they could be the cause of the central hill.

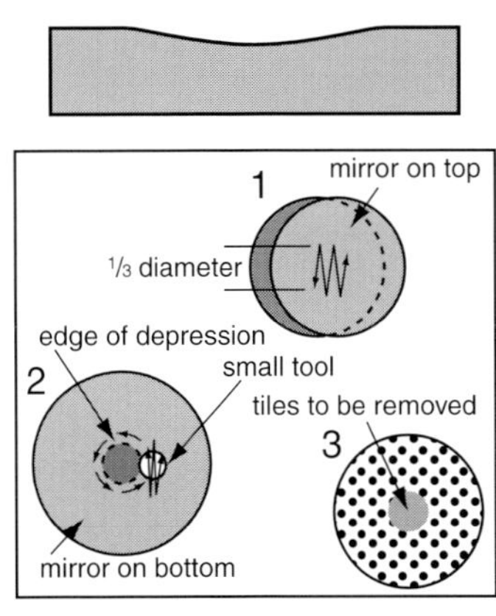

Defect no. 4: Central Depression

Description of the defect: The radius of curvature of the central zone is too short. Probable causes: too much time spent working with the mirror on top, or strokes that were too long.

Recommended treatment: If the depression is not deep (*1*), work with the mirror on top, making $^1/_3$-diameter strokes. After retouching, the zones near the edge might be higher. You must then correct the defect with the tool on top. For a 2- to 3-mm deep depression (millimeters of longitudinal aberrations measured using the Foucault test) (*2*), work much longer and trim the center of the tool. To recover a central defect deeper than 10 mm (it happens!), (3) work only the zones at the side of the depression using a small tool. Press hard and make W-shaped strokes while moving around the mirror. After this retouching, reform the damaged surface using a full-size tool.

Duration: About 30 minutes per millimeter of depth.

Defect no. 5: Raised Zone

Description: A zone whose radius of curvature is too large. This defect is usually caused during the retouching of a central hill using circular strokes.

Recommended treatment: With the mirror on top, place the crown to be retouched 25 to 30 mm from the edge of the tool. Make small, circular strokes, making sure not to let the defect move beyond the edge of the tool. If the defect is situated outside of the 70% zone, retouch it with your thumb or a small polisher.

Duration: Work for about 10 minutes, lifting slightly on the mirror disk if the disk is heavy.

Note: It is important to know where the defect is found in order to retouch the right zone. To visualize its placement, you can draw the crown that you'll retouch with a marker on the back of the mirror.

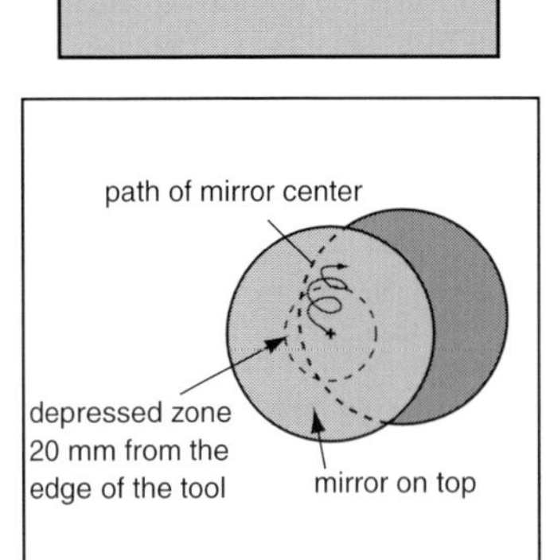

Defect no. 6: Depressed Zone

Description: A zone whose radius of curvature is too short. This defect is caused by repetitive circular strokes of a tool over the same zone of the mirror.

Recommended treatment: Work on the central part of the mirror, placing it on top of the tool. First, make strokes like those shown in the middle diagram of Figure 13.10B, and then make small circular strokes, placing the defect area 20 mm outside of the edge of the tool (see the figure to the left). If the defect is outside of the 70% zone, begin by lowering the center (as in Defect no. 3), then working the edge, with the tool on top, using $^1/_4$ diameter W-strokes. You could also lightly scrape the tool at the corresponding zone.

Duration: Work for 20 to 30 minutes.

Note: Avoid changing the type of stroke. Avoid making figure-eight strokes.

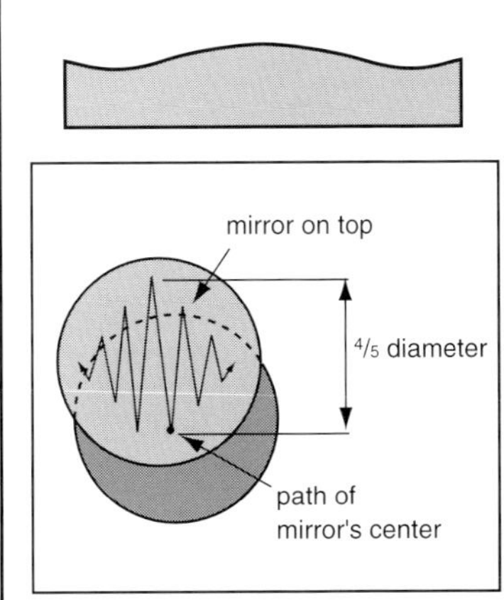

Defect no. 7: Elliptical Mirror

Description: The form of the mirror is between a sphere and a parabola. In the Foucault test the gray zones are larger than the windows of the Couder screen.

Recommended treatment: Make normal parabolization strokes with a maximum amplitude of 4/5 diameter. Work with the mirror on top.

Duration: Work in 20- to 30-minute steps and test the evolution of the mirror in the Foucault test.

Note: If it is impossible to deform the mirror sufficiently and if the elliptical form persists, it may be wise to remake the polisher with softer and/or thicker pitch.

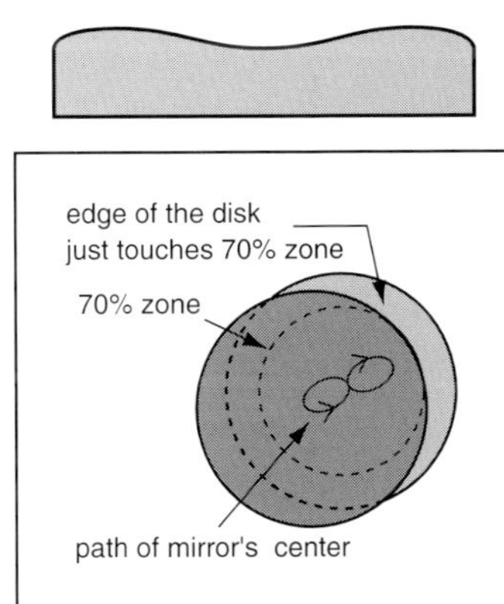

Defect no. 8: Hyperboloidal Mirror

Descripition: The deformation of the mirror is too great. Probable causes: polishing lasted too long, or the pitch is too soft.

Recommended treatment: Work alternately with the mirror and tool on top, with 1/3-diameter strokes in a figure-eight pattern. The edge of the mirror should touch the 70% zone of the tool at the end of each stroke.

Duration: If the deformation is deep ($b < 1.5$; see Section 5.4.5: $b = 0$ for a sphere and $b = -1$ for a parabola), change abrasives. Use cerium or zirconium. Work 30 to 40 minutes per millimeter of extra depth.

Note: You can retouch this defect with a small tool (1/4 of the mirror's diameter). Work on the 70% zone, with small-amplitude circular strokes to avoid leaving tracks (which would be visible only in the Foucault test) on the mirror.

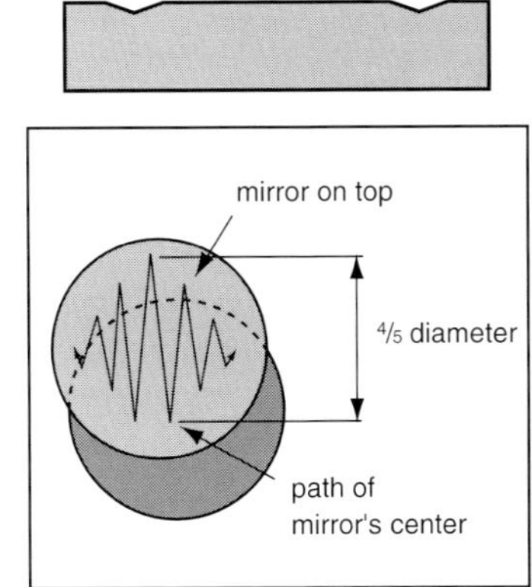

Defect no. 9: Groove

Description: The mirror is depressed over a thin circular zone. This defect may be the result of a botched thumb-retouching or a local polishing defect.

Recommended treatment: Work with a thick, soft tool. Make normal parabolization strokes with the mirror on top to prevent further deformation of the mirror.

Duration: Depends on the depth of the groove. If it shows high contrast in the Foucault test, you will have to work for an hour or more.

Note: When the polisher loses a pitch tile, it must be replaced by another tile of the same hardness. A softer or harder tile can be the cause of a groove.

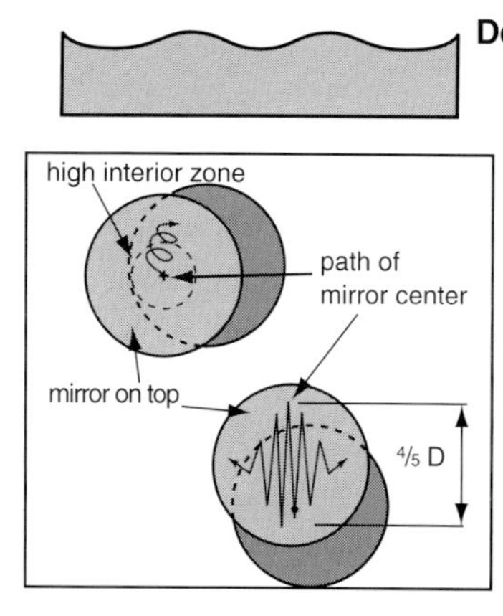

Defect no. 10: Central and Peripheral Depressions

Description: The mirror is worked too much at the center and not enough at the edge. In the Foucault test, the edge of the mirror appears null.

Recommended treatment: Work with the mirror on top making circular strokes over the raised interior zone. Then make strokes as you would to parabolize the edge with the mirror on top. Do not press on the mirror with either stroke.

Duration: Work for fifteen minutes with each type of stroke.

Note: This kind of defect occurs when the polisher is too thin or the pitch is too hard. In this case, the edge of the mirror does not change enough and normal parabolizing strokes are ineffective.

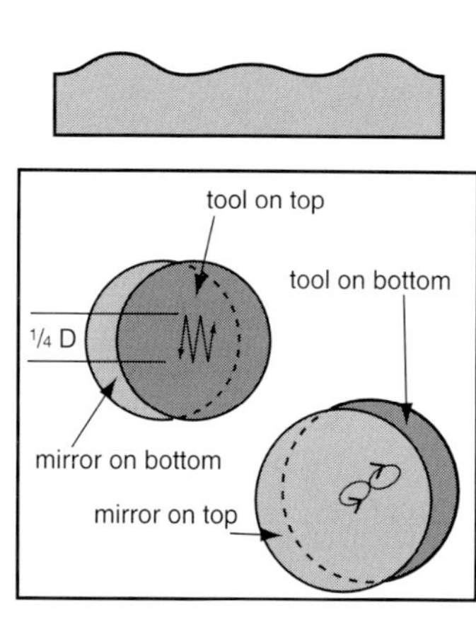

Defect no. 11: Raised Center and Zone

Description: The center of the mirror is not sufficently worked during polishing, and a zone of the mirror is raised. In the Foucault test, the mirror appears flat at the center.

Recommended treatment: Begin with the tool on top, making 1/4-diameter W-strokes. This correction will correct the edge and the center. Next, work with the mirror on top, making 1/3-diameter figure-eight strokes. This action is localized on the raised zone.

Duration: Work for at least a half hour, inverting the mirror and tool evenly.

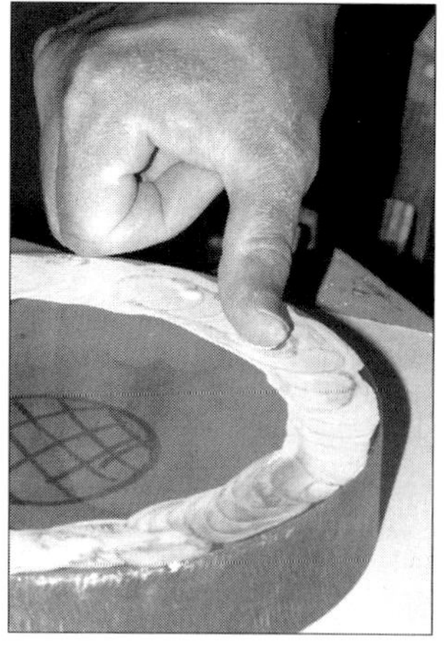

Fig. 14.3 Retouching with your thumb. Using your thumb or a finger you can correct zonal defects that are difficult to fix with a full diameter lap. Your thumb, as a polisher, is equivalent to a small pitch tool. Using moderate speed and pressure corrective action is slow, and you can test your progress with great precision. However, do not move too fast or press too hard; you might create a groove in the mirror. The best way to master this technique is to begin gradually and perfect your technique by analyzing your results under varying conditions. In general, thumb retouching is best done by making small circles first in one direction and then in the other. Use a very creamy mixture of zirconium oxide polishing compound.

Dealing With a Bright Peripheral Zone

The turned-down edge of my mirror is so bright in the Foucault test that it makes measuring the peripheral zone of the mirror almost impossible.

To eliminate this problem, glue a thin band of paper around the edge of the exterior window on the Couder screen. This band should make the exterior zone smaller by only a few millimeters. It will allow you to test the outside of the mirror with greater precision—until, of course, you retouch the turned-down edge.

ation and in planning and executing corrective actions. Again, it cannot be overemphasized: it is essential to keep a notebook to record the kind of stroke, the polishing compound, the hardness of the pitch, etc., as you work. The observations made in the Foucault test are noted after these descriptions. After it is filled with the details of several dozen retouchings, this notebook will be a valuable asset as you seek a solution to any future problem.

15 Making The Secondary Mirror for a 250-mm *f*/4.3 Newtonian Telescope

The secondary mirror for the 250-mm Newtonian telescope is an optical flat with either an octagonal or elliptic shape, 70 mm wide (minor axis) by 100 mm long (major axis) and 10 mm thick. This mirror is ground and polished using a circular blank. Once it is fully polished and corrected to a plane, it is cut to its final shape.

The glass can be Pyrex, a crown glass like BK7 or Zerodur. Don't use plate glass or old glasses of dubious origins; they are likely to have internal strain which will distort your mirror when it is cut. The grinding, polishing and testing procedure for this mirror is the same as that used to make the secondary for the 130-mm mirror (see Chapter 9 beginning on page 125).

15.1 Cutting an Octagonal Mirror

The optical dimensions of the mirror are 70 mm by 100 mm. Protect the optical surface with two coats of shellac and heavy wrapping paper. Cut out the octagonal shaped diagonal with a tungsten-carbide or diamond saw leaving an extra 2- or 3-mm area around the outside edges.

15.2 Cutting an Elliptical Mirror

This operation is done with a drill press equipped with a "biscuit cutter":

> The cutter is a tube of soft metal; e.g., brass or mild steel. The inside diameter should be about 1.5 mm or 2.5 mm larger than the mirror's minor axis; the tube's wall thickness should be about 1.5 mm or 0.75 mm. The tube is fitted to the shoulder of a supporting piece that permits mounting in the drill press and is slotted at the bottom for better access of the abrasive.[1]

Fig. 15.1 A biscuit cutter installed on a drill press. Rotation speed: 250 rpm.

The face of the mirror should be protected by a 4- to 5-mm thick piece of window glass bonded together with arcanson[2] with a piece of tissue paper in-

[1] Jean Texereau, *How to Make a Telescope*, 2nd ed. (Richmond, VA: Willmann-Bell, Inc., 1984) p. 115 (page 121 in early printings).

Fig. 15.2 The elliptical mirror is ready to be retouched.

serted between them. To make the ellipse, cut it at a 45° angle (See Figure 15.2). You can glue the mirror to an old right angle glass prism, or on a thick glass plate wedged solidly against a wood block that has been cut at a 45° angle which in turn is firmly bolted to the drill press table. Advance the biscuit cutter slowly through the glass, frequently adding #120 carbo and water. The biscuit cutter should be removed from the glass frequently in order to avoid heating the glass (you risk melting the arcanson). To protect yourself from the possibility of flying glass, wear gloves and protective eyeglasses. When the mirror is cut out, take the assembly apart by slowly warming it up in an oven and then chamfer the diagonals' edges with a fine stone in order to smooth out any rough edges and to reduce the risk of chipping.

Testing the mirror after cutting will generally show a low-amplitude raised edge, accentuated at the thin edge. This defect can be ignored if it is smaller than 2⁄10 of a fringe. If it is larger, you can retouch it with your finger or even block the mirror and work it according to the method described for the secondary of the 130-mm telescope (See Chapter 9 on page 125).

[2] Arcanson is a yellow glue made by mixing one part white beeswax with six parts rosin. Melt the ingredients in a pot and pour the mixture into rectangular molds. When it has cooled, arcanson is hard. To make a glue, place the two pieces to be assembled on a hot plate and cover the surfaces with the arcanson. The arcanson melts in contact with the heat and can be spread out easily. Allow the assembly to completely cool before cutting the mirror. To remove the arcanson, dunk the pieces for a few minutes into acetone, rubbing them delicately with a cotton ball (wear latex gloves). Use extreme care, acetone is volatile and flammable. It is best to do this out-of-doors.

16 Mechanical Parts for the 250-mm *f*/4.3 Newtonian Telescope

You have now arrived at the point where the optics are finished and aluminized. All that's left to do is for you to make the telescope itself. It consists of four parts: the optical tube assembly, fork, base and right ascension tangent arm.

16.1 The Optical Tube Assembly

The tube is made primarily from plywood. All the wooden parts are glued together with wood glue. Where metal contacts wood or other metal and bolts are not used, epoxy cement makes the bond. Nails are only used to hold the pieces together during assembly. The secondary cage and the mirror box are connected by eight truss tubes, 18 mm in diameter and 810 mm long. Each side of the mirror box supports two of these tubes. The secondary cage—the octagonal assembly at the top of the telescope—supports the spider and the eyepiece holder and is removable. Contact between the truss tubes and the head is made with 1-mm thick pieces of rubber. The rubber helps to dampen tube vibrations.

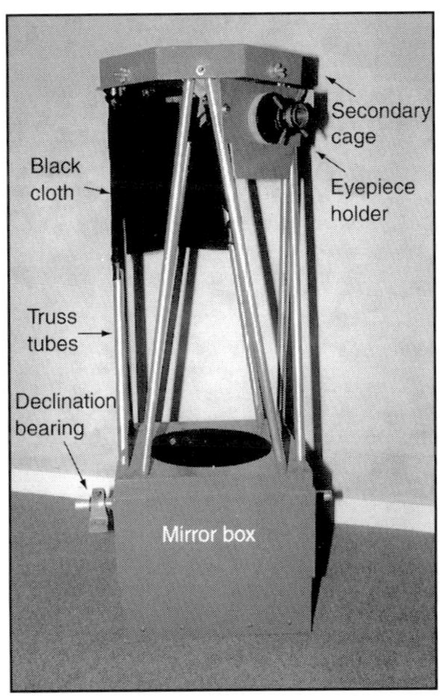

Fig. 16.1 The telescope tube.

To darken the field of view a black cloth is attached to the truss tubes opposite the eyepiece holder.

> **Preserving Wood**
>
> Before assembling the different elements of the tube, coat each of the panels with linseed oil. With a large brush, this operation is easy to do. Let the wood absorb the oil for three or four weeks. The panels will have an orange color. Work and gluing will proceed without any problem. This protection of the wood is most useful if the telescope is exposed to humidity or dew.

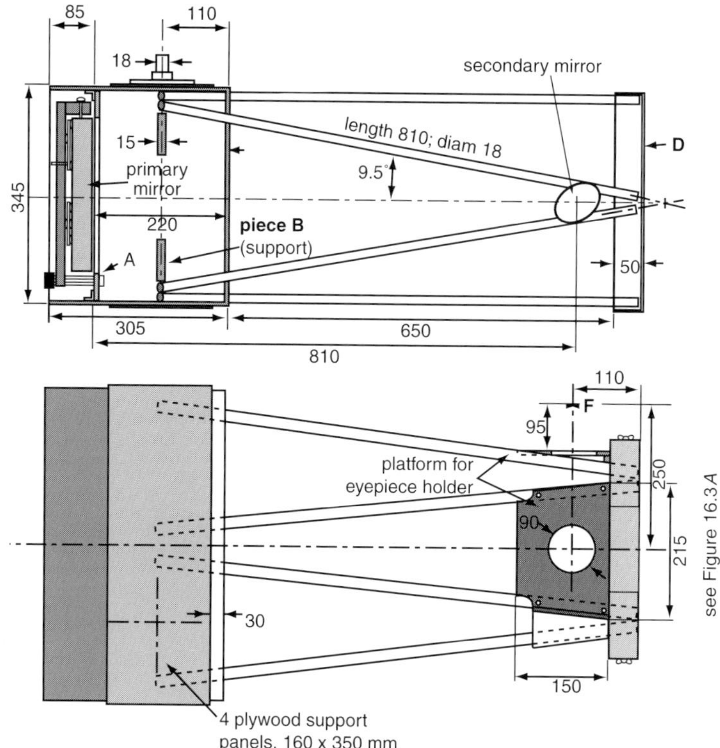

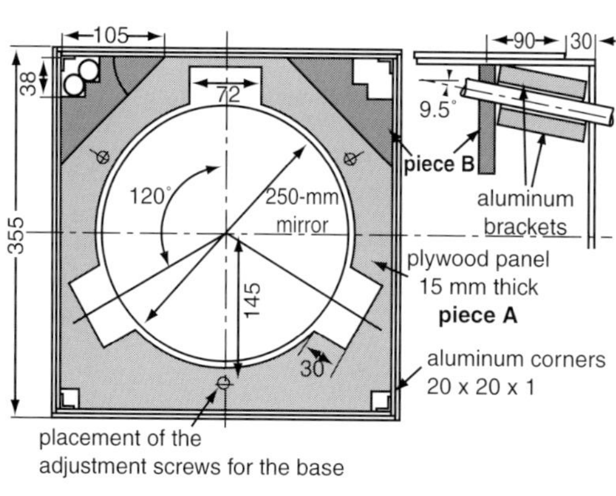

Fig. 16.2 Top: Profiles (from above and 45°) of the truss tube assembly. **Left:** Views of the front of the mirror cell support (piece *A*) and the truss-tube attachments. **Below left:** Half-view of *C*, the diaphragm positioned at the entrance of the mirror box. **Below right:** Half-view of the face and a cutaway view of secondary cage (plate *D*) with the spider assembly.

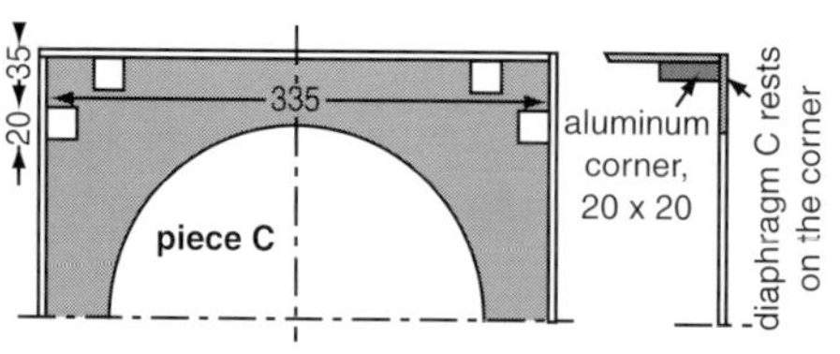

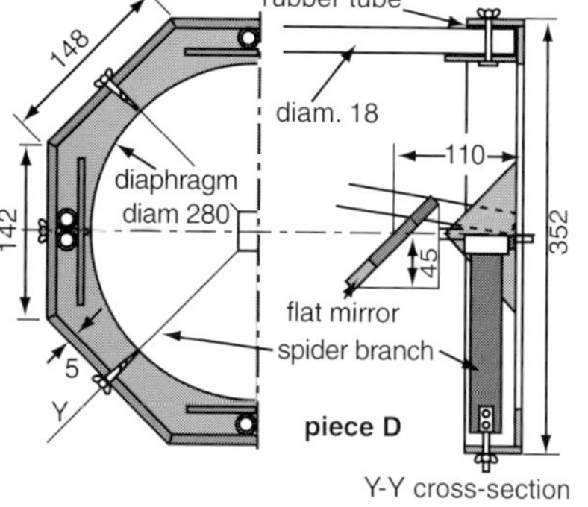

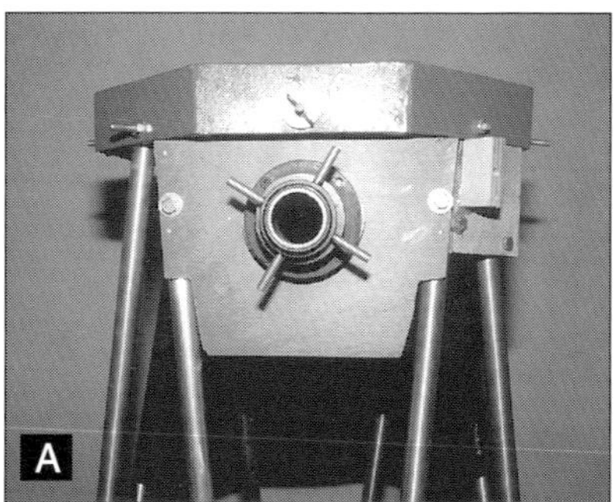

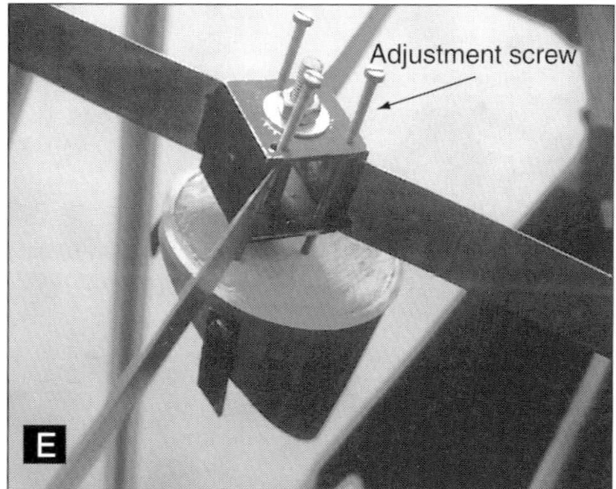

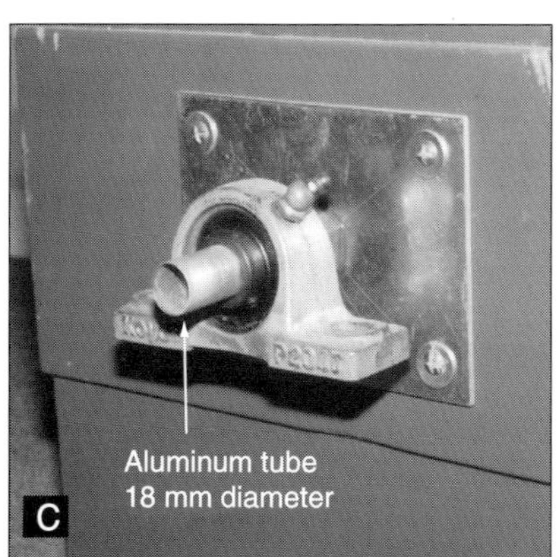

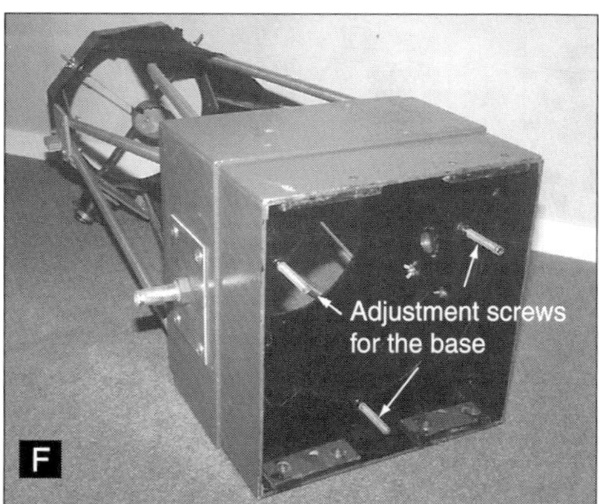

Fig. 16.3 Some views of the tube. A and B: Eyepiece holder (commercial). **C:** Declination axis. The tube is attached to the fork through pillow block ball bearings. **D:** View of the front of the tube. Notice the triangles that attach the head of the tube to the truss tubes. **E:** Detail of the spider and the secondary mirror support (the mirror is not mounted). **F and G:** View of the back of the tube and the placement of the primary mirror cell.

The Primary Mirror Flotation Cell

Fig. 16.4 Unlike the 130-mm telescope, the cell for the 250-mm mirror has nine pressure pads supported on the floating triangles. Each triangle is cut from a 3-mm thick aluminum plate. A hole at the center of gravity of each triangle allows it to pivot on a ball bearing (or the round part of an acorn nut). A screw fixed below each triangle prevents it from turning in respect to the mirror cell. The photographs below illustrate the important parts of the mirror cell. The cell is made of 15-mm thick plywood.

1. Flotation triangles. Views from above and below.
2. View from above the mirror cell. The placement of the ball bearings that hold the flotation triangles are marked *1*, *2* and *3*. The screws at *4*, *5* and *6* are lateral pressure points for the mirror (leave 1 mm of play around the edge of the mirror). At *10* are three 5-mm diameter holes to receive the anti-rotation screws that restrain the triangles.
3. View of the mirror cell with the flotation triangles installed.
4. The mirror is placed in the cell. In *A*, *B* and *C*, attach security pads to prevent the mirror from falling forward.
5. The mirror is mounted in the mirror cell.

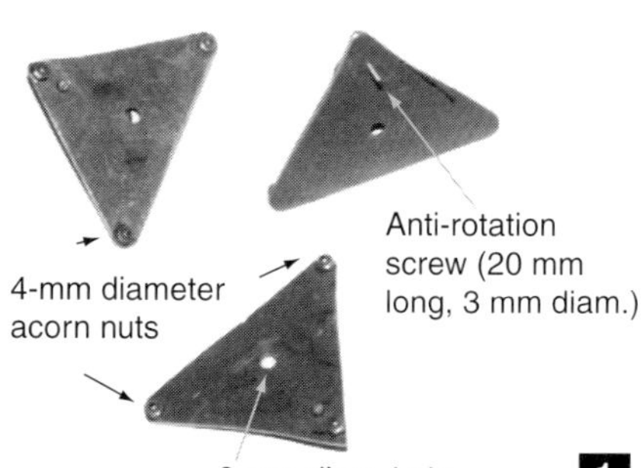

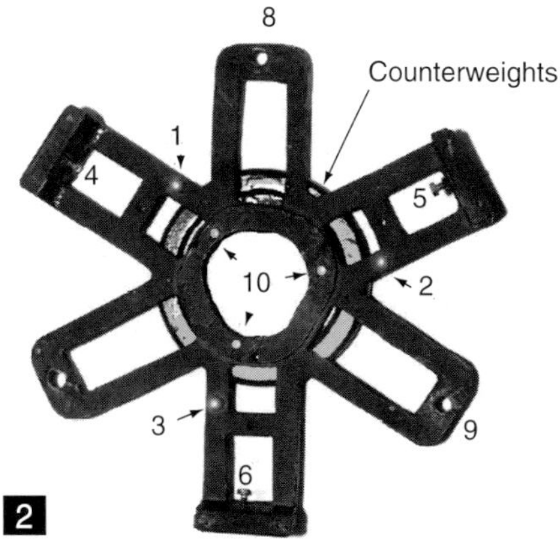

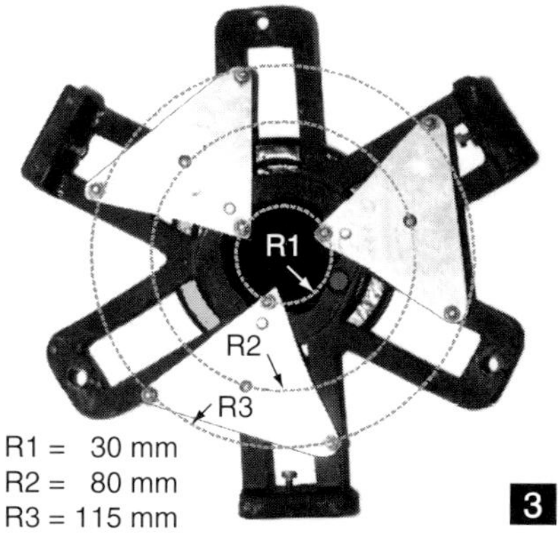

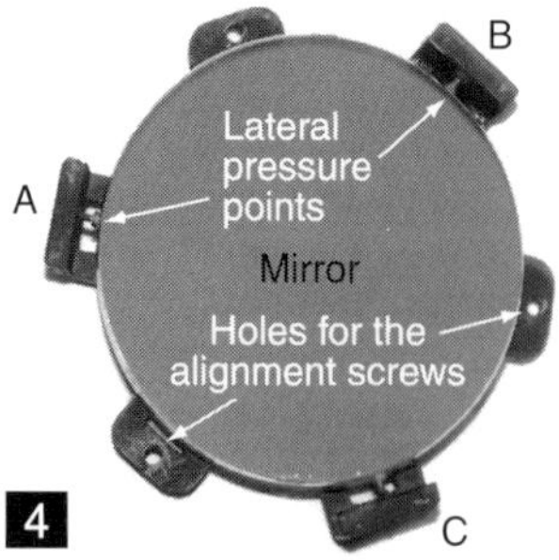

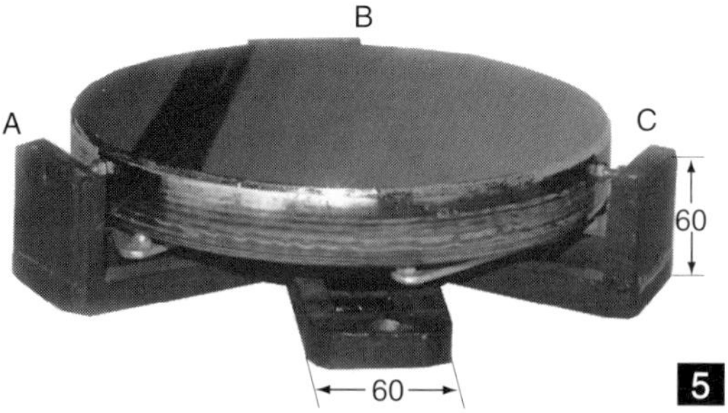

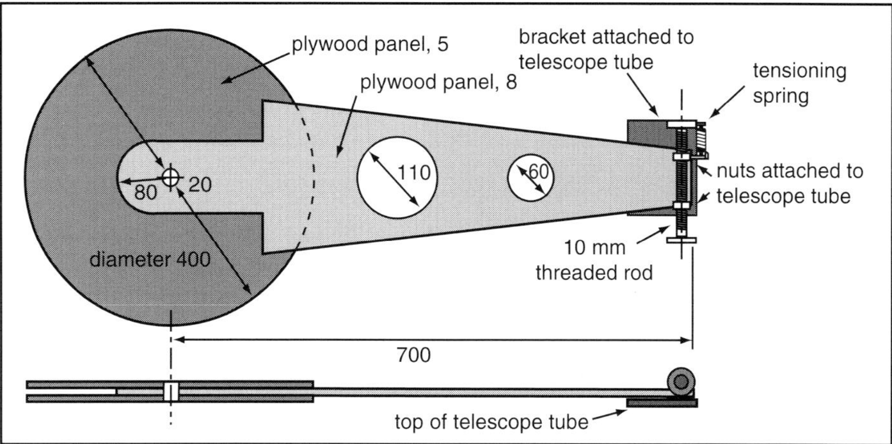

Fig. 16.5 Declination tangent adjustment arm. This assembly is mounted along side the declination axis, between the fork and the telescope tube, on the eyepiece-holder side. The two 400-mm disks act as an adjustable brake. When the brake is tightened, turning the knob at the end of the arm will slowly move the tube toward the north or the south.

16.2 The Fork

The fork is situated between the base and the tube of the telescope. It is attached to the upper end of the right-ascension axis, a 90-mm diameter, 570-mm long steel tube which projects down into the base. Contact between the fork and the base occur at a metallic plate, part *F* in Figure 16.6 on page 184, which is glued with epoxy cement and screwed to the interior face of panel *G*. At 90 mm from the top end, the diameter of the steel axis is cut down from 90 mm to 80 mm (see Figure 16.9 on page 187). The ridge formed from this recess rests on part *D*, an aluminum plate with an 80.1-mm diameter hole in the center. A centrally located bolt screws into the top of this axis and, when tightened, simultaneously locks plates *D* and *F* tightly together.

If the fork and the steel axis are not perfectly aligned (no play is allowed), then the mounting will be unstable and large (low-frequency) vibrations will disturb the instrument when it is used. The interior of the fork is hollow, which allows it to be lighter without loss of rigidity. To dampen high frequency vibrations, you can inject polyurethane foam into the hollow parts of the fork before attaching the last panel (plate *C*).

16.3 The Base

Because of its rigid pyramidal structure, the base of the telescope is able to support about 160 lbs. It is built using 15-mm thick plywood. To maintain rigidity, the base does not fold or disassemble.

The following plans are designed for a telescope used at 45° declination. The mount's north and south "feet" are not adjustable; the alignment of the base in re-

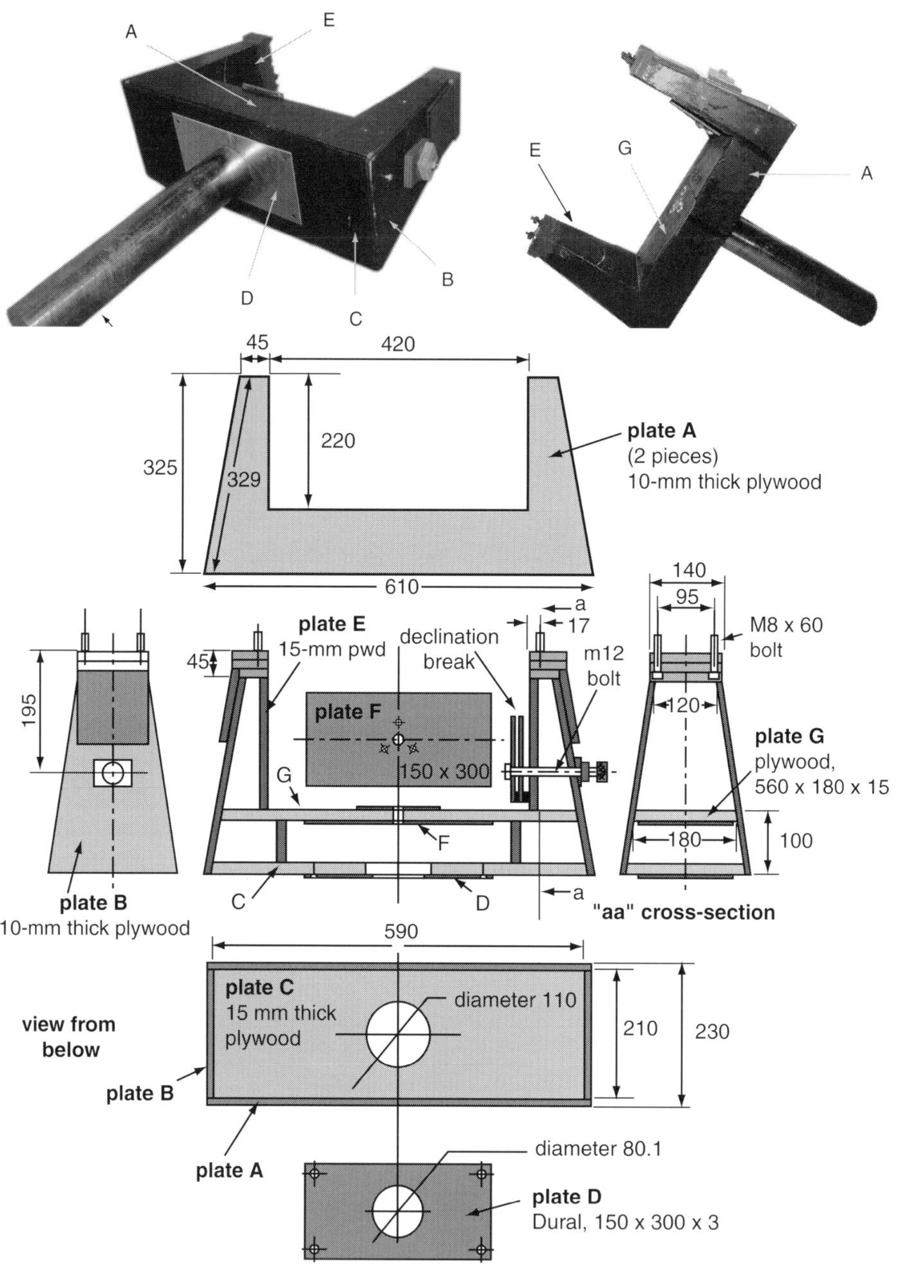

Fig. 16.6 Making the fork: plans and mounting.

lation to the meridian can be effected with nudges, and the alignment of the declination axis is done by adding triangular shims under the feet of the telescope.

16.4 Mounting the Telescope

Figure 16.11 on page 187 shows the 250-mm telescope fully assembled. If the tube is severely unbalanced, add or remove some counterweights. If it is slightly unbalanced, tighten the declination clamp to keep the telescope from moving. Counterweights can also be attached to the fork arms to counteract an imbalance if the tube is slightly off-center in the fork or if you install a heavy camera, finder scope or eyepiece.

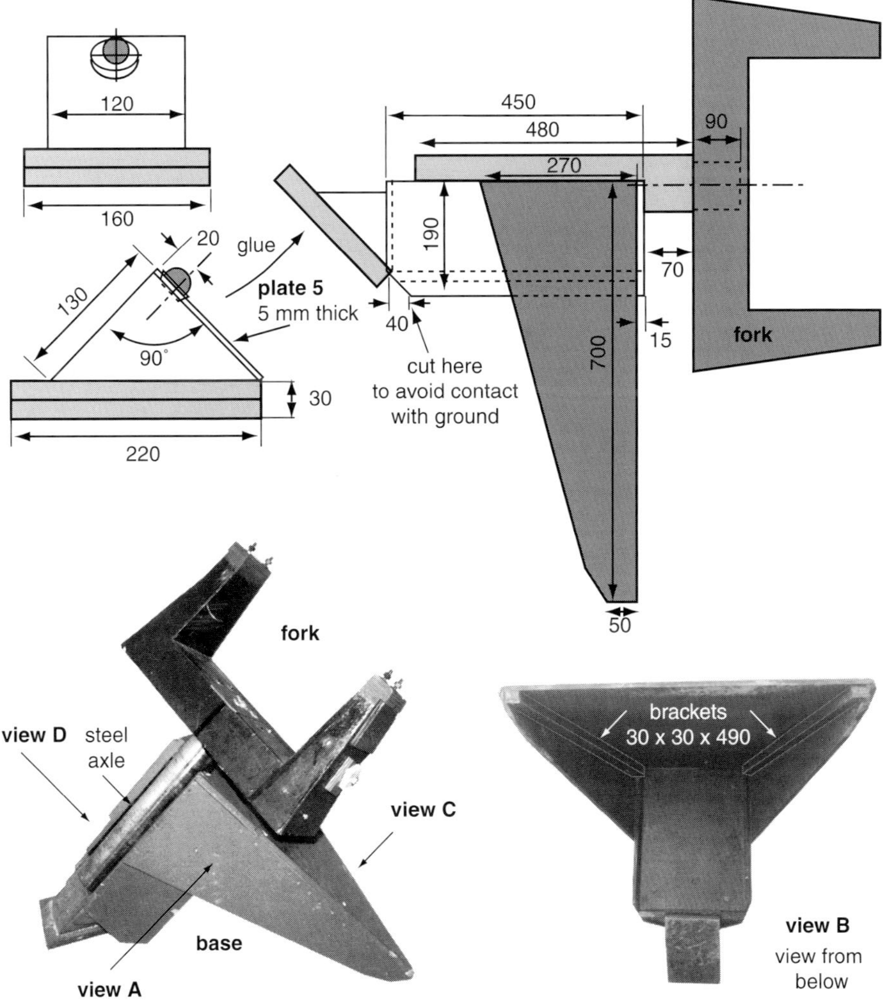

Fig. 16.7 Plans for the base.

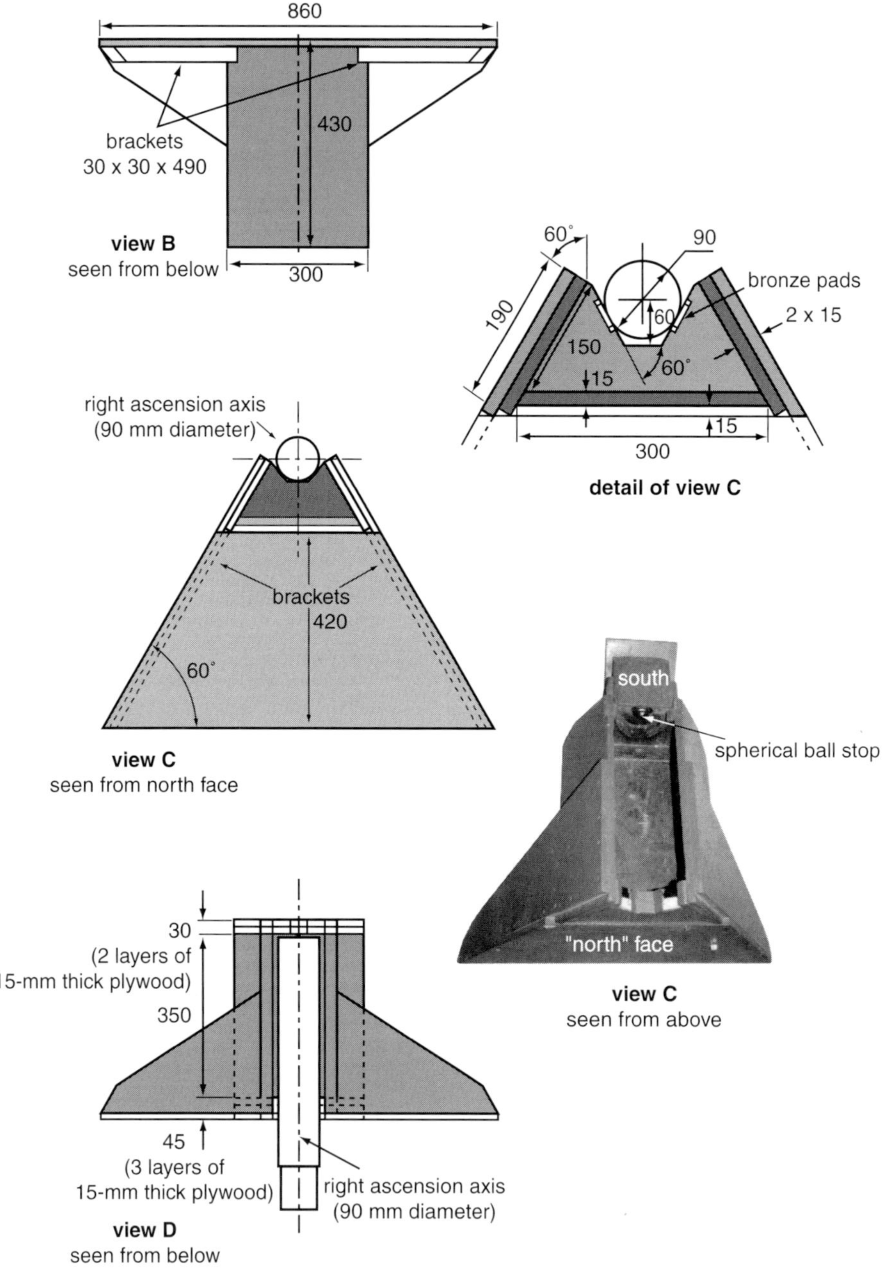

Fig. 16.8 Plans for the base.

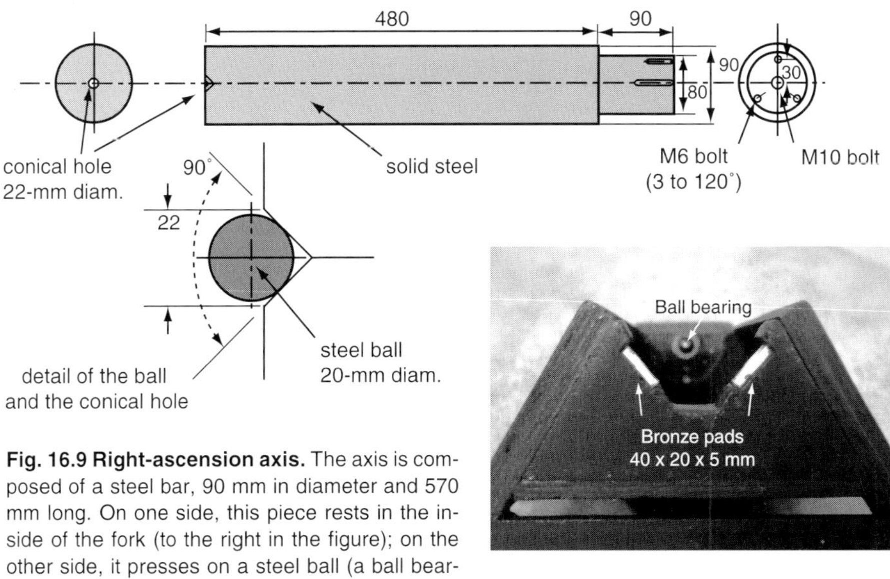

Fig. 16.9 Right-ascension axis. The axis is composed of a steel bar, 90 mm in diameter and 570 mm long. On one side, this piece rests in the inside of the fork (to the right in the figure); on the other side, it presses on a steel ball (a ball bearing), to the left in the figure.

Fig. 16.10 The Photograph shows a view of the front of the base. In the fore, the bronze pads hold the steel axle; in the back the ball bearing on its support.

Fig. 16.11 The 250-mm telescope, assembled. Easily transportable, this telescope has been used throughout France at public observing sessions during star parties and the "Night of stars" in Paris.

17 A Clock Drive

Fig. 17.1 This three hour photograph was taken with a rigidly mounted camera pointed in the direction of the North Star. The resulting star trails demonstrate the rotation of the night sky. Note that stars further from the North Pole seem to move faster and describe longer arcs.

One complete rotation of the celestial sphere takes about 23 hours, 56 minutes and 4 seconds, which is called a sidereal day. Stars seem to rise from the east, culminate in the south, and set in the west. Some stars never set; their movement describes large arcs around the north and south celestial poles. In the northern hemisphere, the axis of rotation of the sky is located close to a moderately bright star called Polaris (see Figure 17.1). When viewed through a telescope, the movement of a star becomes increasingly more noticeable with increased magnification (at the celestial equator this motion is the largest with stars moving 15° per hour, or ½ degree—the diameter of the Moon—in 2 minutes). Movement of the celestial sphere must be compensated for by a clock drive if your observations call for high magnification; it's even more important in astrophotography, where tracking must be very precise or the stars will show trails.

The clock drive is attached to the telescopes at two places: to the right-ascension axis, and to the north panel of the base (see Figure 3.4 on page 19 and Figure 17.2 on page 190). It works on the following principle: a motor, equipped with a geared reducer, turns a screw to which a "tractor" nut is attached. This nut advances as it is turned by the motor (the tractor nut moves 2 mm for every 2 mm movement of the screw). A thin, flexible metal band connects the nut to a triangular smooth sector gear. Turning the nut moves the sector gear, which in turn is attached, with a friction clutch, to the right-ascension axis of the telescope.

To cause the right-ascension axis to rotate exactly with the celestial sphere (one complete rotation in a sidereal day), you must adjust the speed of the motor as a function of the radius of the sector gear of the step of the screw and of the reduction ratio of the motor/gear box.

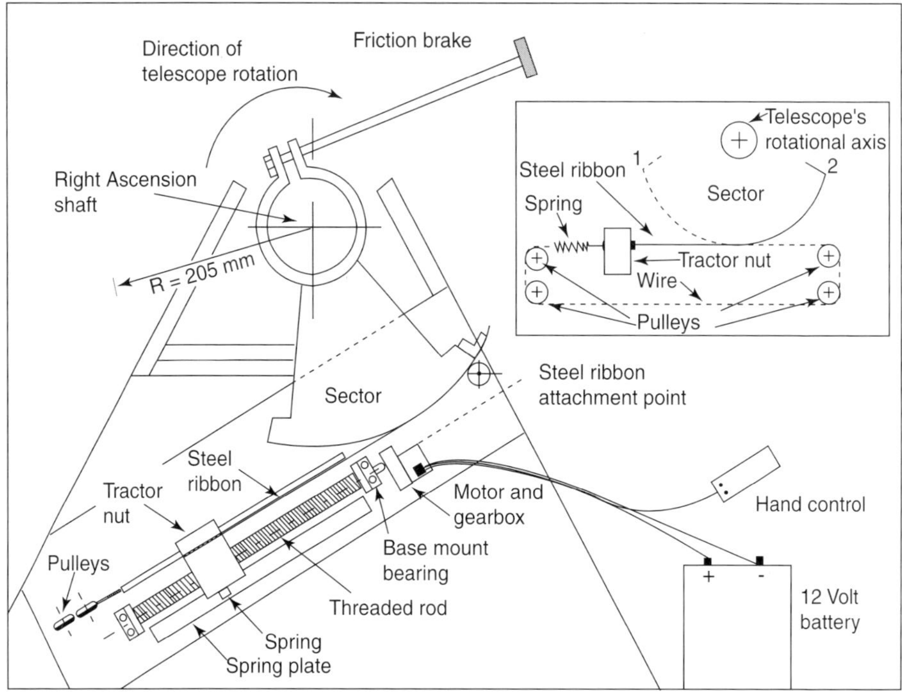

Fig. 17.2 Equatorial drive layout (see also Figure 3.4 on page 19). The break applies just the right amount of friction to the declination axis to allow you to turn the telescope manually without changing the tension on the metal band that is attached to the smooth sector. To maintain tension on the metal band and to compensate for the weight of the smooth sector, the nut is held on the side opposite the ribbon by a wire that passes through a system of pulleys and is attached to the smooth sector on the other side (point 1 in the inset). Point 2 is where the ribbon is attached to the smooth sector. The spring provides a constant source of tension.

For the clock drive described here, the smooth sector has a radius of 205 mm. The perimeter of a circle of this radius is the same as the distance traveled in a day, one turn, 1,288 mm in 24 hours, or 53.67 mm per hour. To keep things compact, we are using a segment or sector of this circle. As will be shown below, you will be able to track an object for the better part of 4 hours before it will be necessary to "rewind" the clock mechanism.

The screw is a 14 mm diameter threaded rod 200 mm long (its usable length in this application). The thread has a 2-mm pitch. It can easily be found in hardware stores in brass, stainless steel and ordinary steel. Since the threads are created by rolling the blank rod through a set of precision dies, they tend to be quite accurate. Non-rusting stainless steel is the preferred material for this application.

Now working backwards from the tractor nut to the motor we can determine how fast our motor must run to make our telescope track. The tractor nut must move 53.67 mm per hour and can run a total distance of 200 mm which means that the tractor nut will run end-to-end along the rod in 3 hours 44 minutes

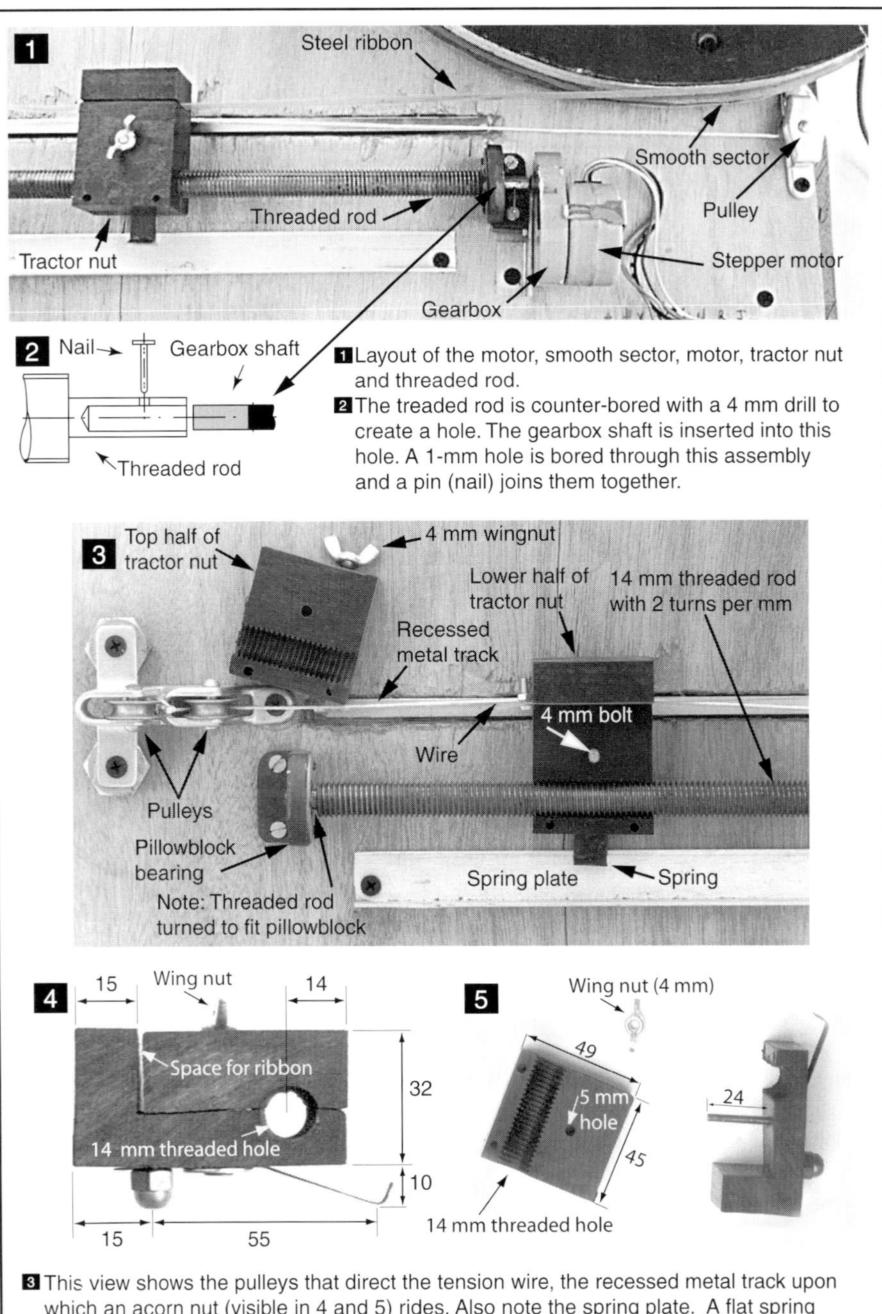

1 Layout of the motor, smooth sector, motor, tractor nut and threaded rod.

2 The treaded rod is counter-bored with a 4 mm drill to create a hole. The gearbox shaft is inserted into this hole. A 1-mm hole is bored through this assembly and a pin (nail) joins them together.

3 This view shows the pulleys that direct the tension wire, the recessed metal track upon which an acorn nut (visible in 4 and 5) rides. Also note the spring plate. A flat spring (more visible in **4** and **5**) puts tension on the acorn nut. Together, they prevent the traction nut from rotating around the screw.

Fig. 17.3

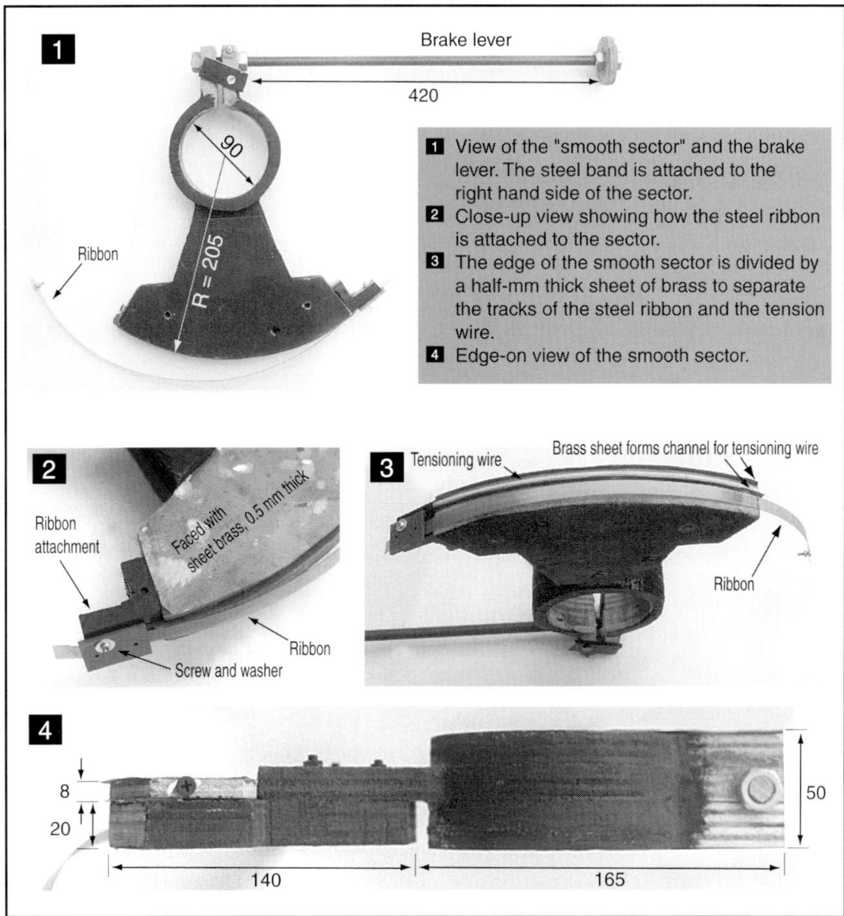

Fig. 17.4

(200mm/53.67 mm = 3 hours and 44 minutes). Therefore, it takes the tractor nut 2 minutes and 14 seconds to travel 2 mm along the screw which is one turn of the screw. The motor's gearbox has a reduction ratio of 500 to 1. Therefore the motor must turn 500 times faster than the screw, which works out to be 3.73 turns per second, or 224 turns per minute.

17.1 The Motor

The motor used in this application is a stepper motor and is powered by a 12 V battery. It has an electronic hand control (described in the next chapter) that allows it to be adjusted to turn at the desired speed (224 rpm into a 500 to 1 reduction gear box). The motor used in the author's telescope is made by SAIA, model UDB20-

N02RN; it has a12 volt input, consumes about 75 watts and steps 15° each time a pulse is received. This particular motor has a 1 to 500 reduction gearbox.[1]

17.2 The "Smooth Sector"

The "smooth sector" is built up from two sheets of plywood and two thin sheets of brass (see item 3 and 4 in Figure 17.4 on page 192). Two sheets of plywood, one 8 mm thick, the other 20 mm thick, are screwed together (but not permanently bonded) and fashioned into the sector which has a 250-mm radius of curvature along one edge. Exactly 205 mm opposite that edge a 90 mm hole is bored to accept the declination axis. Cutting the curve on the sector can be done with a jigsaw or router (with a radius of 205 mm); minor irregularities are then smoothed with a wood rasp and finally finished with a file and sandpaper. Once the curve has been smoothed the sector is separated and sheet brass is placed on both sides of the 8 mm plywood sheet. These brass sheets must extend about 3 to 5 mm above the smooth curve of the sector. The will act a a channel for the tensioning wire. The resulting composite sector gear should be assembled with contact cement and several wood screws. The brass edges should be smoothed so that all sharp edges have been removed, and there is no danger of cutting your hands in the dark when you make adjustments.

17.3 The Screw-Tractor-Nut Assembly

The screw is made from a 14-mm threaded shaft with a 2-mm step (the threads are spaced 2 mm apart). It is cut to a length of 260 mm which will allow a final usable length of 200 mm; see Figure 17.3 on page 191. It should be as straight as possible. Each end of this threaded shaft is turned down in a lathe to a diameter of 6 mm, which is the same as the interior diameter of the pillow block bearings into which they will be inserted. These bearings can often be found at model making shops, and they bolt directly to the mounting plate. The height of the screw relative to the plate on which it is mounted can be adjusted by changing the height of these bearings with metal shims: in our instrument, we raised the bearings by 1 mm with aluminum shims.

The end of the screw that will couple to the motor should be counterbored to produce a 4 mm diameter hole so that the gearbox shaft can slide into it. Once assembled both screw and the gearbox shaft are cross-drilled and then pinned together; see insert 2 in Figure 17.3. To get the nut in starting position on the screw, remove the nail and turn the screw by hand (this may take a minute or two).

The tractor nut is made from phenolic, nylon or other high density, easy-to-machine plastic. The nut's dimensions are shown in inserts 4 and 5 of Figure 17.3. To make it, begin by cutting out a rectangular block, about 50 x 50 x 100 mm, with

[1] Based upon our experience we recommend that you select a motor with more torque. Candidates would be motors manufactured by SAIA such as the UDB10-15-A50C which has a 6 volt input (but run at 12 volts), or UFD10-9.5-K25C which runs with a greater range of input pulse frequencies.

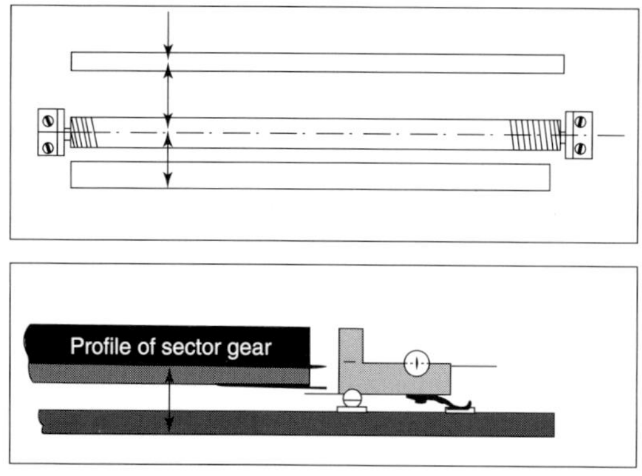

Fig. 17.5 Screw and guide assembly. Top, looking down: Guides are placed on both sides of the screw (the tractor nut is not shown). An acorn nut rides on the 30 mm wide guide. A spring rides on the 35 mm wide guide. Both guides are made of aluminum. **Bottom: Looking edge-on.** The acorn nut on the bottom of the "smooth sector" rests on the 30 mm guide. The spring puts upward pressure on the tractor nut to insure constant contact between the tractor nut and the screw. Note that the upper half of the tractor nut is not shown.

a hacksaw. Next, mark the locations where the holes will be drilled. Always begin by drilling the holes and tapping the screw thread; only after all of this work has been completed should the nut be split. The difficult task is to make the 14-mm threaded hole. The hole should have a finished diameter of 12 mm. It must be smooth and straight—it is best to use a drill press to drill all the holes in this assembly. Begin by using a 3- or 4-mm drill bit, and then progressively enlarge the hole with a series of bits, each 2 to 3 mm larger than the last, until it is 10-mm in diameter; then finish with a 12-mm bit. Threading is done using a 14-mm tap with a 2-mm thread. It is important to pass the tap completely through the piece in one direction and then repeat the operation in the opposite direction. The quality of the drive rate (and of guiding) depends greatly on the quality of the thread—it is a good idea to make and test several nuts and then use the best one on the telescope.

The tractor nut is constrained by the threaded rod, and two points of contact established by an acron nut and flat metal spring riding on metal plates. See Figure 17.3 on page 191 and Figure 17.5.

The ribbon, a 10-mm wide piece of flexible metal, about 400 centimeters long, is slid into the vertical slit (about 1 mm wide) placed between the two parts of the tractor nut. The two pieces of the nut are joined by a nut and bolt that attache the ribbon.

17.4 Maintenance of the Mounting

The mechanical parts of the mounting must be regularly cared for in order to prolong their useful life. After the optical parts, the motor is the most fragile part of the telescope. You must coat the stainless-steel pieces (the screw and the ball bearings) with silicone oil (which doesn't congeal when it's cold and is available at fishing supply stores) so they don't rust. The motor and the reducer should be protected by a plastic cover, and, if possible, removed after each night of observation. These pieces are expensive and don't take humidity well.

18 Electronic Corrector

18.1 Introduction

The equatorial mount is driven in right ascension by a stepper motor. The input to this motor is a series of electric pulses called steps. In this particular motor an electric pulse causes the motor's armature to turn or step 7.5° (when running in half-step mode). Since this motor has a 500 to 1 reduction gear box, the motor must turn 3.73 revolutions per second. To accomplish this the motor is fed about 180 pulses each second. These pulses are created by a small electronic circuit board, whose plans appear on the following pages.

The circuit presented here is relatively easy to build. While only a few components are required, those that are used are sophisticated stepper motor controllers and drivers. We provide the artwork for a single-sided circuit board, drilling templates and part layout diagrams. The circuit board fits completely within a small hand control box which is powered by an external 12 volt deep-discharge battery (see Figure 18.1). The components are readily available from large electronic distributors. In the parts lists presented here catalog numbers have been pro-

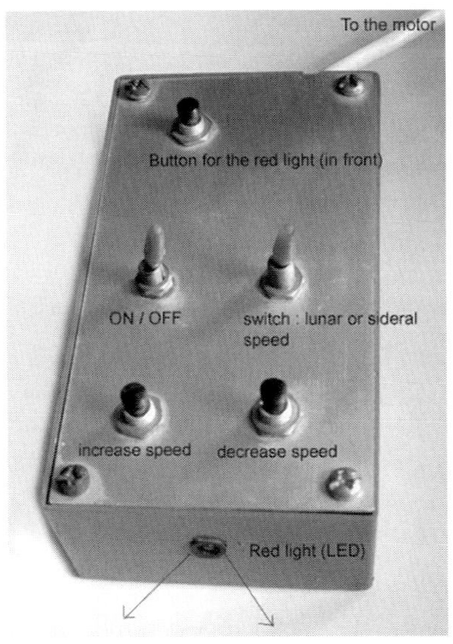

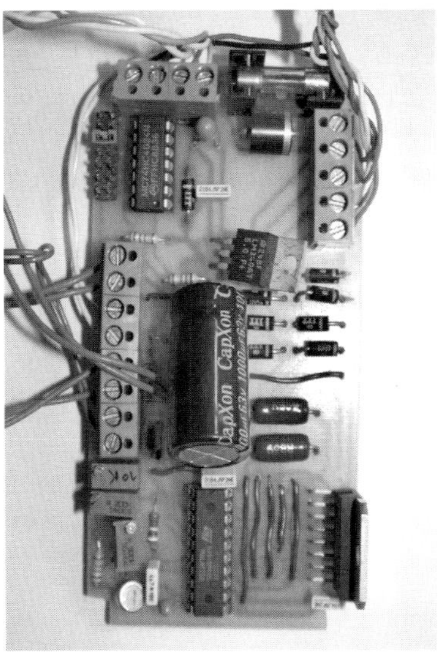

Fig. 18.1

vided for DigiKey and Mouser. Both of these businesses take orders by phone and over the Internet.[1]

18.2 Circuit Overview

- The schematic for the controller is shown in Figure 18.2.
- The thick line that runs around the top and sides of the schematic diagram represents a group of connections. These connections can be traced by their labels (for example, I1, I2, I3, ECL, RST...).
- The control circuit is built up on a single-sided printed circuit board. A full sized parts layout template, circuit board and drilling template are provided in Figure 18.3 on page 198. Circuit board making supplies can be purchased from electronic parts stores or over the web. One such place is Circuit Specialists, Inc.[2]
- Input voltage (+12 VDC) is notated VS and is from a heavy-duty 12-volt battery—it is connected to CON1A on the circuit board. Diodes (D1, D2) prevent damage to the circuit should you inadvertently reverse the battery connections. The stepper motor is run a 12VDC. U2 reduces the voltage to 5 volts (VSS) to power the control logic circuitry.
- U1 is a two-phase bipolar stepper motor controller designated L297 (made by Thomson). Several different versions of this chip are available. You need to order the L297 version. Note that the L297A is a step-pulse doubler which the circuit and motor described here *will not handle*.
- U3 is a dual full-bridge driver (Thompson L285N) designed to interface integrated circuits like the L297 to inductive loads such as stepper motors. This integrated circuit is available in several configurations: make sure that you purchase the L285N version which is described as "Multiwatt Vertical".
- JMP1 is the 7-pin dual in-line jumper block that sets the binary divisor for the clock frequency. By placing a jumper block on one of the seven sets of pins, you can select a band of output frequencies that match the motor and the other variables in your system (radius of the sector, thread pitch of the drive screw, and the motor's gearbox reduction ratio). For example, placing the jumper at Q1 divides the clock frequency by 2, Q2 by 4, and so on up to Q7, where the frequency is divided by 128. The telescope motor, which we specify in this book, requires a 1 to 32 ratio (the jumper block is set to Q5). To double the speed of the motor, set the jumper block at Q6 (1 to 64 step).
- P1 is a potentiometer that varies the amount of current (at 12 volts) that is applied to the stepper motor. This control is adjusted to supply sufficient cur-

[1] Digi-Key Corporation, 701 Brooks Avenue South, Thief River Falls, MN 56701. Phone: 800-344-4539. http://www.digikey.com/. Mouser Electronics, 1000 North Main Street, Mansfield, TX 76063. Phone: 800-346-6873. http://www.mouser.com/.

[2] Circuit Specialists, Inc., 220 S. Country Club Drive #2, Mesa, AZ 85210. Phone: 480-464-2485. http://web-tronics.com.

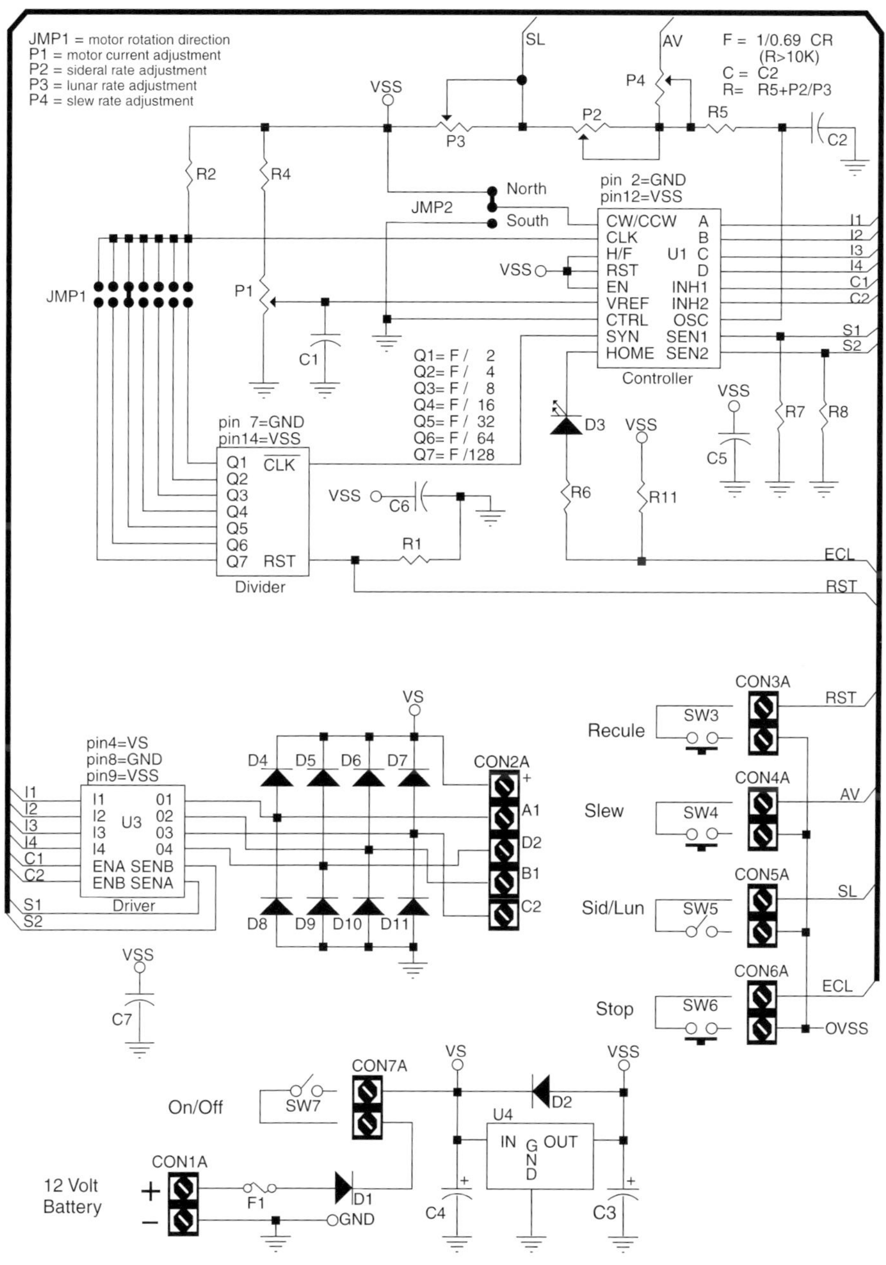

JMP1 = motor rotation direction
P1 = motor current adjustment
P2 = sideral rate adjustment
P3 = lunar rate adjustment
P4 = slew rate adjustment
F = 1/0.69 CR
(R>10K)
C = C2
R= R5+P2/P3
Q1= F / 2
Q2= F / 4
Q3= F / 8
Q4= F / 16
Q5= F / 32
Q6= F / 64
Q7= F /128
pin 2=GND
pin12=VSS
pin 7=GND
pin14=VSS
pin4=VS
pin8=GND
pin9=VSS
Controller
Divider
Driver
North
South
Recule
Slew
Sid/Lun
Stop
On/Off
12 Volt
Battery

Fig. 18.2

Stepper Motor Controller Parts List

Quantity	Reference	Designation
2	SW5,7	Single pole single throw toggle switch
3	SW3, 4, 6	Normally open push button switch
1	U2	74HC4024, 7-stage binary ripple counter
1	U1	L297 stepper motor controller
1	U3	L298N dual full-bridge driver (multiwatt vertical)
1	U4	78L05 3 terminal positive 5 volt regulator
1	C3	1000μF 16VDC radial electrolytic capacitor
1	C1	1μF 35VDC tantalum capacitor
1	C4	10μF 16VDC tantalum capacitor
3	C5, 6, 7	100nF 63VDC disk ceramic capacitor
1	C2	4.7nF 63VDC metallized polyester film capacitor
7	CON 1,3, 4, 5, 6, 7	Weidmüller 2 position terminal blocks, 5.08 mm [DigiKey 1967120000]
1	CON 2	Weidmüller 3 position terminal blocks, 5.08 mm [DigiKey 1967130000]
1	JMP2	Jumper for dual in line header
1	JMP2	7 pin dual in line header [DigiKey WM6814]
2	ACC1, 2	Jumper
2	ACC4, 5	Fuse clip [DigiKey F063-ND]
1	F1	5x20 mm 2A fast-acting fuse
1	D3	Red LED
1	ACC7	LED lens/holder for a T-1 LED, [DigiKey L3002-ND]
1	P1	20K 25 turn Pot. Vertical adjustment. [DigiKey 3292X-203-ND]
1	P2	200 K 25 turn Pot. Vertical adjustment. [DigiKey 3292X-204-ND]
1	P3	10K 25 turn Pot. Vertical adjustment. [DigiKey 3292X-103-ND]
1	P4	470K 25 turn Pot. Vertical adjustment. [DigiKey 3292X-504-ND]
1	R7, 8	1 Ohm 3 watt ceramic resistor
1	R6	150 Ohm W 5% resistor
1	D2	IN4002 100V 1A silicone diode
8	D4,5,6,7,8,9,10,11	UF4002 100V 1A fast recovery diode
1		Machined 14 pin DIP IC socket
1		Machined 20 pin DIP IC socket

Fig. 18.3 Facing Page Top: Parts layout diagram showing the position of the parts on top of the circuit board (side opposite the copper foil). The parts labeled jumpers are wires connecting two points of the circuit that could not be done with a foil trace.
Facing Page Middle: Paths of the printed circuit seen from the side of the parts.
Facing Page Bottom: Plan for the piercing of the plate. Each sign corresponds to a hole with a different diameter, determined as a function of the part.
Important: These three diagrams are actual size and seen from below. The printed circuit must be reproduced at the same dimensions as shown on the left page to fit into the circuit's box.

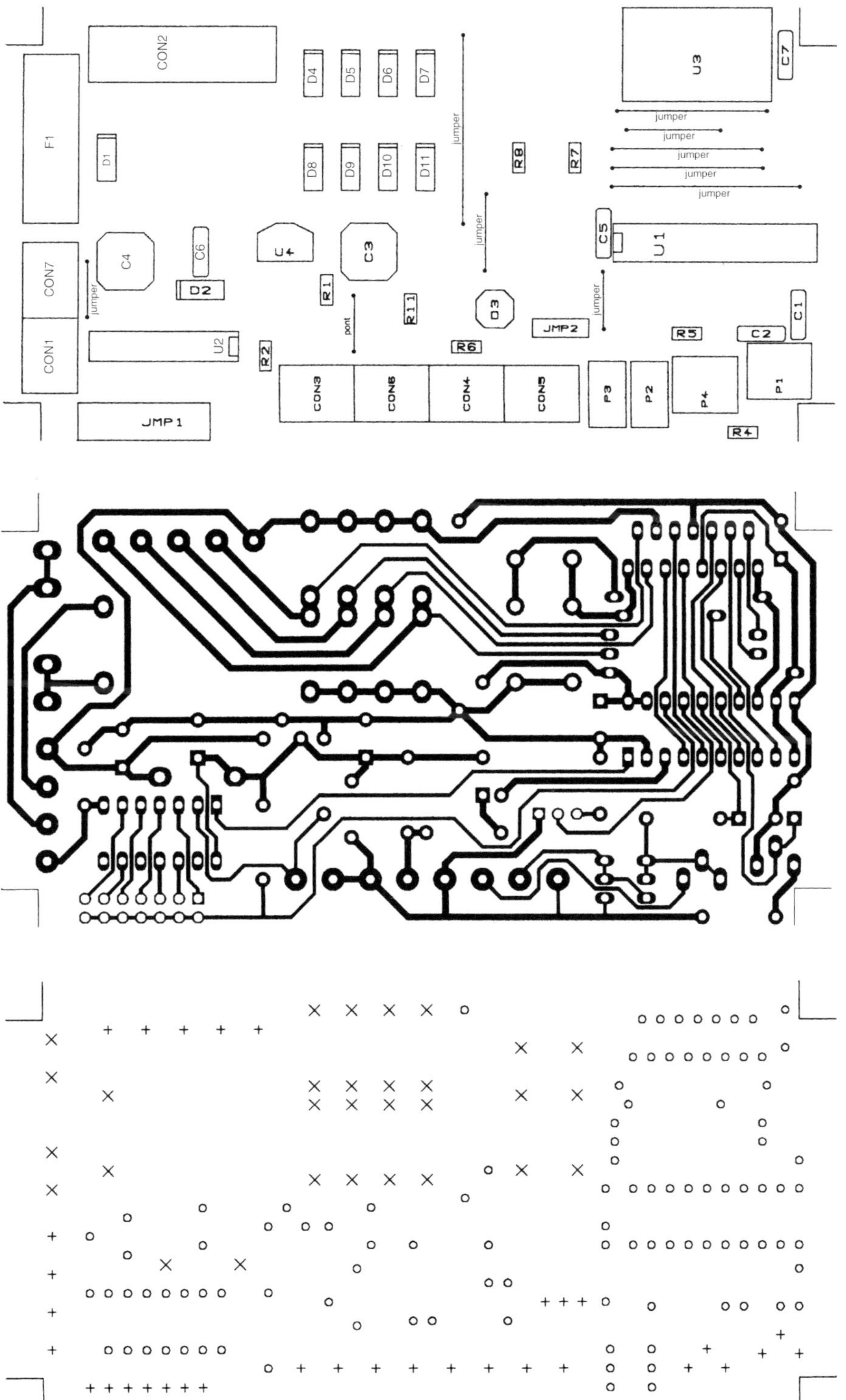
CON2
D4
D5
D6
D7
U3
C7
jumper
jumper
jumper
jumper
jumper
jumper
F1
D1
D8
D9
D10
D11
R8
R7
jumper
C5
U1
CON7
C4
C6
U4
C3
jumper
D2
R1
R11
D3
C1
jumper
pont
JMP2
R5
C2
CON1
U2
R2
R6
CON3
CON6
CON4
CON5
P3
P2
P4
P1
JMP1
R4

rent to just run the stepper under load. Make this adjustment first without the motor installed on the telescope. Start by turning P1 to the maximum (the motor should turn), then decrease the adjustment slowly until the motor stops. Next, very slowly increase the adjustment until the motor just starts to turn. Finally, mount the stepper motor on the telescope, turn on the power and slowly increase P1 until it just drives the telescope. Operating with the minimum amount of current necessary to drive your telescope will extend battery life and insure that the circuit and motor are not subjected to excessive and wasteful heat that could lead to component failure.

- P2 adjusts the sidereal rate clock frequency. You make this adjustment by viewing a bright star close to the celestial equator (first or second magnitude if possible) through a reticile eyepiece (9 mm focal length or 12 mm plus a 2x or 3x Barlow). Let the star move in the field without starting the motor. This way, you can determine the East-West direction and align one of the reticle's cross-hairs with this movement. Next, recenter the star in the field and start the motor. As soon as the telescope begins to move, note the motion of the star in relation to the cross-hair. If the star moves continually toward the east, the motor is moving too slowly. If it moves to the west, the motor is too fast. Movement toward the north or south are caused by misalignment. To adjust the motor's speed, turn P2 and note the effect that movement has on the movement of the star. Continue your adjustments until the star appears to stand still.
- P3 adjusts the lunar tracking speed. These adjustments are similar to those used when setting the sidereal rate using a star except that the Moon is substituted for the star.
- P4 allows for adjustment of the motor's acceleration speed when you press the "advance" button. Turn the potentiometer until the motor misfires. Pull back a bit to the right adjustment. Relaxing and pressing again, check that the motor still turns correctly; if it doesn't, increase again the amount of the potentiometer's adjustment.
- CON 2A is the motor terminal block. If it has 4 wires you have a bipolar motor (see Figure 18.4). With an ohm meter, determine which wires form a circuit—those that do will show resistance, those that do not will show no resistance. Consider the first pair to show resistance to be A and B and the other pair C and D. Next, attach these wires as shown in Figure 18.4. The motor should run, but if it does not, cross A with B or C with D. Similarly, if you have a 5 or 6 wire unipolar motor use an ohm meter to determine which wires are interconnected within the motor. However, in this case you will determine the + wire when your meter reads one half the value of that read when the meter is connected to the other two wires. Connect the motor as shown in the unipolar segment of Figure 18.4. If the motor does not run, you only need to cross A with B or C with D.
- CON3 connects to the motor's stop button.
- CON4 connects to the motor's slew button.
- CON5 is plugged into the speed switch. One position corresponds to sidereal

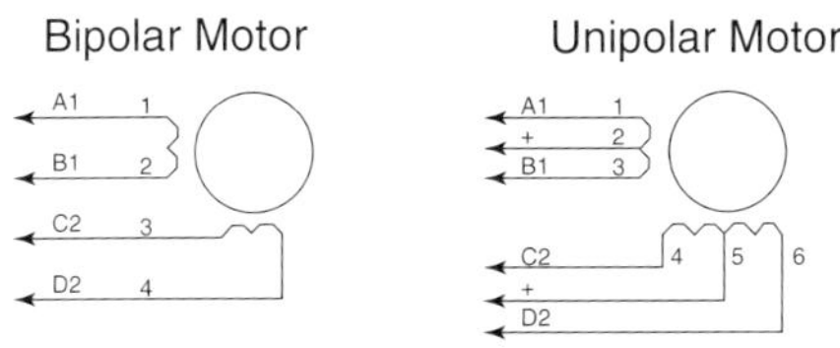

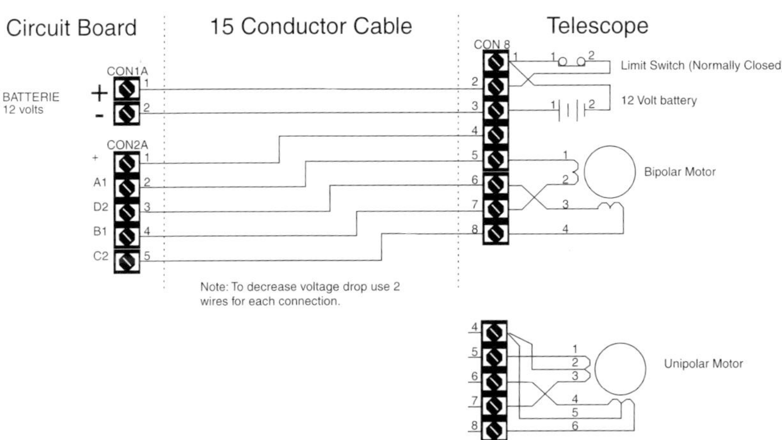

Fig. 18.4

movement (following on a star), the other to lunar motion (a somewhat slower speed).

- CON6 is plugged into the LED button.
- CON7 is plugged into the general start/stop switch.
- 15 conductor wiring is used between the hand control and external circuitry. To reduce voltage drop each connection is made with two wires.

IMPORTANT: during a night of observation, don't forget periodically to check the advancement of the nut on the screw. The nut should never get to the motor mounting, because it could damage the mounting or, worse, the motor (the reducer multiplies the power of the motor by 500!). For security, install a limit switch which opens if the nut gets within 10 mm of the end of the threaded rod; see Figure 18.5.

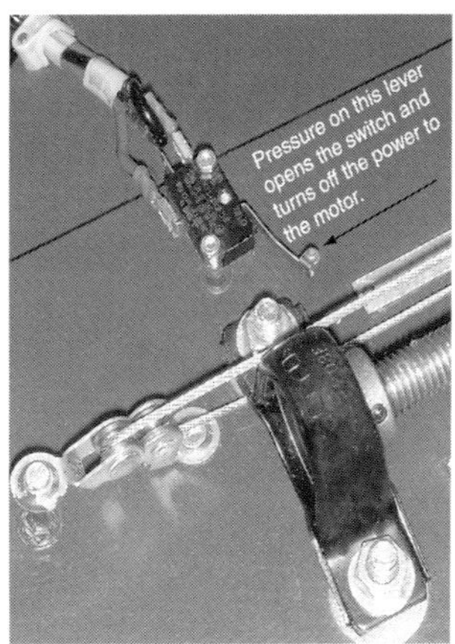

Fig. 18.5 An optional limit switch mounted along side the screw. The advancing drive nut pushes on to the switch and cuts off the power before it jams up. Photo courtesy Hervé Chantegros.

Part 4 A 300-mm *f*/12 Cassegrain-Coudé Telescope

Plans for a 300-mm Cassegrain-coudé (Naysmith) telescope with a focal length of 3,600 mm, using the mounting for the 250-mm telescope in Part 3.

19 Optical Specifications

19.1 Optical Characteristics and Theoretical Performance

19.1.1 Optical Specifications

This telescope has 4 optical elements: a centrally perforated concave primary mirror, a convex secondary mirror, an optical window, and a flat coudé[1] mirror (see Figure 19.1). The primary paraboloidal mirror (M_1) has a 305-mm outer diameter and a 300-mm optical diameter (D_1). Its radius of curvature (R_1) is 1,800 mm, its focal length (f_1) is 900 mm, and its sagitta is 6.25 mm. The mirror has a 45-mm perforation at its center.

The convex secondary mirror (M_2) has a radius of curvature which magnifies (γ) the primary's image by a factor of 4 ($\gamma = 3600/900 = 4$). The distance between the secondary mirror and the focus F_1 of the primary mirror is labeled p. Thus:

$$p = \frac{F_1 + x}{\gamma + 1}$$

where x is the distance between the surface of the primary mirror and the secondary focus F_2 (we set x at 290 mm). In our case, $p = 238$ mm.

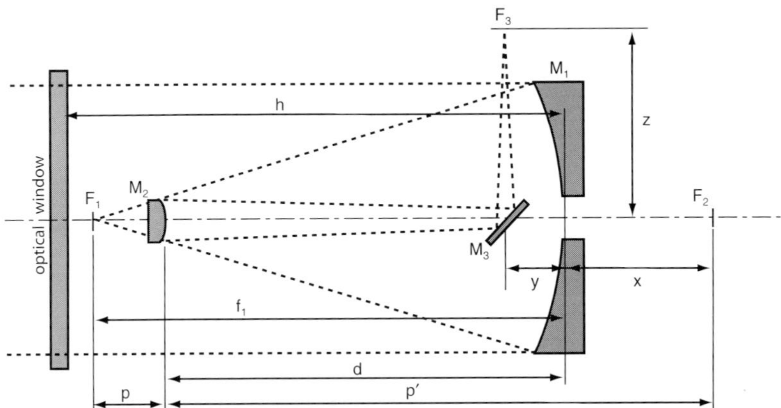

Fig. 19.1 A Cassegrain coudé telescope. Light enters from the left. It travels first through the window and is collected by the primary mirror (M_1), which reflects it toward the convex secondary mirror (M_2). The secondary reflects the light and focuses it at F_2. The third mirror (M_3) is a flat and is placed in front of the primary to reflect the image outside the tube to F_3.

[1] Coudé is French for elbow.

The distance from the center of the primary mirror to the center of the secondary is d, which is calculated:

$$d = f_1 - p, \tag{19.1}$$

where f_1 is the focal length of the primary. For this example, $d = 662$ mm.

The distance between the secondary mirror and the Cassegrain focus F_2 is p'. Knowing the magnification of the secondary mirror is determined by $g = p'/p$, we then have:

$$p' = p \times \gamma, \tag{19.2}$$

and the instrument discussed here is $p' = 952$ mm.

The diameter D_2 of the convex secondary mirror is defined thus:

$$D_2 = \left(D_1 \frac{p}{f_1}\right) + d\tan\alpha \quad , \tag{19.3}$$

where α is the telescope's field of view. We chose a fully illuminated field of view at the Cassegrain focus equal to the diagonal of a 35 mm photographic negative (24 by 36 mm). The field is calculated:

$$\alpha = \arctan\left[\frac{(24^2 + 36^2)^{\frac{1}{2}}}{3600}\right], \tag{19.4}$$

or 41 minutes of arc. The optical diameter D_2 is thus 87 mm and the mechanical diameter is 95 mm. The radius of curvature R_2 of the secondary mirror is given by the formula:

$$R_2 = \frac{2 \times p \times \gamma}{\gamma - 1}, \tag{19.5}$$

or $R_2 = 900$ mm. This mirror is *a hyperboloid*; its coefficient of deformation, b_2, is calculated thus:

$$b_2 = -\left(\frac{\gamma + 1}{\gamma - 1}\right)^2, \tag{19.6}$$

which gives $b_2 = -2.78$ (for the primary mirror, $b_1 = -1$).

The minor axis for the flat coudé mirror (M_3) is calculated exactly like a Newtonian secondary. For this telescope it measures 68 mm. The major axis of the mirror is 95 mm. Its mechanical diameter is 100 mm. It is positioned at a distance of $y = 60$ mm from the surface of the primary mirror. The focus F_3 is at $z = 350$ mm from this mirror and takes into consideration the distance necessary to exit the tube and accommodate a focusing device. Note that $z = x + y$.

This instrument is equipped with an optical window which closes off the

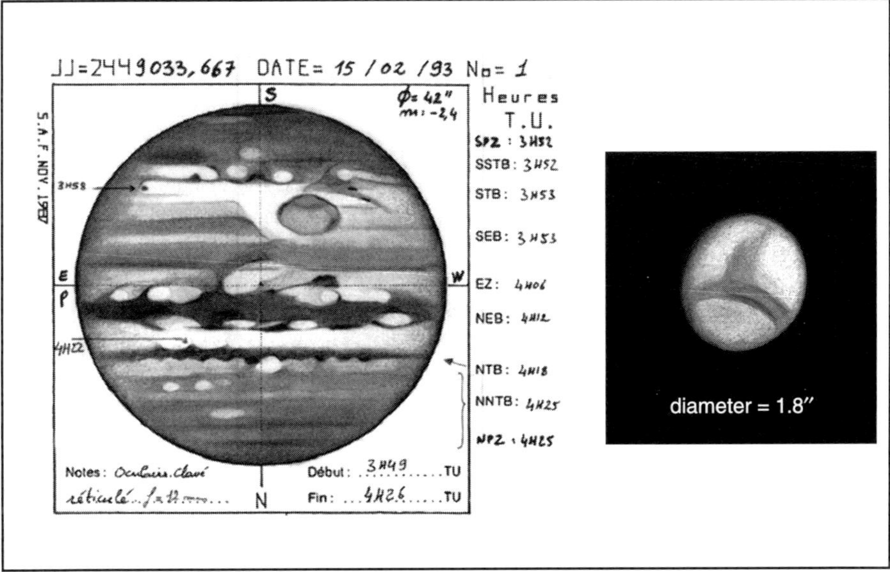

Fig. 19.2 Left: Jupiter observed February 2, 1993, with a 300-mm telescope, magnified 350x. **Right:** Ganymede, the largest of Jupiter's satellites, observed December 29, 1990 at 1000x magnification. To highlight the details, contrast has been strongly enhanced.

front of the tube. It is plane parallel glass 17 mm thick and measuring 340 mm in diameter with a clear aperture of 320 mm. The window is placed 710 mm in front of the primary mirror (*h*).

The support for the secondary mirror includes 150-mm diameter baffles to reduce stray light. These baffles obstruct the primary mirror by 35% and cause a small increase in the brightness of the first two diffraction rings. Baffles are also installed around the flat coudé mirror (see Figure 26.4 on page 256).

19.1.2 Theoretical Performance of the Telescope

This telescope collects 1,840 times more light than the naked eye and has a stellar limiting magnitude of 14.7. At the Cassegrain focus (F_2 in Figure 19.1 on page 205), the diffraction spot has a radius of 8 µm; the resolving power of the instrument is on the order of 0.5 arc-seconds.

This telescope can be used for visual observation, with magnification from 120x (with a 30-mm eyepiece) to 700x (5-mm eyepiece). It is also well-suited for photography, where it fully illuminates a field of 43 mm or 41 arc-minutes. This fully covers a 35-mm frame, which measures 24 by 36 mm.

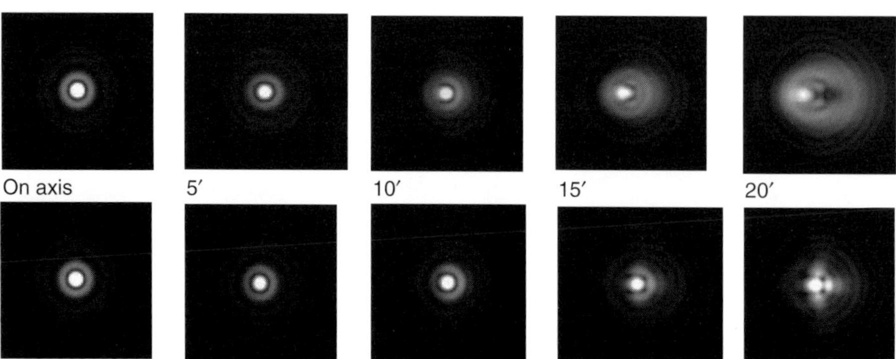

Fig. 19.3 Diffraction images obtained for the 300-mm telescope accounting for the central obstruction (0.35 x *D*). **Top:** The images in the focal plane. They become more flawed as they move farther away from the optical axis; this is due to field curvature. **Bottom:** After compensation for field curvature (by focusing or by addition of a field corrector). It is instructive to compare these images with those for the 130-mm telescope in Figure 6.2 on page 86 and the 250-mm telescope in Figure 12.2 on page 150. Images courtesy of H.R. Suiter.

19.2 Fabrication Tolerances

19.2.1 The Primary Mirror

The primary mirror is paraboloidal and requires a precision of ⅛ wave. To achieve this degree of precision we will use both the Foucault and caustic tests. The mirror will also be rigorously tested for astigmatism.

19.2.2 The Secondary Mirror

The secondary is a hyperboloidal convex mirror and is tested against a concave spherical surface by interference. The precision required is about 1⁄10 of a fringe.

19.2.3 The Coudé Flat

The coudé mirror diverts a narrow light beam (*f*/12) relatively close to the focal plane outside the tube. Calculation of the astigmatism caused by any curvature of the mirror sets the maximum tolerance for a defect of curvature at two fringes over the mirror's 95-mm diameter. The reason this tolerance is so loose is because a given light cone does not cover the entire flat. It will be relatively easy to make this mirror.

19.2.4 The Optical Window

If the two faces of the optical window are not perfectly parallel, it will behave like a small-angle prism and create colored ghost star images. So that these ghost images are invisible to the eye, we want them to be *ten times shorter* than the radius of the diffraction spot at the focus of the instrument. If A is the angle between the faces of the window, the length of the "ghost, or spectrum" measured from blue

($\lambda = 0.486$ μm) to red ($\lambda = 0.656$ μm) is calculated:

$$\Delta D = (n_F - n_C) \times A \ ,$$

where n_F and n_C are the indexes of reflection of the glass for blue and red, respectively, and ΔD is the relative *angle* of the image in F and C light. Their difference, called the average dispersion, is specified by the glass manufacturer. In our case, the telescope window is made of borosilicate crown (BK7) and will present a dispersion of 0.008054 (from the Schott catalog). The maximum length is ρ/10, so:

$$\Delta D < \frac{\rho}{10} \quad \text{or} \quad A < \frac{(\rho/10)}{0.008054} \quad \text{or} \quad A < 12.4 \times \rho \, .$$

At the Cassegrain focus, the radius of the diffraction spot is 0.5 arc-seconds; the angle A should be smaller than 0.5 × 12.4, or about 6 arc-seconds. In mechanical terms, this angle corresponds to a difference in thickness of 10 μm from the thickest edge to the thinnest edge of the plate, over its 340 mm diameter. This represents a "tight" tolerance that can perhaps be eased by a factor of two.

The faces of the telescope window can have a slight curvature which will act like a weak converging or diverging lens, and will not have any noticeable effect because of its large focal length. Color is not detectable if the focal length of the lens is within the following criterion:

$$F > 5.73 \times D^2 \, ,$$

where D is the diameter of the lens in mm.

Assuming a 320-mm clear aperture, the window will have a minimum value of 587 meters for F. If the plate is biconvex, the radius of curvature of the faces will be approximately 587 meters with a sagitta close to 0.025 mm.

19.3 Choosing the Materials

Because of its strict tolerances, the primary mirror should not be sensitive to variations in temperature, which can change its shape and its radius of curvature. It should be made from a material whose coefficient of expansion is almost zero, like Zerodur or ULE (Ultra Low Expansion) silica. For the same reasons, the secondary mirror should be made from Zerodur. However, if there is no alternative, you can make the mirrors from Pyrex or Duran 50, but you should expect the instrument to be much more temperature sensitive.

The coudé mirror can be made from Pyrex. The telescope window, as we have noted, is made from crown glass (BK7 by Schott).

20 Making a Zero-Magnification Finder

This is a practical "heads up zero-magnification" viewfinder. It's made of two glass panes, a small converging lens, and a reticle illuminated by a low-power lamp or light emitting diode (LED). If you do not wish to make this finder, we recommend you purchase a Telrad® or other zero-magnification finder.

The observer views the sky through a glass plate inclined at 45°, which also reflects the image of the reticle focused at infinity by a simple 20 mm diameter 150 mm focal length plano convex lens. The eye, then, sees the reticle projected on the sky. This finder does not magnify images, nor does it allow observation of

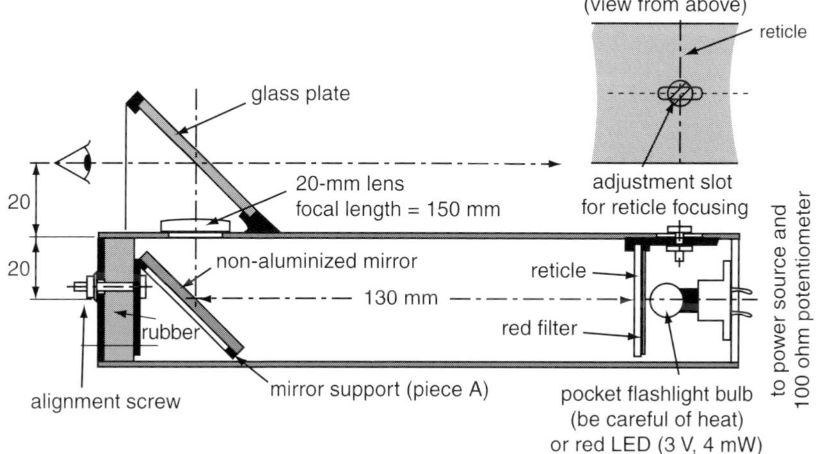

Fig. 20.1 (Above) The reflecting mirror (non-aluminized plate glass) is inclined at 45°. It is glued on piece A, a bent 1-mm thick aluminum plate. The back of the mirror is painted black in order to eliminate stray light reflections. The mirror is aligned by three screws spaced at 120° behind the apparatus. The rubber provides support and resistance.

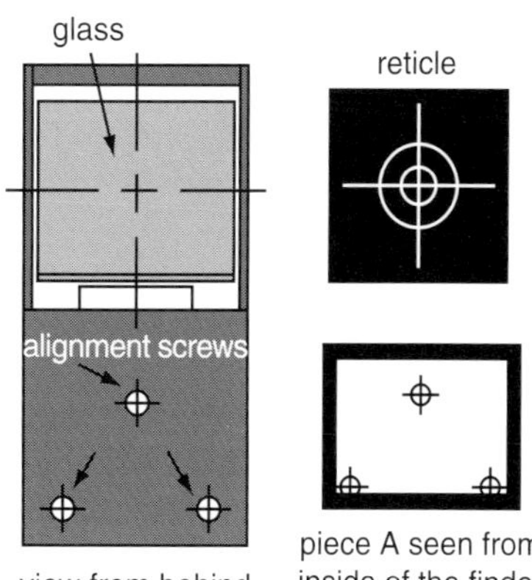

Fig. 20.2 (Right) The reticle is made from a plate of translucent Plexiglas. One of its faces is painted black and then scored with the point of a knife. The illustration shows a cross and two circles of **5 and 13 mm** diameter, which correspond to 2° and 5° fields on the sky.

The 20-mm lens is convex. It is glued directly to the top of the apparatus.

stars dimmer than your naked eye's limit. Once mounted on your telescope and aligned with the telescope's optics, it will enable you to quickly point the telescope to within 2° of your target where you can then switch to your telescope and a low-power eyepiece (40x, 1° field) to further refine pointing.

21 Making the Primary Mirror

Detailed instructions covered earlier in this book are directly applicable to this mirror for rough grinding (see Section 7.2 on page 90), fine grinding (see Section 7.3 on page 92), and smoothing (see Section 7.4 on page 96). However, since the mirror is perforated, the order in which you proceed is important:

1. Perforate the center of the mirror before beginning rough grinding and glue the cutout plug back into the hole thus created.
2. Rough-grind, fine-grind, and begin smoothing the mirror.
3. At the next-to-last smoothing grain, remove the cutout plug.
4. Finish smoothing and polish the mirror without the cutout plug.

21.1 Perforating the Mirror and Putting the Cutout in Place

21.1.1 Why Perforate the Mirror before Grinding?

In a normal Cassegrain configuration the primary mirror is centrally perforated, and the incoming light is reflected back from the secondary mirror through this perforation to the observer's eye. In the coudé variation, the flat is often supported by a fixture that is mounted through this perforation. We do this to avoid unnecessary spider diffraction.

Perforation must be done before rough grinding for two reasons. First, chips and cracks can appear around the edge of the central hole if you perforate the mirror last; however, perforated at the beginning, they disappear during rough grinding. Second, perforation relieves tensions in the glass that can deform a finished mirror. It is easy to see that this kind of event could be disastrous if it occurs at an advanced stage of work.

In general, it is not necessary to test the internal stresses of the mirror before perforating it. Pyrex, Duran 50, and Zerodur glasses are generally homogenous and present no problems if the work is done correctly. If you're using a thick mirror, older glass, or even ordinary glass, you should first test the temper of the glass using polarizing filters and a "wave" plate (see Figure 21.1 on page 214).

21.1.2 Perforating the Mirror

Proper perforation can be done by the glass seller or by a company that specializes in glass work if you provide precise dimensions and tolerances. However, you

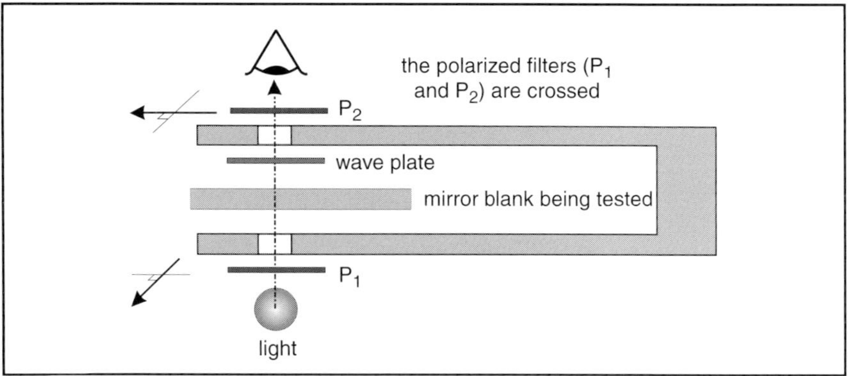

Fig. 21.1 Suggested mounting for testing the internal stresses in a mirror, using polarized filters (ones from a camera work well). For a wave plate, use a sheet of mica or, lacking that, a layer of Scotch tape stuck on the filter P_2. Testing proceeds as follows: **(1)** Before putting the test mirror in place, orient the axes of the polarized filters to completely extinguish the lamp. **(2)** Introduce the wave plate in front of filter P_2 and observe the color of the light (it will be purple if the materials are high-quality). **(3)** Introduce the mirror disk between P_1 and P_2 without changing the orientation of the filters. A change in the color indicates that the glass has internal stress. A change from purple to blue is acceptable. If the color changes to yellow or red, do not perforate the glass!

A Perforation Alternative

Perforation is best done before the mirror is rough-ground. Make the perforation, starting from the back, and stop just before you get through to the optical face (at a distance ε from that face). Fill the cut groove with plaster and then rough-grind the mirror as usual. During grinding, the perforated center will be released. You only need to scratch out a little plaster and fill that groove with lacquer to protect it from moisture. This method has the advantages of not causing chips in the optical face of the mirror and of producing a cutout that is perfectly centered.

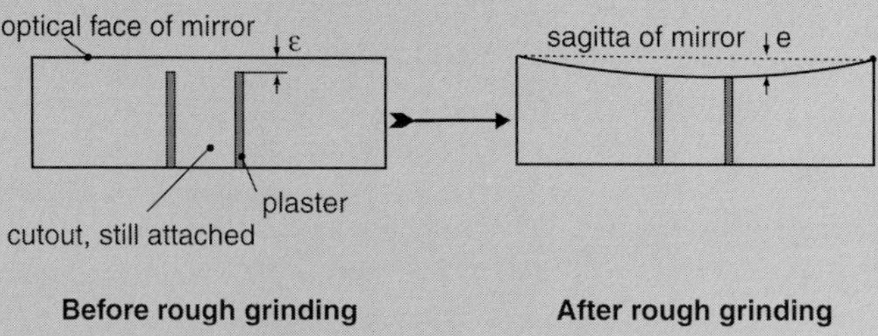

Fig. 21.2 As shown, *e* is the sagitta of the mirror, and ε is between ½ and ¾ of *e*.

must be sure that the plug is returned, because it must be plastered back into the mirror before rough grinding begins.

You can also perforate the mirror yourself. You'll need a heavy-duty drill press, a kilogram of #120 carbo, and water, thick cloth gloves, protective goggles, and a thick shirt. Work proceeds as follows:

1. Using arcanson (see Footnote 2 on page 178), to bond 4-mm thick pieces of plate glass to both the face and the back of the mirror.
2. Mark the center of the mirror with a precision on the ½ mm. The center must be determined carefully; you can't fix a perforation error.
3. To perforate the mirror, use a 43-mm (exterior diameter) "biscuit cutter"; the diameter of the hole created by this tool will be about 45 mm (see Section 15.2 on page 177). When a number of similar mirrors are going to be made, clubs sometimes buy a diamond-edged trepan tool. Drilling is somewhat easier with a diamond-edged trepan, as you do not have to feed it carbo—just water and soluble oil. However, for just one or two mirrors such a tool would be hard to justify.
4. Perforation should be done slowly; it can take more than two hours to complete. Keep the drill and disk well lubricated throughout the process.
5. When you finish the perforation, remove the glass plates from the disk using a solvent or by gentle heating.
6. Make a 1- to 2-mm chamfer at 45° with a fine carbo stone or a 70-mm brass ball covered with #180 carbo.

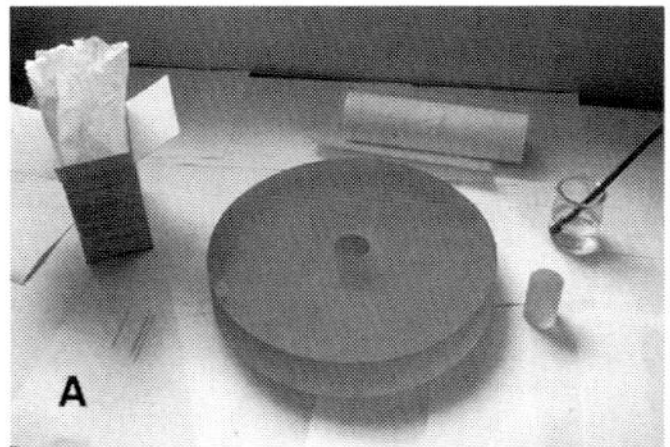

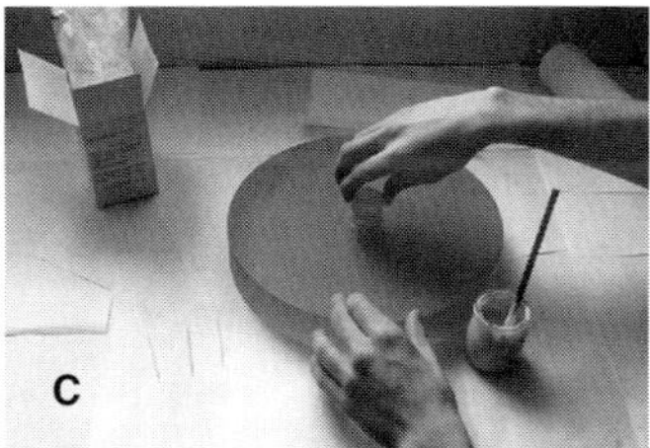

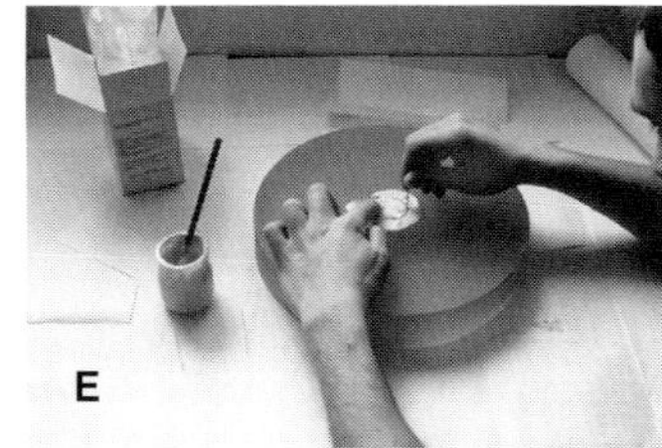

Fig. 21.3 The steps for reattaching the cutout (*A*). Cover the inside of the hole and (*B*) the edges of the cutout with plaster. After sliding the cutout into the hole (*C*), use a toothpick to fill in the rest of the groove with plaster (*D*) and (*E*). Photos: © Karine Lecleire.

21.1.3 Reattaching the "Biscuit"

The cutout biscuit must be cemented back into the hole in the mirror (see Figure 21.3). Use a thin mixture of plaster, and let it dry for at least a day. The dried plaster should be recessed about 1 or 2 mm below the mirror face and then sealed with varnish or lacquer. When you go to plaster the biscuit back into the mirror, do not shim it with toothpicks or matches. Things like this will absorb water and expand and create strain.

The plaster should remain completely dry so the plug does not come unglued during work. At each change of abrasive, carefully re-scrape the plaster and recoat it with varnish. In the unlikely event that the cutout works lose during grinding or smoothing, it must be replaced exactly in the same position to avoid making a high or low step at the center of the mirror. To do this, place the tool face up and thoroughly clean it. Next, cover the face of the tool with a thin plastic film (e.g., Saran Wrap®) and place the mirror, face down, on top. Insert the plug with the concave surface down until it touches the tool. Press on the cutout (a kilogram of pressure does the trick) and pour the liquid plaster around it. Let it dry for two or three days, scrape the plaster and varnish the groove before beginning work again.

21.2 Rough Grinding, Fine Grinding, and Smoothing

Zerodur takes longer to work than either Pyrex or Duran 50. It will take between 8 and 10 hours of rough grinding with #80 carbo to rough out the 300-mm mirror. It will take about two hours less if the disk is made of Pyrex.

During fine grinding and smoothing, make W-strokes of exactly ⅓ diameter. Strokes that are too long cause a bubble to form and can end up causing a hyperboloidal shape during smoothing (a turned down edge several centimeters wide); you would have to return to #180 carbo to eliminate it. With strokes that are too short, the accumulation of emery in the groove surrounding the cutout can cause a central depression of several microns. This problem gets worse when you remove the cutout at the second-to-last grain during smoothing. The defect would be seen when viewing the area with light reflected at a low angle of incidence and must be eliminated before polishing begins.

Fig. 21.4 A critically important moment in smoothing: gently placing the mirror on the tool.

21.3 Removing the Biscuit after Smoothing

Remove the biscuit from the primary mirror before changing to next-to-last abrasive grade during smoothing (#400 carbo or W3) in order to relieve any strain created by the plaster (see Figure 21.5). The removal of the cutout during polishing can cause a strong astigmatism in the mirror, which is very difficult to repair.

21.4 Polishing

Fully polishing a 300-mm Zerodur mirror requires about 12 hours of work. You will need to make two different polishers: the first, with large, hard[1] pitch facets,

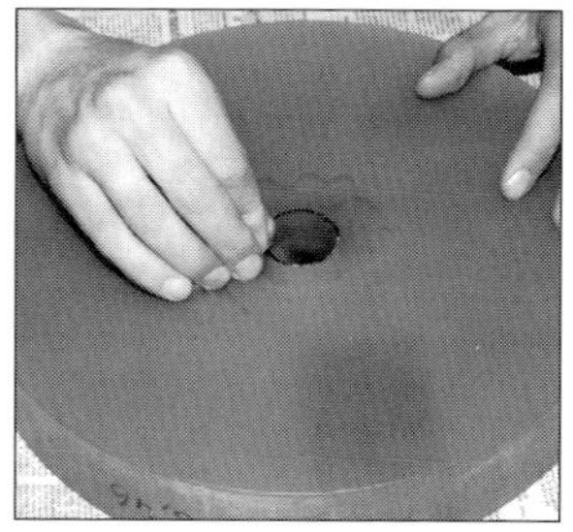

Fig. 21.5 To remove the biscuit, moisten the plaster with lukewarm water. Next, use a metal needle to perforate the plaster with tiny holes until it becomes fragile and you can detach the cutout. This operation takes up to an hour if the plaster is hard (use a thimble to protect your finger as it presses on the needle). Move slowly so that you do not scratch or dig into the glass with the needle. Dévé (1945) suggests dissolving the plaster using a solution of carbonate and soda.*

*Charles Dévé, *Optical Workshop Principles* (London: Adam Hilger, 1945).

60 mm on a side, will be used to fully polish out the mirror while maintaining a spherical form. The second tool, with 30-mm pitch facets made from softer pitch, will be used to parabolize the mirror.

21.4.1 Making the Polishing Tool

To make the first polisher warm the tool and then wrap a masking tape dam around its outside edge and carefully fill it with melted pitch to about 5 mm depth (see Figure 7.10 on page 103). When cooled, cut V-shaped grooves into the pitch using a 5- to 10-mm soldering iron tip or a single-edged razor blade (remember to cut the lap off-center, see Figure 21.6 on page 218). Once the lap is cut, remove the pitch from the central part of the tool to create a circular hole 2–3 mm wider than the diameter of the cutout. Pressing the mirror into contact with the lap is done in a tub of warm water. Trim the squares if the pitch spreads into the V's or to within 2–3 mm of the central hole and/or the outside edge of the tool.

21.4.2 Beginning Polishing

Place a piece of thick felt cloth or thin foam rubber padding of the same diameter as the mirror on the work post. This cushion will adapt to and press evenly over the back of the mirror, and reduces the likelihood of astigmatism caused by unequal distribution of the mirror's weight. When working with the tool on top, turn the mirror about ⅛ of a turn on the work post every ten minutes, and turn it an extra ⅓ turn every hour (this way, over time, the mirror is never in the exact same position twice and any small residual support problems are averaged out).

Polishing begins by charging the lap with a somewhat thick mixture of cerium oxide and water making ¼-diameter W-strokes, alternating mirror on top and tool on top. With the mirror on top, occasionally add water through the central hole. Check the shape of the polisher and trim the tool about every hour, or as soon as the pitch spreads near the hole or the outside edge of the lap. It is important to do this in order to prevent a turned-down edge around mirror's central perforation

[1] The right hardness can be obtained by heating the pitch to evaporate the violate elements if it is too soft or by adding turpentine if it is too hard.

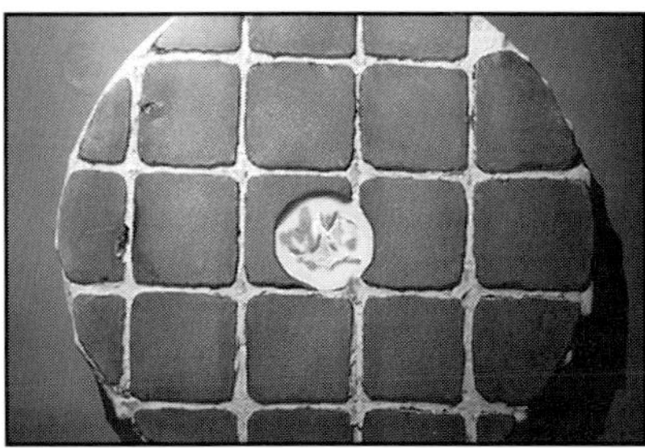

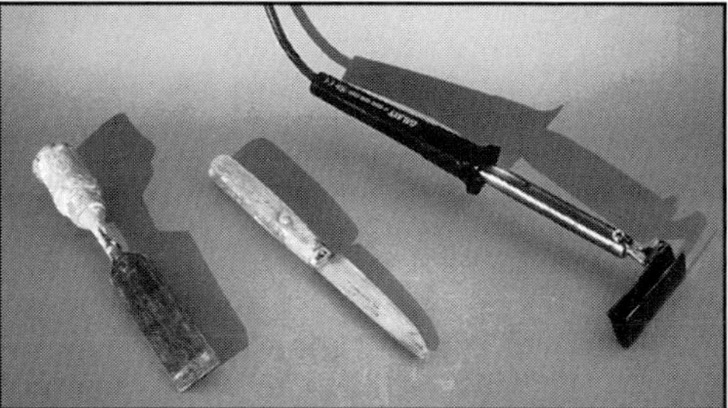

Fig. 21.6 Left: 300-mm polisher with 60-mm squares. The pitch in the center has been trimmed because of a central hole in the mirror. Note that the lap has been cut off-center. **Right:** Some tools that come in handy when making a polisher: a chisel, a knife, and a soldering iron. Photos: © Arnaud Gaillon.

and its outside edge!

The immediate area around the mirror's perforation may remain gray for an hour or two after polishing begins. The depression is created by the accumulation of abrasive around the central hole during the final stages of abrasive smoothing. Move the mirror slightly to the side (about ⅕ of the diameter) when it is on top. The slight variation of the curvature of the mirror caused by these strokes allows the recovery of the center. Slight remaining grayness around the perforation can be eliminated by a parabolization stroke such as that shown for "Stroke C" in Figure 13.10 on page 163, which concentrates wear over the central region of the mirror. However, it is best not to have this problem to begin with; it can be minimized by carefully extending the duration of the final dozen fine grinding wets.

Those who wish to make a telescope window for this instrument can test the window with the primary mirror while it is spherical.

21.4.3 Parabolizing the Mirror

Once the primary is fully polished out you are ready to begin parabolizing. Start by making a new 7- to 8-mm thick lap with pitch softened with turpentine or linseed oil. This lap should have 30-mm pitch facets with approximately 10-mm channels.

> *Note: To make a 400-mm or larger pitch lap, use pitch squares 50 mm on a side with 20-mm channels. Large squares allow for a better contact than small 20- or 30-mm squares used for 200- and 250-mm mirrors. Moreover, the surface will be smoother, more regular, and less subject to rippling.*

Still using cerium oxide and your new lap, parabolize the mirror from the center outward (see Section 13.3.1 on page 163) with the mirror on top. From time to time, work with the tool on top with slow ¼-diameter strokes for 1 or 2 revolutions around the work post. These strokes are used more for smoothing the overall surface rather than for advancing parabolization. During parabolization, carefully control the amount of polishing compound you use since an excess can cause a turned-down edge. Work for the first hour without pressing at all; pressure should not be more than the weight of the mirror.

Test the overall form of the mirror with the Foucault test, without the Couder screen. Note the value of the difference between the central zone and the edge of the mirror. The mirror is slightly deformed (one or two mm at most), compared to the theoretical 12½ mm of longitudinal aberration desired for the final parabola (in the case where the Foucault tester or the source is fixed and the knife moves).

If you detect a too-deep central depression, make smaller strokes. If there is a turned-down edge, trim the tool to the exact diameter of the mirror minus just a few mm for a bevel.

Test your progress every half-hour. When the difference between the central zone and the edge of the mirror approaches 8 mm, use a Couder screen (see Section 5.4.6 on page 61) and measure the aberration for each zone. At the same time, change polishing compounds; the end of parabolization will be done with rouge. Wear is slower and the surface will be smoother.

22 Testing the Primary Mirror

22.1 Foucault Test

Testing the 300-mm mirror is done with the same Foucault test apparatus used for the 250-mm mirror. Each zone of the Couder mask corresponds to about 1.5 mm of longitudinal aberration Λ_C. The null is not observable over the entire area of a mirror because the mirror is strongly deformed. To make measurements, look for a simultaneous extinction at the center of each pair of windows.

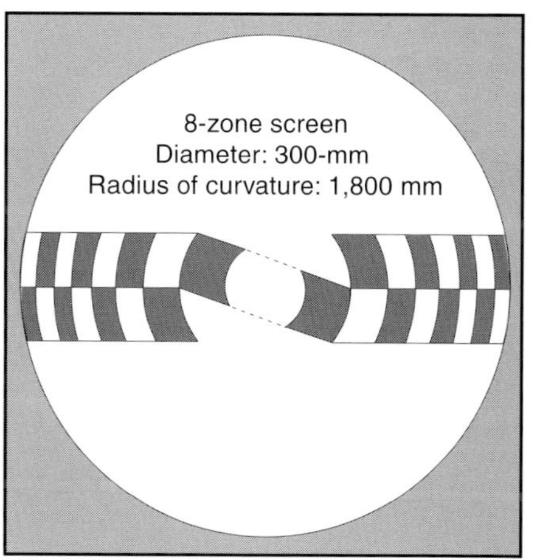

Fig. 22.1 An 8-zone Couder screen made for a 300-mm mirror. The optical diameter of the mirror is 300 mm, and its radius of curvature is 1,800 mm.

The center of the first window is obstructed by a 50-mm diameter disk to occult the central hole in the mirror (45 mm diameter) and the eventual turned-down interior edge.

In mm: $hx(\text{hole}) = 25$;
$hx(1) = 53$; $hx(2) = 75$; $hx(3) = 92$; $hx(4) = 106$;
$hx(5) = 118.5$; $hx(6) = 129$; $hx(7) = 140$;
$hx(8) = 150$.

The calculation of hm and hm^2/R is done as with the 250-mm mirror, except for zone 1, where $hm(1) = [hx(\text{hole}) + hx(1)] / 2 = 39$ mm and $hm^2/\text{R}(1) = 0.84$.

22.1.1 Retouching the Parabola

The retouching methods for this mirror are similar to those described for the 250-mm mirror (see Section 8.3 on page 116). However, there are a few important differences:

- Because of the mirror's weight you must limit your strokes so that the central perforation does not reach the edge of the tool; otherwise, you will create a severe central depression.
- The most common defect is, without a doubt a turned-down edge. It is easily created when you parabolize by wearing down the glass near the outside edge. It is best controlled working with the tool on top, lightly charged with polishing compound, using short W-strokes.
- To see how the retouching is progressing, observe the movement of the polishing liquid at the edge of the mirror just after the tool passes over it. If the liquid does

The Wire Test

You can test the telescope mirror by replacing the Foucault knife-edge with a taut wire with a diameter larger than or equal to the size of the slit. On the mirror, you'll see a dark circular ring and vertical bar that indicates, respectively, the height of the zone under measurement and the centering of the wire at the optical axis of the mirror. Make these measurements using a Couder screen; notate the depth of a zone when the shadow of the wire passes simultaneously through the center of the right and left window. This method, which works with paraboloidal mirrors, does not work for spherical mirrors because the shadow of the wire spreads out over the whole surface of the mirror and does not allow measurement of the average depth. The size of the shadow determines the precision of the measurement; a thin, straight wire permits you to point to a zone more exactly than a wire that is spread or bent. Adjust the size of the windows to match the wire's shadow; this way, you can benefit from more windows than you can use with the knife-edge. The methods of test reduction and calculating the form of the wavefront are the same as they are with the knife-edge.

The wire test cannot completely replace the Foucault test because it does not show the quality of the mirror surface or defects like a turned-down or raised edge. Neither is it very useful on spheres or mildly corrected mirrors. On the other hand, the wire test does allow you to detect and measure astigmatism in paraboloidal mirrors by observing the shape of the shadow rings.

This method is described in more detail in Section 23.3 beginning on page 233.

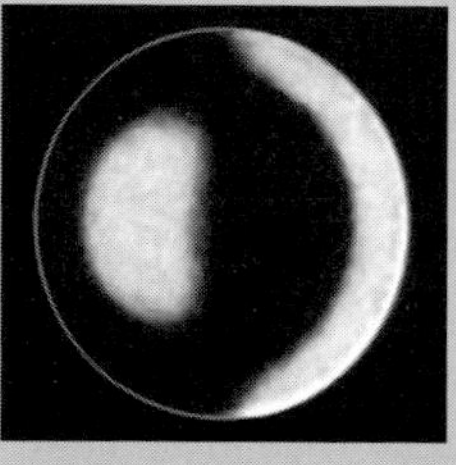

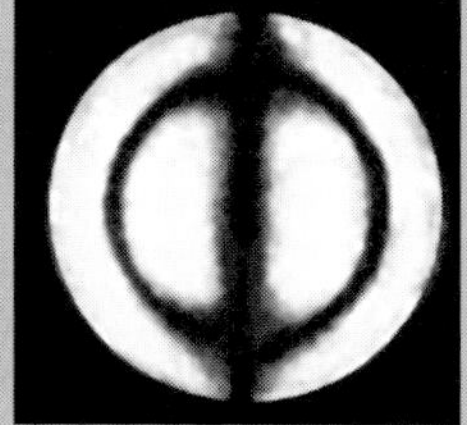

Fig. 22.2 These images illustrate the similarity between the Foucault test (top) and the wire test (bottom).

not run off from the edge of the mirror (as it does in smoothing), continue polishing; the tool is still not working on that zone.

It is possible to finish a 300-mm *f*/3 mirror using the Foucault test. Still, it can be useful to use a complementary, less subjective test—the caustic test, invented by the astronomers R. Platzeck and E. Gaviola in 1939.

22.1.2 Why the Caustic Test?

In 1859 Leon Foucault publicly described his test at the center of curvature of a mirror, which has since been used in making thousands of excellent mirrors. Despite its great usefulness, certain disadvantages show up in applying the test to nonspherical surfaces, such as the paraboloid, that make it difficult to use and which set a practical limiting accuracy of zonal measurement at about two millionths of an inch [on the wavefront]. Some improvement results when the zonal mask is discarded, testing

Electronic Foucault Testing

As the focal ratio of the mirror diminishes, the number of windows in the Couder screen increases and measurements become difficult to make—the windows are too thin and there's more parasitic diffracted light. One can avoid these problems by using a video or CCD camera positioned where you normally locate your eye behind the Foucault knife-edge. To eliminate the diffraction at the edges of the test windows, place the Couder screen on the TV monitor. This method, proposed by the American amateur Ken Ramsey (*Sky & Telescope*, April 1994), allows you to use Couder screens with 10 or 12 windows to test highly deformed mirrors and to detect irregularities that are imperceptible with a normal screen. Moreover, you eliminate the subjective factor of personal perception, and it is possible for several people to simultaneously view and evaluate the mirror.

If you have access to a CCD camera you can take a succession of exposures, stack the resulting images and measure the light intensity of each of the Couder screen's windows using an image processing program such as Richard Berry and James Burnell's Astronomical Image Processing for Windows (AIP4WIN) software which accompanies their book: *The Handbook of Astronomical Image Processing* (Willmann-Bell, Inc.).

> then being done by observing shadow edges and the overall shape of the surface (Everest, Gaviola). However, shadow edges are never very sharp and the overall curve, such as the doughnut, can be studied only in the case of moderate focal ratios. Hence a zonal mask is needed for a short-focus paraboloid and for any other [strongly] aspherical surface.[1]

The caustic test is most useful for fast and large mirrors where the deformation from a sphere is great. On the other hand, it does not offer greater precision than the Foucault test for mirrors with weak parabolas. Use the more-difficult caustic test only when the situation demands it.

22.2 The Caustic Test

22.2.1 Principles of the Caustic Test

A light source is placed at the center of curvature (C) of a paraboloidal mirror (see Figure 22.3 on page 224), and it illuminates two small regions a and b, at height h, on the surface of the mirror. In the Foucault test, we have assumed that the center of curvature of a zone in the Couder screen corresponds to a point of intersection P of the light coming from this zone with the optical axis.

In reality, it is otherwise; the paraboloidal mirror has a radius of curvature that is larger at the outer edge than it is at the center, and thus the reflected light rays converge beyond point P, respectively at A and B, symmetrical in relation to the optical axis. All the points of the paraboloidal mirror obey this rule, and the

[1] I. H. Schroader, *Amateur Telescope Making*, 2nd ed., vol. 1 (Richmond, VA: Willmann-Bell, Inc., 1996), p. 251.

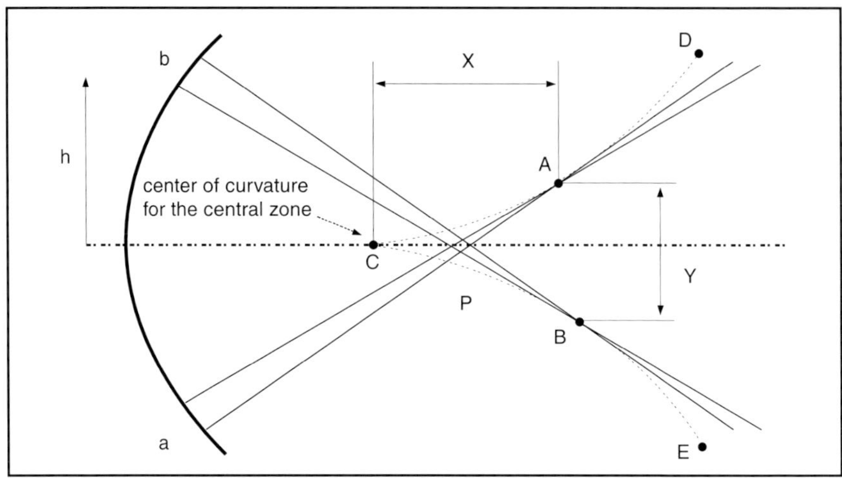

Fig. 22.3 Geometry of the caustic test. The caustic "horn" is shown with a dashed line.

respective foci form a figure that is called *caustic*. Its shape is similar to the bell of a trumpet (the curvature is shown as a dotted line in Figure 22.3). The caustic is the external envelope of the light beams reflected by the mirror. The distance between the pointed extremities of the caustic, at C, and the projection of A and B is labeled Y. For a perfect paraboloidal mirror, the terms X and Y can be calculated as follows:

$$X = \frac{3h^2}{2R} \quad \text{and} \quad Y = \frac{2h^3}{R^2}, \tag{22.1}$$

where R is the radius of curvature of the central zone, and h is the height of the measured zone. (Consider here that the entire test apparatus moves on both axes. If the source is stationary, then double the values of X and Y.) Divide the mirror into a certain number of zones for which you calculate the theoretical values of X and Y. Then, for each zone of the mirror, position the test apparatus on the optical axis at the calculated point X, and measure the corresponding distance Y (the size of the caustic figure) using an eyepiece with a graduated reticle. The difference between the theoretical and measured values of Y allow you to determine the form of the wavefront reflected by the mirror.

22.2.2 Adapting the Foucault Tester for the Caustic Test

The Foucault tester must be modified in order to perform the caustic test. Replace the knife-edge with a 9-mm focal length eyepiece with graduated reticle and a barrel diameter of 24.5 mm. The eyepiece enlarges the image of the slit returned by the mirror. The magnification is given by the formula:

$$M_{\text{eyepiece}} = \frac{1}{4f}$$

where f is the focal length of the eyepiece in meters. In our example, the 9-mm eyepiece magnifies 28x. The reticle is vertically oriented (it can be a straight line or a cross) at the same height as the slit source. The transverse movement platform is equipped with a micrometer readable to a micron (which is not trivial).

22.2.3 Making a Test Mask

Make a circular mask from thin, but rigid, paper board or even plywood, with small round holes cut with a punch or a Fostner drill bit. These zones have a diameter of around $R/100$, where R is the radius of curvature of the central zone of the mirror ($R = 1800$). See Figure 22.4 on page 226. Make an even number of these holes, spaced symmetrically across the central diameter of the mask. The holes closest to the two outside of the mask should be tangent to the edge of the mirror. For a perforated mirror, the interior holes are also tangent to the edge of the central hole (for the 300-mm mirror described here, leave a margin of 5 mm around the perforation to compensate for the interior turned-down edge that might be caused in polishing).

After you cut out the mask, make small removable shutters, numbered on their backs, to cover the windows in the screen. These are attached over top of the windows with pieces of adhesive tape; you should be able to remove and reattach them easily, without having to touch the mirror or its support.

22.3 Performing the Caustic Test

22.3.1 Alignment

Place the mirror in the testing support without the caustic mask and adjust the measuring apparatus at the mirror's center of curvature and light the lamp. The image of the slit should project into the eyepiece. If necessary, adjust the mirror support screws to center the slit image in the eyepiece. Next, narrow the slit to 3⁄100 mm and check for vertical alignment. The reticle of the eyepiece should be rotated to parallel the slit image. Next, align the longitudinal axis of the test apparatus with the optical axis of the mirror. For this, make a circular screen of the same diameter as the mirror with three holes in a line, one at the center and the two others at the edges of the mirror (see Figure 22.5 on page 227). When you place the screen in front of the mirror, you'll see three different images of the slit in the eyepiece. Move the tester along the optical axis of the mirror, and you'll see that the two lateral images move either closer together or further apart, depending on which direction you move the apparatus. At the point where the light beams from the outside zones intersect with the optical axis, the images will be superimposed with each other and the central image, and they will separate again if the platform is moved. If the central image moves laterally, you will need to adjust the orientation of the entire apparatus as well as the transverse displacement screw in order to align the tester.

When the platform is aligned with the optical axis of the mirror, the two im-

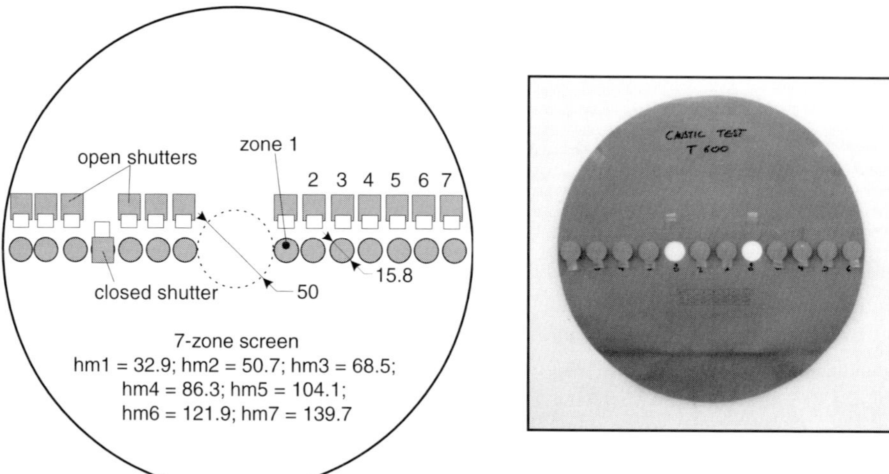

Fig. 22.4 Left: The test screen for the 300-mm mirror has 7 pairs of windows, each 15.8 mm in diameter (R/114, which is close to R/100). The first hole is located 25 mm from the center (perforation); leave 2 mm between each window up to the edge. All measurements indicated in mm. **Right:** 6-zone screen for a 600-mm mirror.

ages will be symmetrical in relation to the optical axis (shown by the central image), whatever the longitudinal position (the images should be on opposite sides of the reticle).

The last adjustment is to determine the position of the center of curvature of the central zone of the mirror (point *C* in Figure 22.3 on page 224), also called the caustic point (see the box on page 228). This position should be known to within 2/100 or 3/100 mm.

22.3.2 Procedure

Note that in this procedure the source and the eyepiece move together.

1. Using Equation 22.1 (page 224), calculate the theoretical values of *X* and *Y* for each zone of the screen. To each value of *X*, add the value read from the micrometer at the caustic point (this zone will serve as a reference, and all the values of *X* have this point for their origin).
2. Place the 7-zone caustic screen in front of the mirror and open only the left window of zone 1 (light interference is produced if complementary windows are opened together). Position the apparatus at the theoretical value of *X* (including the addition of the *caustic point*) corresponding to that zone. At the eyepiece, center—carefully—the reticle on the visible image of the slit and record the value Y_l on the transverse displacement screw.
3. Without moving the tester or the mirror, close the left window and open

the right one. Center the image again on the reticle, adjusting only the transverse axis. Record Y_r, the value read from the screw.

4. Repeat the above three steps for each of the other zones.

Note: Apart from the zone at about 0.4 height (and depending upon the mirror), you can open the left and right windows at the same time, because the two images are completely separated and will not interfere with each other.

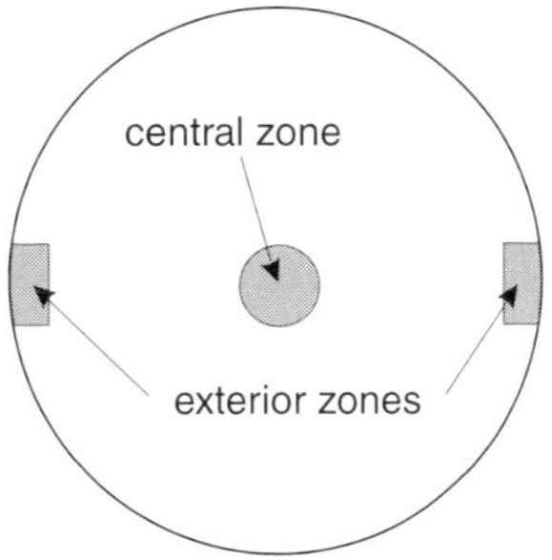

Fig. 22.5 Measurement screen with three perforations.

22.3.3 Interpreting the Measurements

Make a test sheet similar to that shown in Figure 20.6 and record the theoretical values of X and Y along with the measured values of Y_l and Y_r. Calculate Y_m, the difference between Y_l and Y_r.

Next enter into this table the measured values of Y_m for each zone of the caustic screen, and compare them to the theoretical values of Y. Calculate the reduced measurements corresponding to this data thus:

$$\text{Reduced measurements} = m \times (Y - Y_m)\frac{R}{2h_m},$$

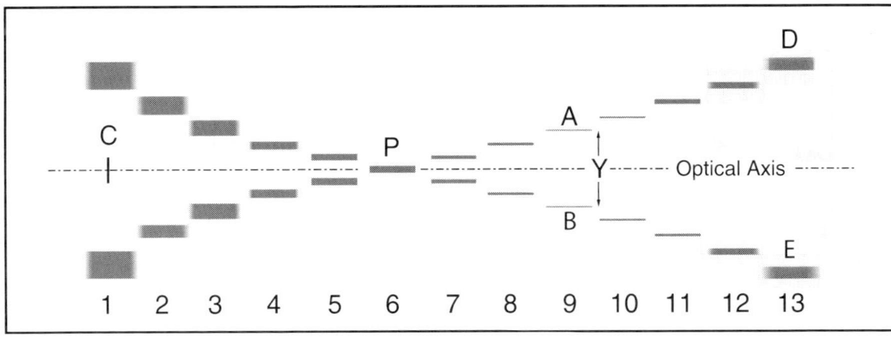

Fig. 22.6 Images of the slit obtained for a peripheral zone of the mirror (windows *a* and *b* from Figure 22.3 on page 224) by moving the eyepiece along the optical axis. Image no. 1 is seen at the center of curvature for the central zone of the mirror (point *C*). There are two images because the light beams coming from *a* and *b* have not yet intersected. Image no. 6 is seen at the intersection of the light coming from *a* and *b* and the optical axis (point *P*); the two images are superimposed but the image is not sharp. It is at this point that the Foucault knife-edge is positioned to obtain the null at zone *h*. The two sharpest images from the slit are at image no. 9 (the caustic test permits you to visualize the real focus of the zones on the screen). These light beams are found away from the optical axis at points *A* and *B*. Measure the distance *Y* by which *A* and *B* are separated. After that point, you can observe the two images of the slit at points *D* and *E* (image no. 13), at the edge of the test. **The image of the slit** is sharp when the eyepiece is at the caustic point.

Finding the Caustic Point

You begin without the caustic screen. Moving the tester along the optical axis, observe the appearance of the light spot in the eyepiece. When the apparatus is at the caustic point, you will see a clean image of the slit on a milky background. If you bring the apparatus nearer the mirror, the central image will get blurry; if you move further away from the mirror, the image will be bordered by interference fringes and then be doubled. The caustic point corresponds to the first appearance of the clean image of the slit as you move away from the mirror. This measurement is difficult and requires some experience: *it's the principal source of errors in the caustic test.*

Another solution is to place the Couder screen in front of the mirror and to cover all but one pair of holes (at height 0.5). At the eyepiece, you can determine the longitudinal position as the images of the slit coincide with the optical axis at point *P* (at this point, the solitary image of the slit is bordered by high-contrast interference fringes). Knowing the depth of the zone measured in the Foucault test (you can use the theoretical value H_m^2/R of the 0.5 zone if the mirror is better than ¼ λ), you can calculate the position of the center of curvature for the paraxial zone of the mirror by a simple subtraction. Example: the H_m^2/R of the 0.5 zone is 10 mm, and the *micrometer travel is* 12 mm. *If the source and the eyepiece move together*, the exact center of curvature of the central zone of the mirror—the caustic point—is located at 12 – (10/2) = 7 mm travel on the micrometer.

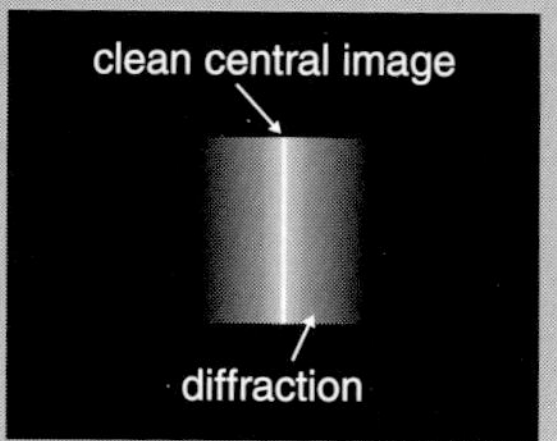

Fig. 22.7 The image of the slit is sharp when the eyepiece is at the caustic point.

Advantages and Disadvantages of the Caustic Test

Advantages

- The caustic method is impersonal: the perception of the observer doesn't matter much in the measurements.
- The caustic test is more precise than the Foucault test if the caustic test screen has more windows than the equivalent Couder screen.
- Measurements are not influenced by diffraction at the edges of the test windows.

Disadvantages

- It's difficult; the principle source of errors is in finding the caustic point.
- It takes much longer than the Foucault test.
- The caustic test requires a moving platform that is perpendicular to the optical axis and precise down to a micron, and such an instrument can be rather expensive.
- The caustic test does not allow the observation of the overall surface quality of the mirror, the detection of a turned-down or raised edge, nor the presence of ripple—defects smaller than the size of a test window are imperceptible.

Test Sheet		
Caustic test	Diameter	300
300 mm f/3	Radius	1800
	Zone D	15.8
	Frac R	113.9
	Separation of zones	17.8
	Tester motion entry	2 moving

ρ = 0.002013 mm

Zone		1	2	3	4	5	6	7
Hm [mm]		32.9	50.7	68.5	86.3	104.1	121.9	139.7
Theoretical X		0.902	2.142	3.910	6.206	9.031	12.383	16.263
Caustic point		12.201						
Theoretical X+point		13.103	14.343	16.111	18.407	21.232	24.584	28.464
Theoretical Y		0.022	0.080	0.198	0.397	0.696	1.118	1.683
Measured Y_l		24.990	24.960	24.902	24.798	24.695	24.535	24.150
Measured Y_r		25.008	25.037	25.100	25.196	25.393	25.675	25.830
Y_{tot}		0.018	0.077	0.198	0.398	0.698	1.140	1.680
Yth-Ym		0.0040	0.0034	0.0004	-0.0012	-0.0016	-0.0219	0.0030
Reduced		0.218	0.122	0.011	-0.026	-0.028	-0.323	0.038
Constant k		-0.066						
Reduced - k		0.284	0.188	0.077	0.040	0.038	-0.257	0.104
TA_f		0.0026	0.0027	0.0015	0.0010	0.0011	-0.0087	0.0040
TA/ρ		1.29	1.32	0.72	0.48	0.54	-4.32	2.01
Slopes u x 1 million		-2.88	-2.95	-1.62	-1.06	-1.21	9.66	-4.49
[1 000 000 = nm/mm]								
Hx [mm]=	24.0	41.8	59.6	77.4	95.2	113.0	130.8	148.6
Coords	0.0	-51.3	-103.8	-132.6	-151.6	-173.2	-1.2	-81.1

max	0.0	nm
min	-173.2	nm
PTV	=1/3.2	wave

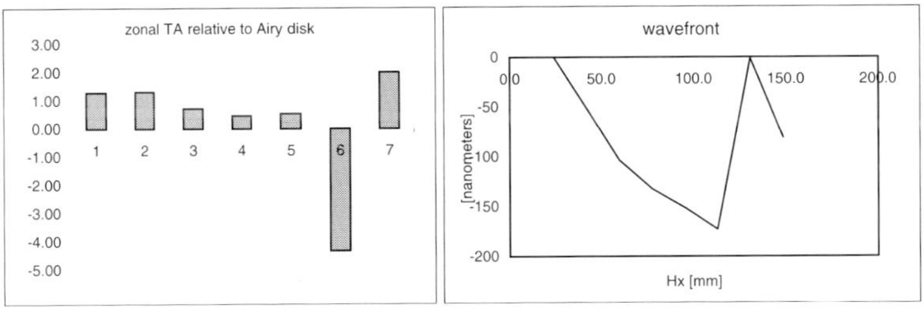

zonal TA relative to Airy disk
3.00 2.00 1.00 0.00 -1.00 -2.00 -3.00 -4.00 -5.00
1 2 3 4 5 6 7

wavefront
0 -50 -100 -150 -200
[nanometers]
0.0 50.0 100.0 150.0 200.0
Hx [mm]

Fig. 22.8 A sample caustic test sheet made during polishing. It is not representative of a finished mirror. Created with an Excel spreadsheet available for download at www.will-bell.com. Courtesy of H.R. Suiter.

where R is the radius of curvature at the center of the mirror, (m = 1 for fixed-source tester or m = 2 for moving) and h_m is the average height of each zone. Using these reduced measurements, calculate the reduced longitudinal aberrations and the transverse aberrations TA_f of the mirror using Equations 5.9 and 5.10 (on page 66). Other calculations and the graph of the profile of the mirror are obtained as they are in the Foucault test.

The type of retouching needed is deduced from the graph of the transverse aberration or of the wavefront profile (to know how much glass to remove). After every correction, measure the value of R; recalculate the theoretical values of X and Y before every new set of measurements.[2]

[2] An Excel spreadsheet program, developed by H.R. Suiter, is available for download at www.willbell.com which will allow you to plug in your data and automatically reduce it.

23 Testing for Astigmatism

23.1 What is Astigmatism?

A mirror is astigmatic if it has different radii of curvature across two diameters. In other words, the mirror has two distinct foci. Astigmatism is normally inherent in many optical systems, especially in off-axis images. It can also be caused by strain in an improperly annealed blank. None of these forms of astigmatism are the subject of this chapter. Here we will address astigmatism caused by improper technique and how to avoid it. This discussion is especially important for large mirrors such as the 300-mm Cassegrain mirror described in this book. This and other mirrors beyond 130 mm are increasingly sensitive to how they are supported and worked during the optical fabrication process. During polishing you must test for astigmatism to make sure that if any can be measured, that it is within acceptable limits—if this condition is beyond acceptable limits, steps much be taken to correct it, or it will have significant negative performance consequences.

The shorter focal length of an astigmatic mirror is called the *tangential focal length*, *T*, and the longer focal length is called the *sagittal focal length*, *S*. The distance *L* on the optical axis that separates *T* and *S* is called the *astigmatic distance*. According to A. Couder,[1] the maximum tolerable astigmatic distance *L* is:

$$L = 1.45 \times \lambda \times \left(\frac{F}{D}\right)^2,$$

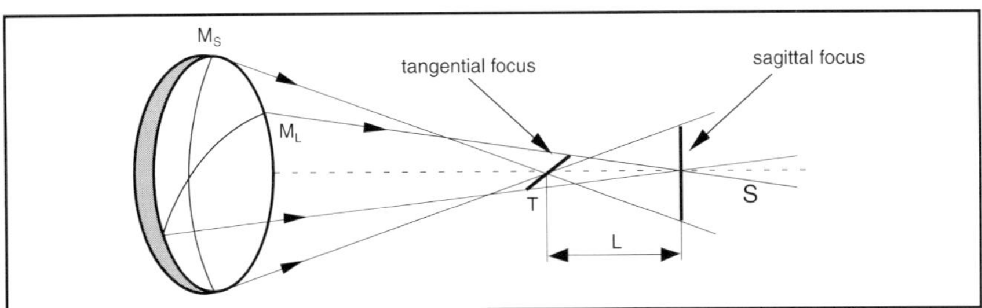

Fig. 23.1 In this diagram, the image of a star is stretched horizontally when it is formed at *T*. This image is found at the focus of the vertical meridian M_S (with the shorter focal length). At the sagittal focal length *S*, the image is stretched vertically and is situated at the focus of the horizontal meridian M_L (with the longer focal length). Diagram provided by S. Debruyne.

[1] A. Couder, recherches sur les deformations des grands Miroirs employés aux observatoires astronomiques (thesis, 1932).

where λ is the wavelength of light, F and D are the focal length and diameter of the mirror. This maximal length corresponds to an error of $\lambda/5.5$ on the wavefront.

23.2 How to Detect and Measure Astigmatism

1. Using the artificial star test or the wire test (see Section 23.3) determine if the mirror is astigmatic.
2. If the mirror is astigmatic, measure the length of the astigmatism and calculate the difference of the sagitta of the curvatures.
3. If this difference is larger than $\lambda/5.5$, determine the orientation of the meridian with the shorter focal length in order to retouch the mirror. If the difference is less than $\lambda/5.5$, the astigmatism can be ignored, and the mirror is considered good.

23.2.1 Using the Artificial Star Test

At the center of curvature of the mirror, place a point light source—the artificial star. The artificial star is obtained by replacing the light source of the Foucault or caustic test apparatus with a sheet of aluminum foil pierced by a very sharp needle to make a tiny hole. Examine the reflected image of the source with a powerful eyepiece (at least 10 mm focal length) to confirm that you can comfortably see the reflection of the star and that the image is not too bright or too dim, in which case, make another hole sized appropriately and retest. Recall that the magnification is given by

$$G_{\text{eyepiece}} = \frac{1}{4f},$$

with f the focal length of the eyepiece in meters.

You can also create an artificial star with a polished steel ball bearing placed at the mirror's radius of curvature. Light this ball from a distance with a bright source (the sun, a lamp). The reflection of the light on the ball simulates an artificial star. Visibility of the reflected image is determined by the brightness of the source.

23.2.2 Observations in the Artificial Star Test

In a non-astigmatic mirror, the image of the source is round both inside and outside focus (the center of curvature—see Figure 23.2). Consider an astigmatic mirror with the diameters oriented as they are in Figure 23.2. At the tangential radius of curvature, the image of the star is stretched horizontally. At the sagittal radius, it is stretched vertically. At the circle of least aberration, the star looks like a cross (see Figure 23.3, frame 5).

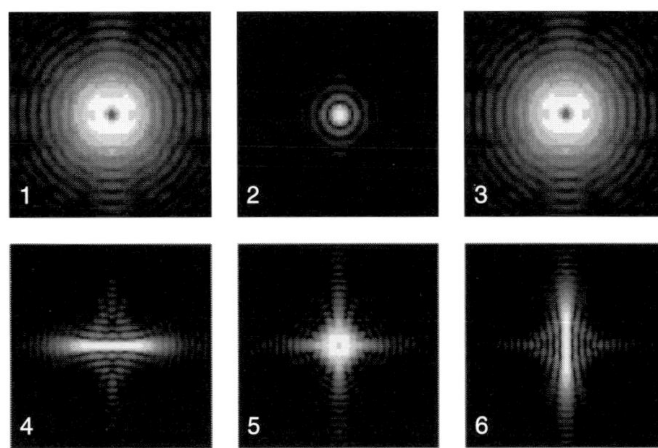

Fig. 23.2 Top: Appearance of the artificial star during testing of a spherical mirror that is not astigmatic. In *1*, the image is observed in front of the center of curvature *C*. In *2*, the diffraction spot is perfectly circular at *C*. At *3*, the image is observed behind *C* and is similar to image *1*. **Bottom:** The mirror is astigmatic on the order of 1.5λ. The tangential image is horizontal (*4*), and the sagittal image is vertical (*6*). At the circle of least aberration (*5*), the star appears as a cross. Contrast is lowered in this figure to show dim structure.

23.2.3 Measurement and Correction

Measure the length of the astigmatism, *L*, separating the two centers of curvature *S* and *T*. Next, calculate the difference between the sagitta of the principal diameters, ε, using the equation:

$$\varepsilon = \frac{1}{16} \times L \times \left(\frac{D}{R}\right)^2 ,$$

where *D* is the diameter of the mirror and *R* is its radius of curvature. *This formula is valid only if the source and the eyepiece move together.* If the source is stationary, double the value of ε.

If ε is less than 0.18 λ, or λ/5.5, the astigmatism is negligible.

If ε is between 0.18 λ and 0.25 λ, the mirror has astigmatism that will be slightly visible during observation, but still technically tolerable.

If ε is greater than 0.25 λ, you'll need to retouch the mirror during polishing. Retouchings are centered on the meridian with the shorter focal length (M_S). To find this meridian, observe the image of the artificial star at the tangential radius (the closest to the mirror); the meridian M_S is always perpendicular to the axis along which the image is stretched, whatever the orientation of the mirror on its support.

23.3 The Wire Test

The wire test lets you measure the astigmatic length with more precision than the artificial star test on paraboloidal or strongly deformed mirrors only (not on spherical mirrors). It also allows you to localize any eventual polishing retouchings.

23.3.1 Using the Wire Test

You'll need a Foucault tester equipped with a rotatable slit that will move with no longitudinal play and which can be easily interchanged with the regular slit light

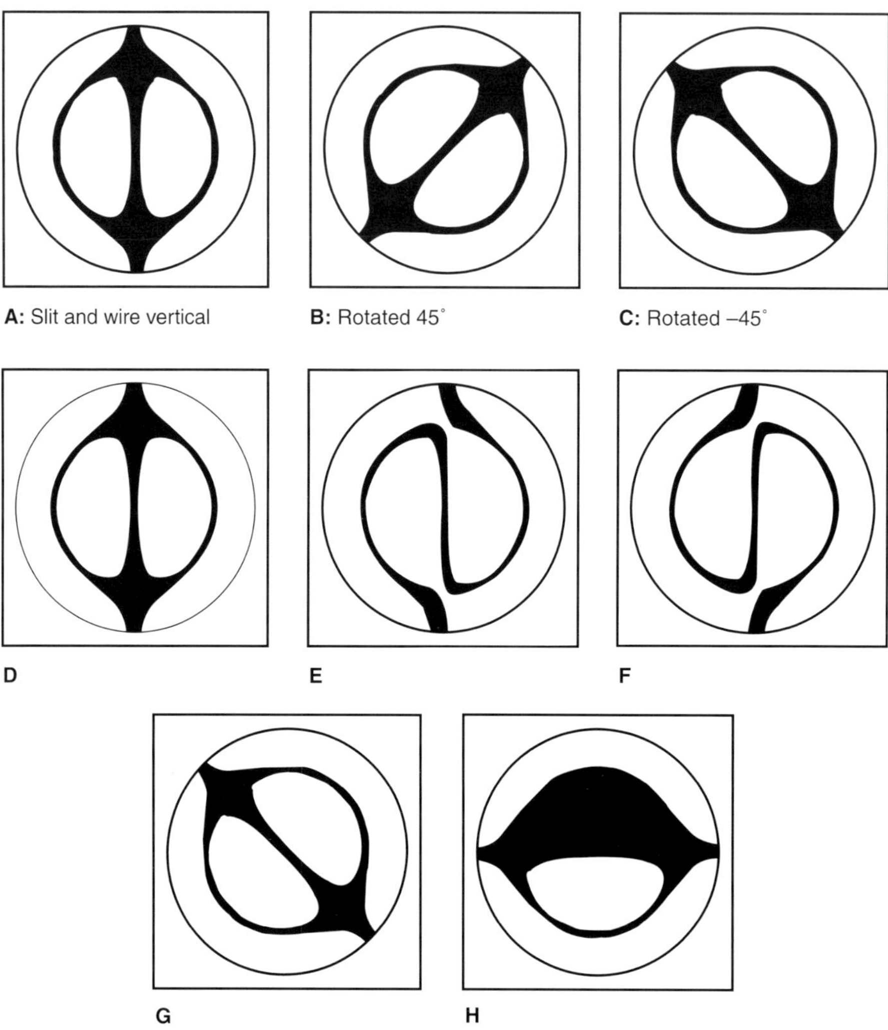

Fig. 23.3 Along the top row are three images of a paraboloidal mirror without astigmatism. The black ring is circular and maintains the same diameter, whatever the orientation of the wire and slit. Illustrated by S. Debruyne. In **D**, the wire and the slit are parallel over one principal meridian of the mirror (*ML* or *MC*). In **E** and **F**, the wire and slit are not aligned with a meridian. Illustrated by S. Debruyne.

source. The knife-edge of the Foucault tester is replaced with a wire, 1/10 mm thick (a section from a long strand of hair works fine), glued to a turning ring (with no longitudinal play) that attaches to the tester's light source (see Figure 8.2 on page 110).

With the rotatable slit installed on the test apparatus but retaining the knife edge, position the knife-edge at the center of curvature of the mirror's 0.7 zone. Next, without moving the tester, replace the knife-edge with the wire, vertical and parallel, to the slit.

23.3.2 What You Should See

First case: the mirror has no astigmatism.

Observe the lighted disk, which will be crossed by a dark vertical line and a dark circle coincident with the 0.7 zone (Figure 23.3A). When you turn the source and the wire 45° simultaneously, the image observed behind the wire turns in the same direction and by the same angle, and it keeps the same shape (Figure 23.3B and C).

The more exactly the slit and wire are parallel, the better the contrast in the observed image. When the alignment is perfect, the dark figures are outlined by diffraction fringes (see Figure 23.4).

Second case: the mirror is astigmatic.

Without knowing the orientation of the principal diameters, place the mirror on its test support. First, we will try to determine the meridian with the shortest focal length (which is the most important one, because it determines the location of the retouching).

When the slit and the wire are vertical, there are three possible images you might see. In the first case (see Figure 23.3D), the slit and wire are aligned with one of the diameters of the mirror (a rare case). Usually, the dark image visible on the mirror looks like a barred spiral in one of two orientations (see Figure 23.3E and F). By turning the wire and slit in the direction of the spiral (to the left in Figure 23.3E, to the right in Figure 23.3F), you can arrive at a circular image of the dark ring (Figure 23.3G; this image is derived from Figure 23.3E). The dark line in the image will be parallel to the meridian of the mirror with the longest radius of curvature; consequently, it is perpendicular to the meridian with the shortest radius of curvature.

If the original image looks like Figure 23.3H, turn the wire and slit until you obtain a spiral image, then apply the preceding technique.

With a marker, trace the shortest meridian on the back or side of the mirror.

To perform the following procedure, reorient the mirror in its support, with meridian M_S oriented vertically (see Figure 23.1 on page 231).

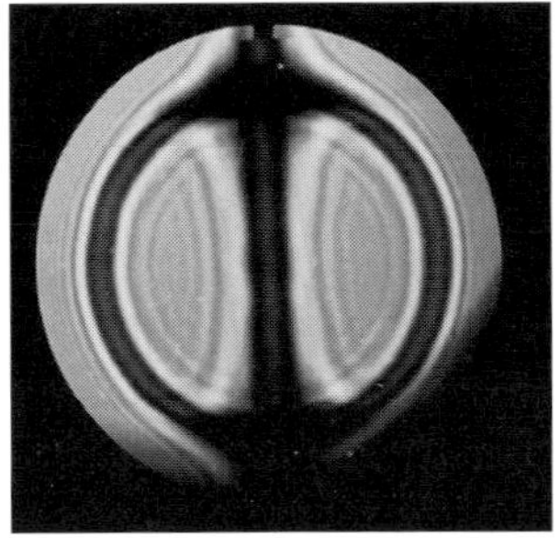

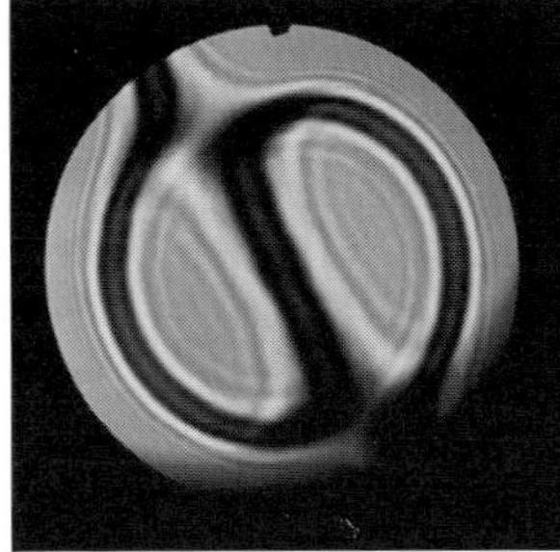

Fig. 23.4 600-mm mirror with a 4,800-mm radius of curvature as measured in the wire test. **Left**, the wire is parallel to the meridian with the longest focal length. **Right,** the wire and the slit are turned 25° to the left. The spiral form appears. The astigmatism is $\lambda/2$ on the wave front. Photographs © D. Vernet and S. Debruyne.

23.3.3 Measurement and Interpretation

1. Using a marker, being careful not to scratch the mirror, trace a circle with a diameter of 70% of the mirror's diameter. This circle generally indicates the zone of least aberration, in reference to the best-fit reference sphere, of the paraboloidal mirror.
2. Orient the slit and the wire vertically, and align the test apparatus with the optical axis of the mirror to obtain the image shown in Figure 23.3D on page 234.
3. Move the support of the slit and the wire (together) to superimpose the dark circular shadow exactly on the 0.7% zone of the mirror. Record the number indicated on the micrometric screw as m_1.
4. Turn the slit and the wire 90° to observe the image shown in Figure 23.4 on page 235. The black circle is larger than the marker-drawn circle on the mirror. Move the test platform toward the mirror to superimpose the circles again. Make a second reading of the micrometer screw (m_2).
5. The smallest distance corresponds with the center of the smallest radius of curvature and vice versa. Subtract the greater value of m_1 or m_2 from the lesser to obtain the astigmatic length L. Calculate the difference of the sagitta of the principal diameters of the wavefront ε using the equation:

$$\varepsilon = \frac{1}{16} \times L \times \left(\frac{2h}{R}\right)^2$$

   where h is the height of the measured zone. In this example, we have arbitrarily picked the 0.7 zone, but we could have picked any other region of the mirror; the interior zones give less precise results, and exterior zones are harder to measure. Remember, this equation is for a moving-source tester. Double it for a fixed source tester.

As in the artificial star test, retouch the mirror if ε is greater than or equal to $\lambda/4$.

23.4 How to Retouch an Astigmatic Mirror

Identifying the meridian with the shortest focal length enables you to plan corrective retouching. If the astigmatism is large ($\varepsilon > \lambda/2$), retouch the mirror locally using a polisher with a sub-diameter lap. If the defect is weaker ($\varepsilon < \lambda/2$), retouch the mirror with a full-size lap.

23.4.1 Using a Sub-diameter Lap

Use a small circular lap, about ⅓ of the mirror's diameter, made of soft, relatively thick pitch squares. On the back of the mirror, trace, with a felt tipped marker, triangular sections centered on the meridian with the shortest focal length (see Figure 23.5). The mirror is placed on a uniformly flat padded surface. Charge the sub-di-

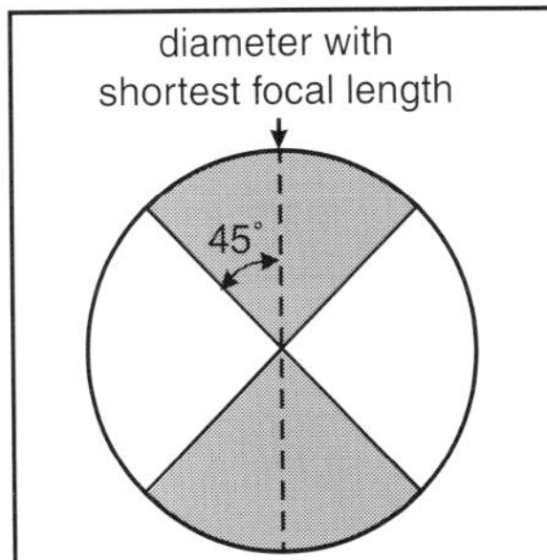

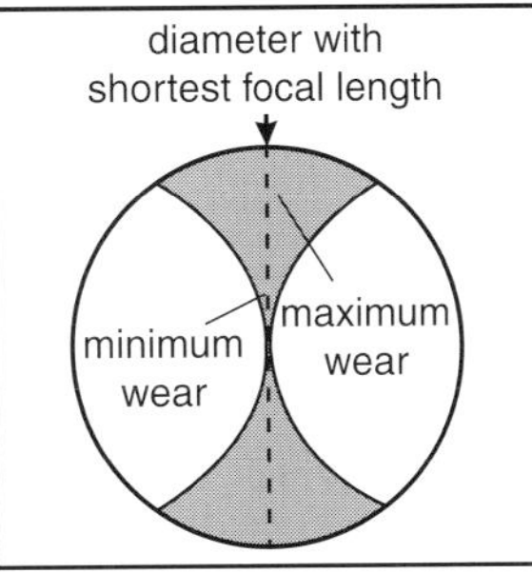

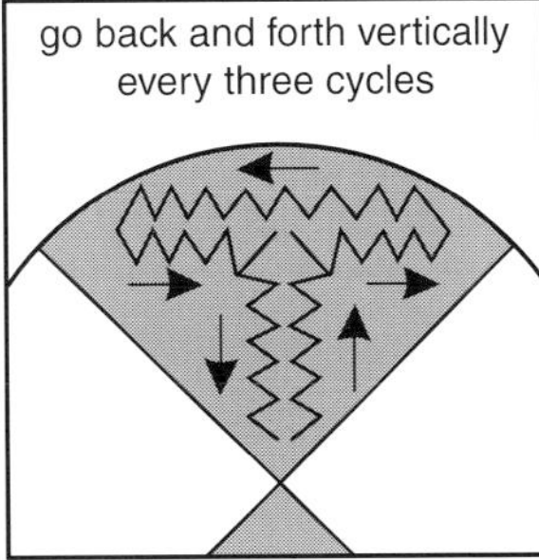

Fig. 23.5 Left: divide the surface of the mirror into two symmetrical work zones centered on the mirror; the tool should only move over these (gray) regions. **Center:** representation of the worked area of the mirror. **Right:** A cycle of T-strokes traced by the center of the tool.

ameter lap with cerium oxide and begin your corrections with T-strokes (see Figure 23.5 Right). These strokes concentrate more polishing action at the edge of the mirror than at the center (theoretically, the tool should never pass over the center of the mirror). The difficulty of the retouching is in maintaining the overall regular form of the mirror. Do not work too fast, because you run the risk of creating irregular defects: deepened zones, astigmatism over three or four diameters, a mirror that is spherical on one axis and paraboloidal on another, etc. Move on to another section of the mirror every 15 or 20 minutes.

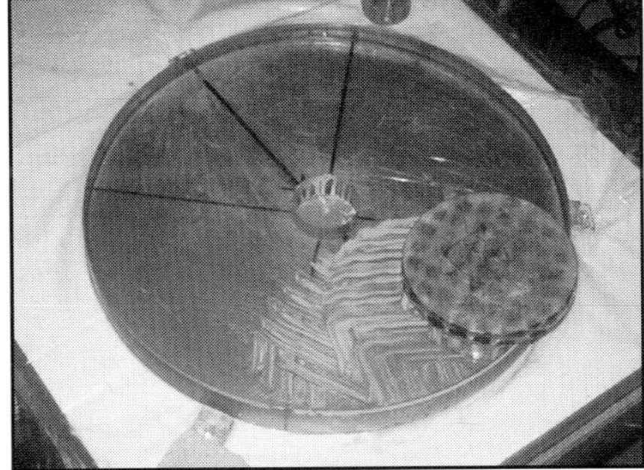

Fig. 23.6 Retouching a 600-mm mirror using a 200-mm polisher. Notice the division of the mirror into triangular sections.

The repeated passage of a small tool over the high zones of the mirror progressively diminishes the astigmatism, but it also introduces rippling on the surface. This is why, each hour, you should polish the entire mirror with a full-size tool. Make normal parabolizing strokes with the mirror on top, turning the mirror frequently. Test the general form of the mirror with the Foucault test, and make exact measurements on each of the principal diameters of the mirror. Measure the evolution of the astigmatism with the wire test.

23.4.2 Using a Full-Size Tool

The fastest method of retouching is to work only with the tool on top, placing a band of thick cloth 20–30 mm wide under the mirror and corresponding with the shortest diameter of the defect. Make large W-strokes with the full-size polisher to maintain the paraboloidal form of the mirror (be careful of turned-down edge). Frequently turn the tool under your hands. It will be necessary to polish this way

How to Avoid Astigmatism

The easiest way to eliminate astigmatism is not to create any in the first place. In order to do this, follow these rules:

- If the mirror is thin, its back should be carefully smoothed flat before beginning rough grinding of the front surface. Irregularities in the mirror's back can affect the polished face during smoothing and polishing.
- Make sure that your support is solid, flat, and allows an even distribution of the mirror's weight. You can put the mirror on a 5- to 8-mm thick felt or rubber pad. Don't fasten the mirror to the polishing post; it should be held in place by the 3 wooden blocks placed at 120° intervals around the edge. There should be some play, do not wedge it tight against these blocks.
- During polishing, avoid letting the tool stick to the mirror; this way you can avoid making irregular zones on the mirror.
- The test support for the mirror should be made to avoid the appearance of astigmatism on the mirror. The mirror should not be squeezed when it is held in place by the screws. It is important to limit the rubbing on the mirror's sides and the bending of the mirror under its own weight. Outside of testing, the mirror should rest vertically on two points separated by 90°, but it can also be held in place by a strap.

Turn the lower disk (mirror or tool) on the post a ¼ or ⅓ turn every 15 minutes. Turn the superior piece in the opposite direction about ¼ turn more often—at least once every 5 to 10 strokes.

How can you improve the precision of the wire test?

The more exactly the following two conditions are met, the more precise the wire test will be:

- The diameter of the wire should be slightly larger than the opening of the slit. In this configuration, you will get high-contrast images similar to those in Figure 23.4 on page 235.
- The thinner the wire is, the more precise the measurement will be (the slit must still be thinner than the wire).

for several hours to wear down, little by little, the irregularities of the mirror and recover its symmetry.

24 Making the Convex Secondary Mirror

The secondary mirror is the most difficult piece to make in a Cassegrain telescope. It is physically small, convex, and not directly testable with the Foucault test. To test it, we use an interferometric method with a spherical concave reference disk that has the same radius of curvature as the convex secondary mirror.

24.1 Procedure

The procedure for making the secondary mirror is as follows:

1. Rough-grind, fine-grind and smooth two glass blanks. The convex disk will become the mirror, the concave one the tool. The sagitta of the mirror should be 1.78 mm over a 95-mm diameter.
2. The back of both the mirror and tool are then ground and polished on a cloth lap. The polish needs to be sufficiently transparent that interference fringes can be seen when the opposite face is in contact with a reference surface.
3. Make the concave mirror spherical. This mirror becomes your reference disk.[1]
4. Make the convex mirror spherical and then, using the spherical concave disc as a reference, use the fringe test to guide your figuring to a hyperboloidal surface on the convex mirror.

24.1.1 From Rough Grinding to Smoothing

The procedure is the same as for the 130-mm mirror (see Chapter 13 beginning on page 153). Important points to note are:

- Use a tool disk (concave) that is larger (by 4 or 5 mm) than the mirror (convex) to avoid a possible edge defect on the concave piece during polishing.
- Be careful to avoid chips, which can form at the edge of the convex piece during rough grinding. Make a 4- to 5-mm chamfer and maintain it during the course of your work. Also avoid making strokes that are too long while pressing on the concave tool.
- At the end of rough grinding, the sagitta should be precise to within 1⁄100 of a

[1] For a highly-deformed mirror (the secondary of the 300-mm telescope, for example), you must make the concave piece hyperboloidal in order to be able to test the convex mirror correctly (see Section 24.1.6, page 242).

With a ring spherometer, the radius of curvature, R_{cx}, is calculated as follows:

$$R_{cx} = \frac{(R_{int})^2 + e_1^2}{2e_1}.$$

For a concave mirror, R_{cc} is calculated:

$$R_{cc} = \frac{(R_{ext})^2 + e_2^2}{2e_2}.$$

If you replace the ring with 3 balls, of diameter *b*, placed at a distance *D* from the feeler, the radius of curvature *R* is:

$$R = \pm b + \frac{D^2 + e^2}{2e}$$

(+*b* for the convex mirror, –*b* for the concave).

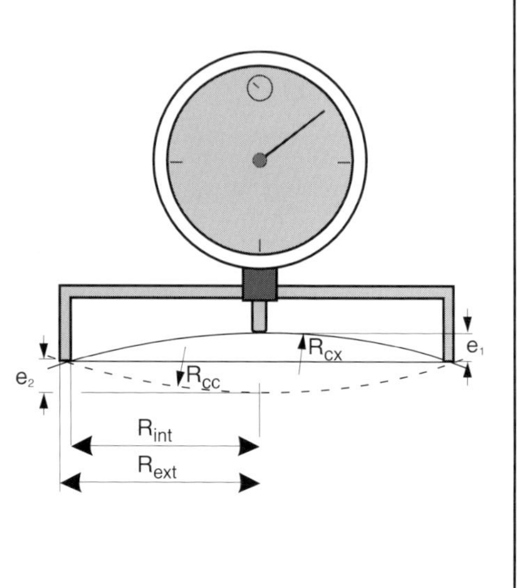

Fig. 24.1 Ring spherometer used to measure the radius of curvature of small-diameter mirrors. See Figure 5.3 on page 50 for a three-point spherometer.

millimeter. The focal length and the values *p* and *p′* (see Figure 19.1 on page 205) are very sensitive to variations in the radius of curvature of the secondary mirror. Measure with a ring type spherometer (see Figure 24.1).

- When the correct sagitta is achieved, make the same number of wets with the mirror on top as with the tool on top in order to conserve the curvature of the faces until the end of smoothing. At this stage, the surfaces should touch at all points and there should be no air bubble between the two disks.

24.1.2 Polishing the Back of the Tool and Mirror

In order to do the interference test, completely smooth the back of the mirror on the back of the tool; then polish them individually by working them on a flat polisher covered with a cloth (place the cloth over the hot pitch) and highly charged with cerium oxide. Polishing will last for 15–30 minutes. The quality of the surface and the flatness of the back has no influence on the form of the interference fringes during testing.

24.1.3 Polishing the Mirror and the Reference Disk

Make two plaster polishing tools: the first, convex, to polish the concave disk; the second, concave, to polish the convex disk. Make the polishers using the method described for the 130-mm mirror. Begin by flash polishing the mirror and the reference disk with their respective polishing tool to make a first interference test.

Work on each disk for 20 to 30 minutes, making ⅓-diameter W-strokes.

After cleaning the mirror and reference disk with a clean tissue or paper towel and a soft brush, place three shims made of cigarette paper at 120° intervals around the edge of the concave reference disk. Place the mirror on top of these shims. The disks are tested by interference using a device like that shown in Figure 9.8 on page 132.

If the curvatures of the two disks are precisely identical, you will see straight, evenly-spaced interference fringes.

If the curvatures are different, you will see concentric circular fringes. By pressing on the top disk with your finger, you can see whether the circles are convex (the rings will expand, and the disks will touch at the center) or concave (the fringes will contract, and the disks will touch at the edges). If the difference in curvature is greater than 7 or 8 fringes, return to the last grade of abrasive used in smoothing, working with the concave disk on top if the fringes are convex (expanding) and vice versa. A concave defect is harder to retouch than a convex one. It is even desirable to begin polishing with a convexity relative to 3 or 4 fringes. At this stage, check the spherical form of the concave reference disk using a Couder screen and a Foucault tester. Retouch any large defects in the concave piece before polishing the convex mirror.

24.1.4 Beginning Polishing

First, do ¾ of the polishing of the convex mirror in order to give it the same radius of curvature as the reference disk. If you observe convex fringes, place the mirror on top in order to "deepen" the mirror (make it more concave). If the fringes are concave, work with the tool on top to "flatten" the mirror (make it more convex or less deep). Polish in 20-minute cycles, making straight or W-shaped strokes with an amplitude no larger than ⅔ to ¾ of the mirror's diameter, with a transverse amplitude of ⅛-diameter. Make sure not to press at the end of a stroke. If this regime creates a deformation in the mirror, smooth the surfaces using ⅓-diameter W-strokes. Polish the reference disk using the same techniques; however, it is not necessary to polish it completely, since it is easier to change the radius of curvature of a piece if its surface is still gray.

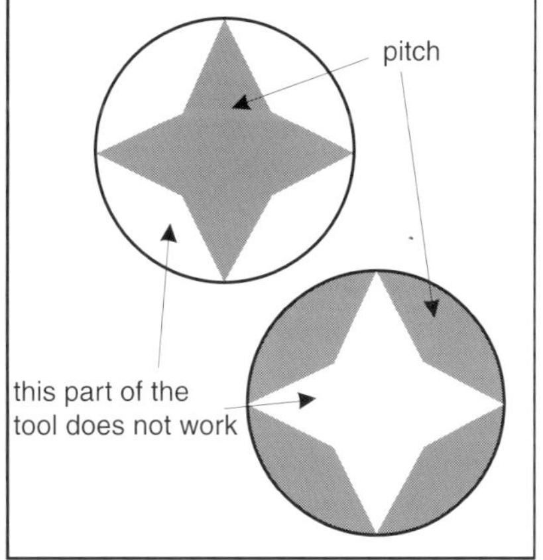

Fig. 24.2 Examples of polishers after pressing a paper mask: the gray zones indicate the pitch surface in contact with the mirror. **Top:** a tool for flattening the curve; **bottom**: a tool for deepening the curve on a convex mirror.

Another method for deepening or flattening the mirror is to remove the pitch from certain areas of the tool. Remove the pitch at the center of the tool to flatten or reduce the curvature; remove the outside to deepen the curvature. You can also press triangular paper masks into the pitch (see Section 24.2, page 241).

24.1.5 Making the Reference Disk Spherical

When the curvatures of the mirror and the reference disk are very close (1 to 2 rings), it's time to make the reference disk spherical. The piece, measured in the Foucault test, must achieve a precision of $\lambda/10$ wave to be useful as a reference for the convex mirror. If you wish to ignore extra-axial aberrations in the mounting, you need to bring the Foucault knife-edge within 10 to15 mm of the slit, either by modification of the tester or by using a beamsplitter or a right angle prism (available from Edmund Scientific or Melles Griot). The spherical reference disk should be checked using a 3- or 4-zone Couder screen. The deformation measured for each zone corresponds to the reduced measurements used in the calculation of LA_C (see page 66).

The concave reference disk finished, complete the polishing of the convex mirror. The observed interference fringes should be straight. At this stage, the convex mirror should have no more grayness.

24.1.6 Hyperbolizing the Convex Mirror

The calculation of the deformation on (see Equation 19.6 on page 206) gives $b_2 = -2.78$. We can calculate an expression for the hyperbolic surface relative to the closest reference sphere using:

$$x = \frac{b_2}{8} \times \frac{r_2^2 y^2 - y^4}{R_2^3} \tag{24.1}$$

where y is the distance to the center of the mirror, R_2 is the radius of curvature of the secondary mirror, and r_2^2 is its semi-diameter. The maximum difference, ε, is

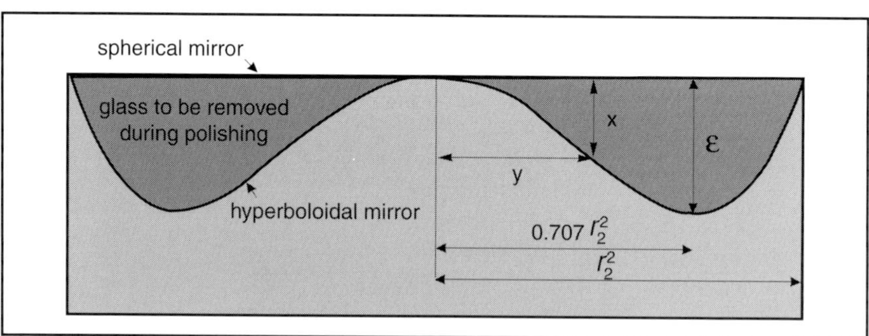

Fig. 24.3 Comparison of hyperboloidal and spherical mirrors with the same radius of curvature (differences have been greatly magnified). The maximum difference is determined for the 0.707 or "70%" zone. When a convex hyperboloidal mirror is tested, the graph of the central fringe should resemble the profile of the hyperboloidal mirror pictured here.

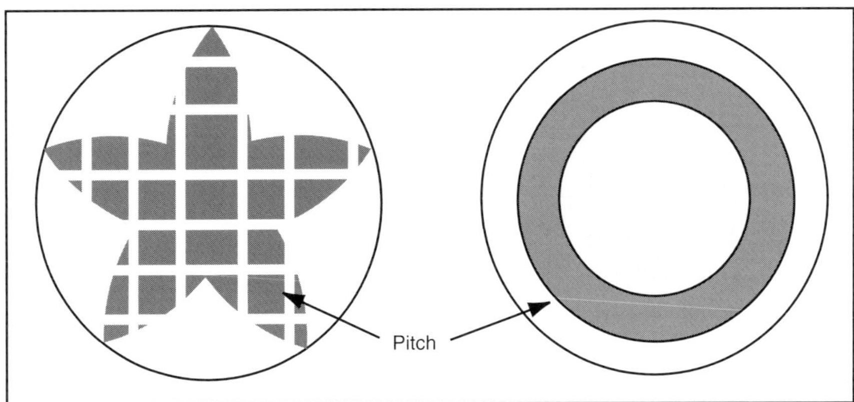

Fig. 24.4 Specially constructed pitch laps will concentrate wear in one particular area on the glass. The figure on the right is a petal lap which will concentrate most of the wear in the center if the strokes are center over center, but since the "petals" extend out to the edge, wear is blended across the diameter of the mirror. On the left is a crown lap. This lap creates wear in a limited area. This lap can work very fast and create zones that are difficult to correct.

determined for the zone of height 0.707. It is calculated using the following formula:

$$\varepsilon = \frac{b_2}{32} \times \frac{r_2^4}{R_2^3}.$$

The standard procedure is to work on the convex mirror using a tool with an area removed that is inversely proportional to the area you wish to correct. You could also use a tool with pitch in the shape of petals or a *crown* tool with a diameter of 70% of the mirror's (see Figure 24.4).

Note: we are talking about the optical diameter. Do not include the mirror's polished outside edge in tests or in calculations if its exterior diameter is larger than its optical diameter.

For weak deformations (<1½ fringes), figuring is not difficult. Make local retouches, each a few minutes long, using a small polisher or even your thumb. To test the convex mirror in reference to the concave spherical, cut a template out of paper; its form is defined by Equation 24.1. Determine the retouches needed by comparing the central fringe with the template.

For large deformations (as is the case for the 300-mm mirror whose maximum difference ε is 1.22 μm or 4.14 fringes for $\lambda = 0.589$ mm) make a series of tools with crowns centered on the 70% zone, but with different sizes. Shape the convex mirror slowly, being careful not to create "stair-step" fringes. To avoid this problem, the edge of the tool's crown should be irregular (e.g., in a sawtooth pattern), and the pressure on the tool should be light.

Test methods for this mirror are slightly different from those for previous examples. When the largest difference ε is greater than 2 fringes, the central fringe is too greatly deformed to permit reliable measurements by matching fringes with a template.

You must, therefore, make a hyperboloidal mirror with the same deformation (b_2) as the convex mirror in reference to the existing sphere. This concave hyperboloidal mirror will be used to test the convex one. When the convex and the concave mirrors have the same deformation (b_2), the interference fringes will appear straight, as they do for flat mirrors. If a defect remains, the fringes will be concentric and regular (their spacing decreases as a function of the square root of the distance from the center of the mirror).

The form of the concave reference mirror can be tested using a Foucault tester and a Couder screen. Note that the theoretical longitudinal aberration *LA* for each zone of the screen is given by:

$$LA = \frac{-b_2 \times h_m^2}{R_2}.$$

So that the radius of curvature at the center of the reference doesn't vary (since it's the same as that of the convex mirror), the reference mirror must be deformed from the edge (see also Section 13.3.2, page 164). The *LA* of the outer zone is beyond that of the central zone, similar to paraboloids, and the above formula applies for a fixed-source tester. As in any Foucault test, you must divide the anticipated *LA* by 2 when using a moving-source tester.

25 Making the Telescope Window

25.1 Why Make a Telescope Window?

The telescope window is a "porthole" with flat parallel faces, made of perfectly transparent and homogenous optical glass. This piece, installed at the top of the telescope tube, seals its opening and greatly reduces air currents inside the tube. Thus, turbulence inside the telescope is much less than in an open telescope. The telescope window is also used to support the secondary mirror, which eliminates the need for a spider, so stars are not as highly diffracted at focus, and images have greater contrast. On the other hand, making or buying a telescope window costs more in time, work and money. You don't need a telescope window unless you're observing planets or stars at high resolution, pursuits which also demand irreproachable optics.

25.2 Perforating and Edging the Telescope Window

So that it can serve as a support for the secondary mirror, we will perforate the window with a 45- to 50-mm diameter hole at its center. Proceed as described in Section 21.1.2 on page 213. Don't reattach the cutout after making the cut; just fill in the hole with some modeling clay.

It is also necessary to true-up the outside of the window with a diamond impregnated grindstone to make it perfectly circular. This operation, called *edging,* can be done by a specialized company, or it can be specified when you order the BK7 disk from the manufacturer. For reasons of material and safety, it's not a good idea for amateurs to do this.

25.3 Smoothing the Faces

25.3.1 Preparing the Tool

Before smoothing, the BK7 plate should have a diameter of 340 mm and be 18 mm thick. To grind the faces, make a plaster or wood tool and cover it with ceramic squares (see Figure 25.1). Use black optician's blocking pitch to attach the squares, and level the squares by pressing them with the tool. Be sure to

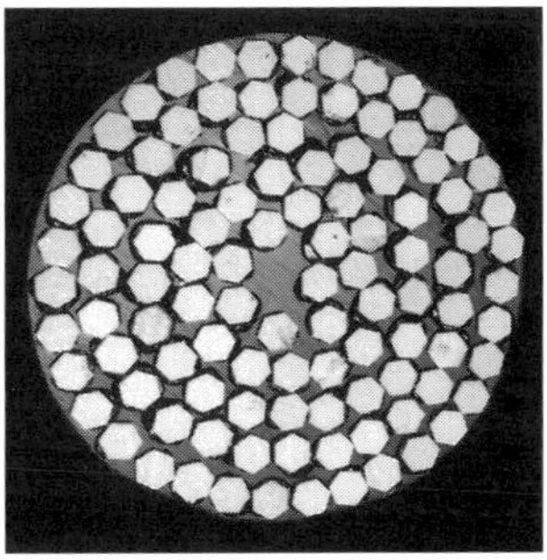

Fig. 25.1 Plaster smoothing tool with hexagonal ceramic tiles.

scrape the edges of the squares to make sure that none of the optician's blocking pitch touches the glass during smoothing.

25.3.2 Testing the Parallelism of the Faces

Use a micrometer graduated in hundredths of a millimeter to measure the thickness of the disk at eight points equally spaced around its edge. Before any measurement make sure to clean both the micrometer's contact faces and the glass; there should be no dust or grit on either. The largest measurement indicates the thickest part of the outside edge; the smallest, the thinnest. The difference between the thickest and thinnest edges is called "wedge". This difference should be less than 1⁄100 mm. Mark the thickest section on the side of the piece with a marker (see Figure 25.2).

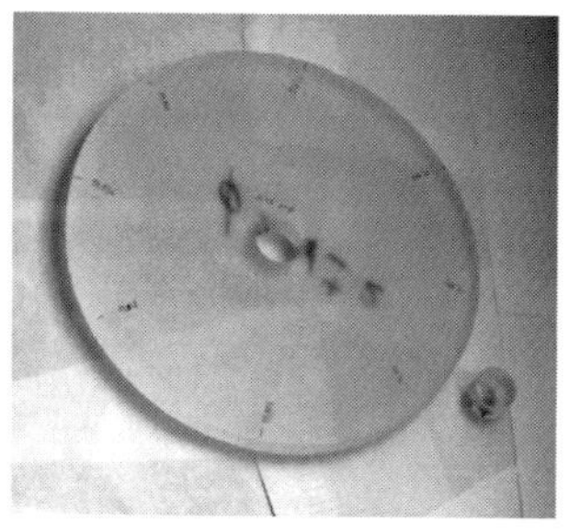

Fig. 25.2 Telescope window photographed before smoothing. Thicknesses are notated around the edges of the piece.

25.3.3 Smoothing

Cover the workpost (whose platform must be completely flat and horizontal) with a thick linen or flannel cloth of the same diameter as the window disk. The cloth should stay dry. During smoothing, when the disk will be worked on from above, make sure to turn the glass about a ⅓ turn on the cloth about every fifteen minutes to avoid astigmatism.

Begin work with W240, using normal smoothing strokes (W-strokes ⅓ of the disk's diameter in amplitude). If the plate has large surface irregularities (the side you are working on has few places where the tool touches the surface), use #360 carbo. Work on the two faces alternately, flipping the disk every hour, both with the tool and the disk on top.

To correct for any wedge, work with the glass on top, pressing at the location of the thickest edge (marked on the side). With the tool on top, make ⅓ diameter W-strokes on the thickest side of the mirror, crossing the strokes at 90° over one another. Finish the surface with two wets using normal centered strokes. The parallelism of the faces must be within the 0.01-mm tolerance before moving on to W1.

25.3.4 Testing the Flatness of the Smoothed Faces

Flatness is checked during grinding using a three-point spherometer equipped with a micrometric head graduated at least in hundredths of a millimeter, (see Figure 5.3 on page 50). Measuring the sagitta of the window is done by "zeroing" the spherometer on a reference flat. The flat's precision can be a half-fringe or better over the diameter of the spherometer. Using W240, the curvature of the faces should be less than a hundredth of a millimeter when measured with a 150-mm diameter spherometer. After a wet or two with W1, you should find (on the spherometer) a maximum sagitta of 2 or 3 μm, and this should be reduced to 1 to 2 μm at the end of smoothing. To improve the precision of the measurement, use a spherometer with a larger baseline or one with documented greater accuracy. Another option is to flash polish the surface and test by interference (see Section 5.5 on page 73). Usually 5 to 10 minutes polishing on a pitch lap is sufficient to see a fringe pattern.

25.3.5 Finish Smoothing

Finish smoothing is done by working alternately with the window on top and on the bottom with normal strokes. Do as many wets as you need; taking your time and working carefully will produce a very smooth, regular surface.

Before proceeding to the polishing stage, coat the interior and exterior edges of the window with shellac. This will help to keep abrasive powders from getting stuck in the roughly ground sides which may cause scratches if they fall off and onto the lap; when finished you can remove the shellac with denatured alcohol.

25.4 Polishing the Faces

The procedure for polishing the telescope window is as follows:

1. Quickly semi-polish the first face
2. Fully polish out the second face
3. Return to the first face and fully polish it out
4. Retouch the two faces as needed

By alternating polishing of the two faces you eliminate strain which was set-up in the glass when it was ground. If you were to fully polish out the first surface before going on to polish the second surface, the glass would deform; this is known as the Twymann effect. To polish the telescope window, make a tool with large facets of relatively hard pitch. As was the case for the perforated primary mirror, trim away the lap's central facet to mirror the central perforation of the window.

Hold the glass directly in your hands, turning it somewhat frequently and making sure not to localize the pressure of your fingers on the glass (which would have the effect of creating zonal defects or a central hole). You can protect the face

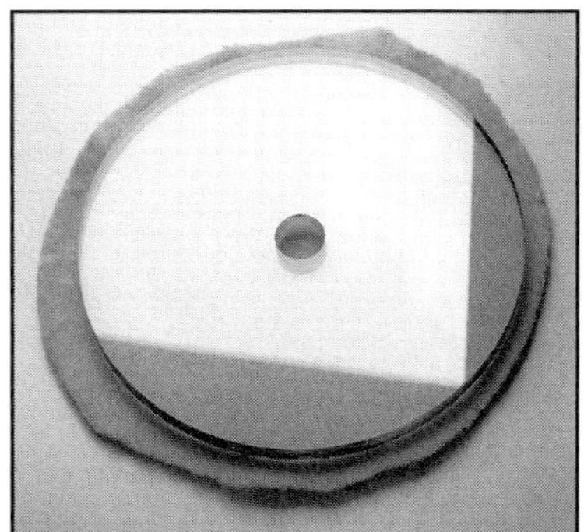

Fig. 25.3 During polishing and smoothing the telescope window should rest on a thick cushion of linen, flannel or felt. Turn the plate about ¼ turn every five minutes to avoid astigmatism.

that's not being polished with bands of contact paper or masking tape.

25.4.1 Semi-polishing the First Face

Work for an hour with the window on top, then another hour with the tool on top. When the window is on the bottom turn it about ¼ turn every five minutes. Work with a tool charged with cerium oxide without exerting downward pressure with your hands. Handle the glass plate as if it were very flexible and easily bent because that is exactly what happens at the wavelength tolerances you are working toward. If the surface has an interior or exterior area that's still very gray after two hours of work, it probably was caused during smoothing. In such a case, it's usually preferable to return to the W6 smoothing abrasive.

25.4.2 Polishing the Second Face

Clean and protect the first face with contact paper or masking tape before beginning to polish the second. Proceed in the same manner and with the same polisher. Begin with cerium oxide, and work until all gray areas have disappeared (8 to 10 hours). At this point, change to rouge and fully polish out the surface.

25.4.3 Testing for Flatness

Flatness of the faces of the telescope window can be tested quickly using a 100- to 150-mm flat reference disk to determine if each face is flat, concave or convex. This summary test tells you whether there's a central hole or a crown. The regularity of the edge is unimportant, because the outside 20 mm of the window is hidden by the mounting flange.

25.4.4 Finishing the First Face and Foucault Testing

When the second face is completely polished, finish the first face using rouge, but do not protect the second face. In the Foucault test, you will test the wave form transmitted through the window. The window is mounted in front of a spherical mirror with a clear aperture at least equal to that of the window (320 mm). Figure 25.4 illustrates the principle of the test. If you do not have another spherical mirror, you could use the primary mirror for the 300-mm telescope before it is parabolized. Its 300-mm diameter does not quite allow a full aperture test of the window, but it's close enough.

First, observe the spherical mirror alone at its center of curvature (C) and find a null. Next, place the window vertically in front of the mirror; the image of the slit on the mirror will move to C'. Light beams cross the window twice; the distance separating C and C' is calculated with the following equation:

$$CC' = 2 \times e \times \left(1 - \frac{1}{n}\right) ,$$

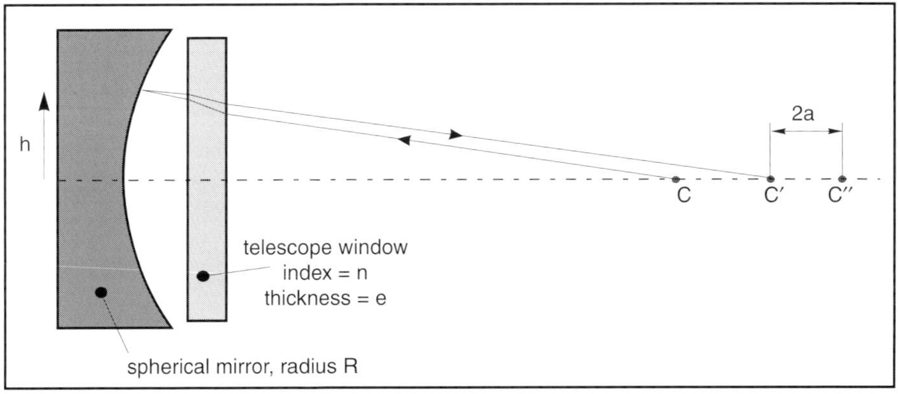

Fig. 25.4 Testing the telescope window in front of a spherical mirror with a Foucault tester.

where e is the thickness of the glass and n is its refractive index. For a BK7 plate 17 mm thick with an index of 1.51680, CC' is 11.58 mm.

When you are visually establishing the null by longitudinal displacement of the knife-edge, you will notice that it's not positioned exactly at C' but at C''. If the window is concave, C'' is beyond C'; on the other hand, if the plate is convex, C'' will be closer to the mirror than C′. The distance separating C' and C'' is labeled $2a$. The focal length F of the telescope window is calculated:

$$F = R \times \frac{R + a}{a},$$

where R is the radius of curvature of the spherical mirror. For example, the window is tested in front of a spherical mirror with a 3,000-mm radius of curvature, and $2a = 10$ mm. The focal length of the window is then 1,803 m, which is acceptable (the calculated tolerances at the beginning of this section give us a minimum focal length of about 590 m).

In this test set-up, the telescope window is placed in a converging light beam, which introduces a slight spherical aberration and gives the impression of the appearance of the null. The distance $2a$, separating C′ and C″, should be increased by the value of the longitudinal aberration introduced by the test set-up. This aberration, Λ, is calculated by the following equation:

$$\Lambda = 2e \times \left(\frac{h}{R}\right)^2 \times \frac{n^2 - 1}{2n^3},$$

where h is the height of a given zone (see Figure 25.4). For a 320-mm spherical mirror with a 3000-mm radius of curvature, h/R is equal (at the edge) to 160/3,000; thus,

$$\left(\frac{h}{R}\right)^2 = 2.8\times10^{-3} .$$

The 17-mm thick BK7 plate, when tested against a 3 m spherical test mirror, has an inherent longitudinal aberration (Λ) of 0.018 mm. This value is on the same order as the errors in the Foucault test, so it will be considered negligible. It is possible for a telescope window to have all of the defects visible in the Foucault test.

> *Note: when testing the telescope window in front of a 300-mm f/3 primary mirror (900 mm focus), Λ will be 0.05 mm. In this case you must begin to account for the spherical aberration in the exterior zones of the piece, outside of measurement, using a Couder screen (calculate and subtract Λ from the heights* h *of the exterior zones).*

25.4.5 Interpreting the Results of the Foucault Test

The test mirror should be perfectly spherical and without observable defects when Foucault tested. The shadows observed in the Foucault test after the interposition of the window are produced by refraction, and they should be interpreted exactly *opposite* as they are for mirrors. If the knife-edge is to the right, you should imagine that an imaginary light is coming from that direction (not from the left) to determine the direction of any defects (see Figure 5.6 on page 52 for these conventions). Since the light beam passes twice through the window, the quantity of glass to remove is the *double* of that on a mirror presenting the same wave error.

It is possible, and even advisable, to measure the longitudinal aberration when Foucault testing the optical window. Use a 5- or 6-zone Couder screen to calculate the average heights (h_a) of each of the Couder screen windows. First, take measurements of the longitudinal aberration on the mirror alone; we'll call these measurements M_M. Second, after interposing the telescope window, and without moving the mirror, take another series of measurements (M_W). Subtract M_M and the value of C″ from M_W. Divide the result by 2 to obtain the value of the window's defects which will be the case when the light passes only once through it in normal use. At this point, the focal length of the window should be within tolerances. The result is notated Λ_R. Knowing the primary mirror's focal length (F) and the radius of curvature (R) of the spherical test mirror, the longitudinal aberration (LA_f) at the focus of the instrument is calculated thus:

$$LA_f = \Lambda_R \frac{F^2}{R^2} .$$

The transverse aberration (TA_f) at the focus of the instrument is calculated using the following equation:

$$TA_f = \Lambda_R \times h_a \times \frac{F}{R^2} ,$$

Fig. 25.5 When both faces are polished, the telescope window is perfectly transparent. © Karine Lecleire.

where h_a is the average height of the measured zone (corresponding to the center of each window of the test screen). Add a constant, k, to the longitudinal aberration values to refine the location of the circle of least aberration. Next, compare the transverse aberration at radius ρ of the diffraction spot (the values of TA_f/ρ should be less than 1). Knowing TA_f lets you determine the slopes and the profile of the wavefront transmitted by the plate while you're making it (see Equations 5.11 through 5.14, starting on page 66). The rest of the procedure is the same as for the paraboloidal mirror.

This test with the Foucault tester can also be done using a paraboloidal mirror, but the procedure is more complicated. The null cannot be observed, and it is necessary to make measurements to know the form of the window disk.

Testing tells you the form and the amplitude of the window's optical defects. Retouching should be done by working alternately on the two faces; the methods described to correct problems on curved mirrors are applicable. Don't forget to place the window disk on a thick, dry cloth and turn it ¼ turn every five minutes. These precautions are very important and help to prevent astigmatism, a defect that often passes undetected in a quick test of the mirror.

In conclusion, it is useful to remember that a good telescope window doesn't have to have flat faces. In our case, the most important quality is the wave front that's transmitted. That has been determined in the Foucault test, which takes into consideration the defects of the two faces (which are added or subtracted from one another), the thickness of the glass and the refractive index of the material (including any problems of inhomogeniety in the glass).

26 The Optical Tube Assembly and Mounting

The plans in this chapter are for an optical tube assembly, which includes the primary mirror cell, the secondary and coudé mirror supports, and the mounting for the window.

The cylindrical telescope tube has an interior diameter of 360 mm and an exterior diameter of 370 mm. It is made to fit on the mounting made for the 250-mm Newtonian telescope (see Chapter 16 beginning on page 179). The lower part of the tube (at the level of the declination axis) is held in a square or octagonal "box." The side bearings and the declination arm are attached to the sides of the box. The eyepiece holder is situated 175 mm from the end of the tube and can be removed to allow access to the interior of the tube.

The coudé mirror is mounted in front of the primary mirror before the mirror cell is installed at the end of the tube. All adjustments to the primary and third mirror are made from the back of the tube.

The secondary mirror is mounted on the telescope window before the window is attached to the tube.

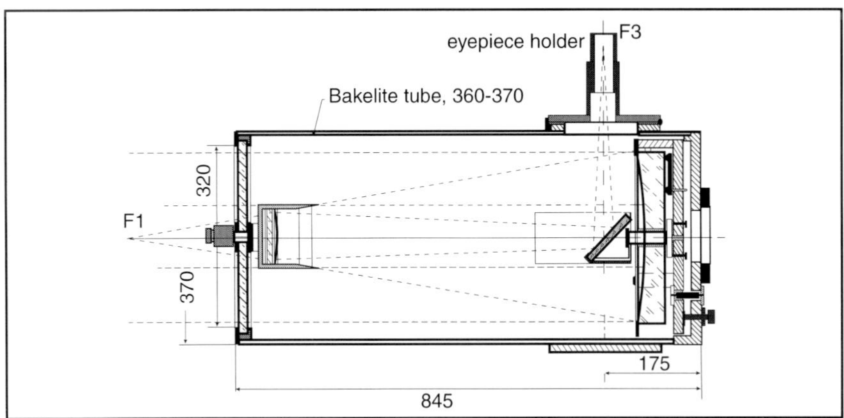

Fig. 26.1 The tube for the 300-mm telescope.

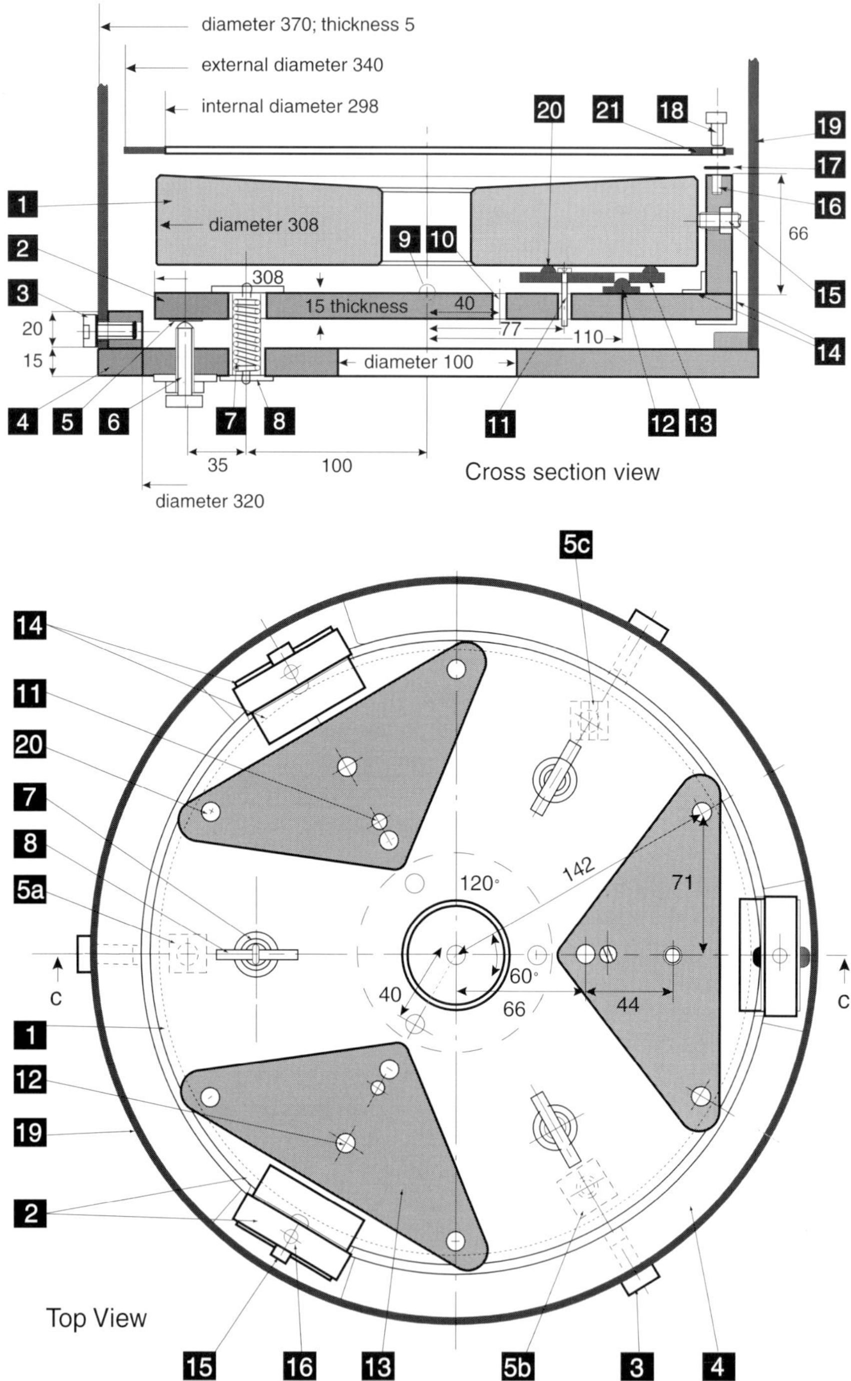

Fig. 26.2 9-Point Primary Mirror Flotation Cell.

Legend for the 9-Point Primary Mirror Flotation Cell:

1. Primary mirror, 305 mm diameter, 50 mm thick.

2. Mirror support, 15-mm thick plywood, 154 mm diameter, with 3 teeth spaced at 120° to hold the lateral mirror supports.

3. Threaded bolt to hold the mirror cell in the Bakelite tube.

4. Bottom of the mirror cell, 15-mm thick plywood, 370 mm diameter, 100 mm diameter perforation at the center for access to the adjustment screw for the third mirror.

5. Flat brass plate (piece *5a*). The other brass plates (*5b* and *5c*) are placed at 120° intervals. These adjustment screws push on these plates. They keep parts *2* and *4* from pivoting independently of each other.

6. Adjustment screw for the mirror cell (x 3): Threaded bolt, 40 mm long, with a locking nut.

7. Compression spring between pieces *2* and *4*.

8. Steel bar, 4 mm diameter, 40 mm long. Six pieces (2 per spring).

9. Central pivot for the third mirror's support (ball diameter 8 mm).

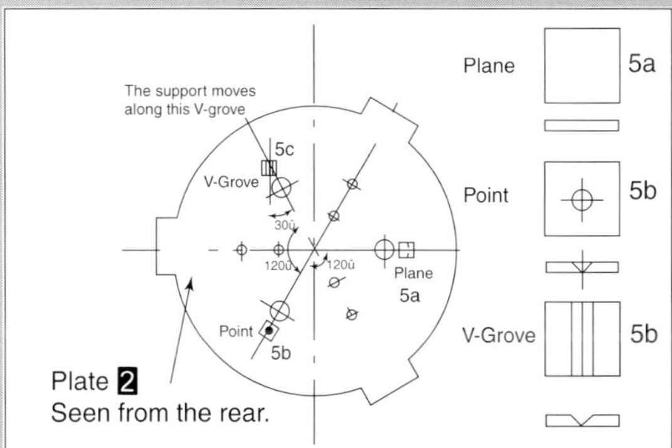

10. Pierced opening, 8 mm diameter, for the third mirror's adjustment screw.

11. Threaded bolt, 30 mm long, holding the flotation triangles in place (x 3).

12. Brass or stainless-steel acorn nut, 8 mm diameter; pivot for the flotation triangles (x 3).

13. Flotation triangles (x 3); 4-mm thick aluminum.

14. Corner reinforcements (6 pieces total).

15. Nylon screw for lateral control of mirror (x 3). Threaded incompletely to stop the screw. Leave a little play (a few tenths of a millimeter) between the screws and the mirror so that they don't strain the mirror.

16. Threaded hole, 6 mm diameter, at the top of the 3 braces of the mirror cell.

17. Washer, 6 mm diameter (x 3). Thickness is determined so that the diaphragm is between ½ and 1 mm above the mirror.

18. Threaded nut screw (x 3) to hold the diaphragm.

19. Bakelite telescope tube. Exterior diameter: 370 mm; interior diameter: 360 mm.

20. Acorn nut or nylon bolt; point of contact for the mirror (3 per triangle).

21. Diaphragm ring, which keeps the mirror from falling over. Brass, 3 mm thick. Interior diameter: 298 mm; exterior diameter: 340 mm.

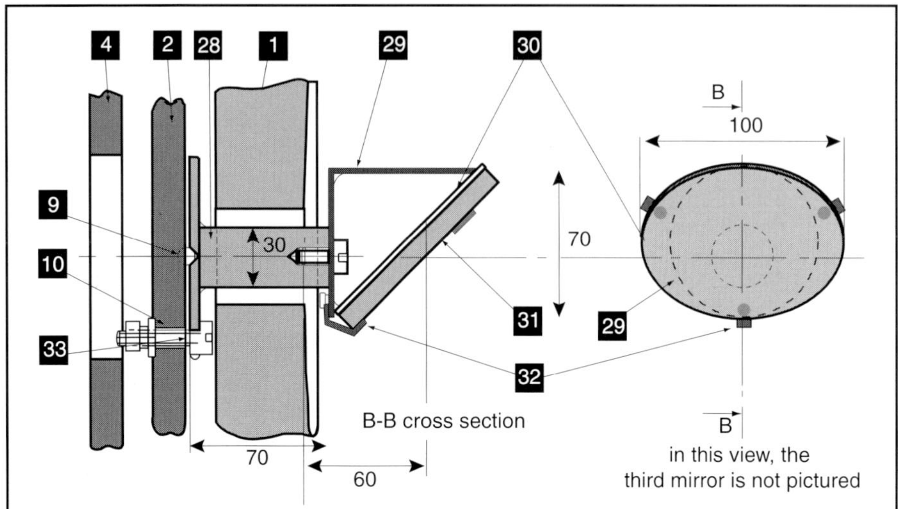

Fig. 26.3 The Coudé Mirror Mounting. The mirror(*31*) is mounted right in front of the primary mirror (*1*) on a cylinder cut at a 45° angle (*29*), mounted at the extremity of the tube (*28*) that passes through the center of the primary mirror. This metal cylinder is mounted on an aluminum plate 3 mm thick and 100 mm in diameter (*30*), held in place by 3 pads positioned at 120° intervals (*32*). Adjustment is accomplished by three locking bolts (*33*), at 120° intervals passing through the holes (*10*) pierced in the primary mirror cell (*2*). The ball bearing (*9*) serves as the pivot point for the adjustments.

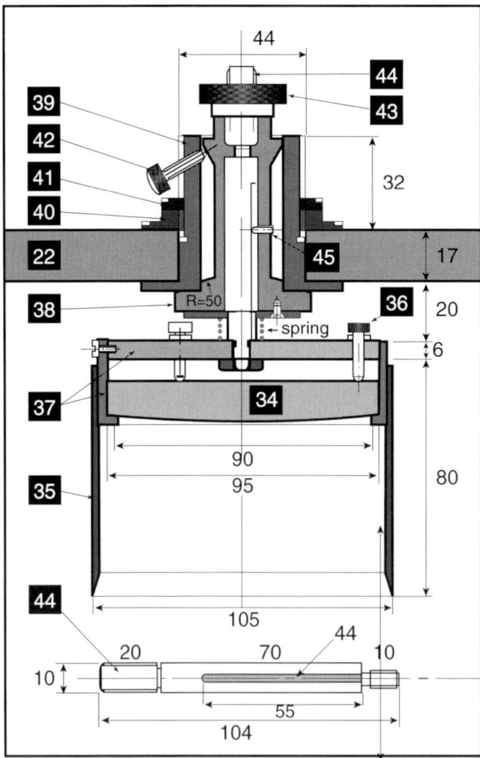

Fig. 26.4 Mounting the Secondary Mirror. The convex mirror (*34*) is mounted in a cell (*37*) and held from behind by three screws at 120° intervals (*36*). The mirror has a little play in its cell to avoid straining. The adjustment system for the mirror is accessed through the hole in the telescope window (*22*). Piece *39*, threaded on the outside, holds 2 nuts which lock together (*40* and *41*) and which retain the support in the hole of without strain. The 3 adjustment screws spaced at 120° (*42*) push against the top of the piece (*38*). The pivoting of the mirror is accomplished by the contact of parts *38* and *39*; these contact points are spherical and were ground against each other with a fine emery. The central axis (*44*), detailed at the bottom of the figure, slides through piece *38* and is controlled by the bolt (*43*) to move the secondary mirror toward or away from the primary mirror. The headless bolt (*45*) is lodged in a hole bored in piece *44*; it keeps this piece from rotating in relationship to piece *38*. The pivoting adjustment system (parts *38* and *39*) is a mechanical solution inspired by the model proposed by Jean Texereau (1984).

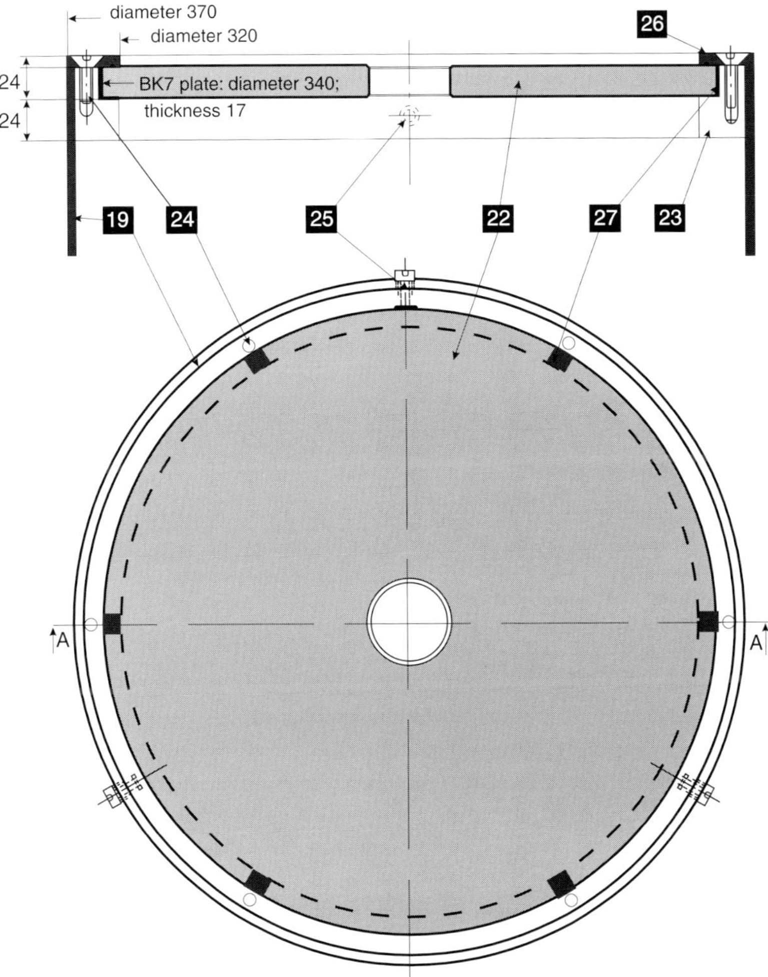

Fig. 26.5 Telescope Window Cell. The telescope window (part *22*) measures 340 mm in diameter and 17 mm thick. Its central hole measures 45 mm in diameter. It rests on a support that is slightly larger (*23*), leaving 1 mm of lateral play. Contact between the window and the mechanical part of the telescope is maintained through the six thick (1 mm) paper pads (*27*) placed every 60°. The telescope window is held in its support by a ring (*26*), with a 320 mm interior diameter and 370 mm exterior diameter. This ring is fixed in the support (*23*) for the window by 6 bolts (*24*) at 60° intervals. A silicone or neoprene washer establishes contact between the ring and the window. The assembly is attached to the telescope tube (*19*) by three bolts (*25*) spaced 120° apart.

Part 5: Use and Care of Your Telescope

27 How to Use Your Telescope

27.1 Aligning the Optics

A telescope is an assembly of optical and mechanical parts that must be aligned before it can be put to use. Optical alignment, also called *collimation*, allows the starlight to follow the correct path from the primary mirror to the instrument's focus. Alignment is very important, and it must be done carefully if you are to obtain clear images.

27.1.1 Aligning a Newtonian Telescope

Collimation of a Newtonian telescope is done in two steps. The first step can be done during daylight, no matter what the state of the sky. The second step fine-tunes this alignment at night using a star viewed at high magnification.

27.1.1.1 Preliminary Alignment

Install the primary and secondary mirrors in their cells. Rack the focuser in without the eyepiece in place and look through the hole in the eyepiece holder and observe the reflection of the primary mirror in the secondary mirror (Figure 27.1*A*). Loosen the central support screw for the secondary and rotate the entire assembly until the mirror's reflection passes up and through the focuser. While you are doing this you probably will have to turn the adjustment screws to center the reflection of the primary in the secondary (see Figure 27.1*B*).

> *Note: the primary mirror must be completely visible in the reflection of the secondary; if you cannot see the primary's outside edge, the secondary mir-*

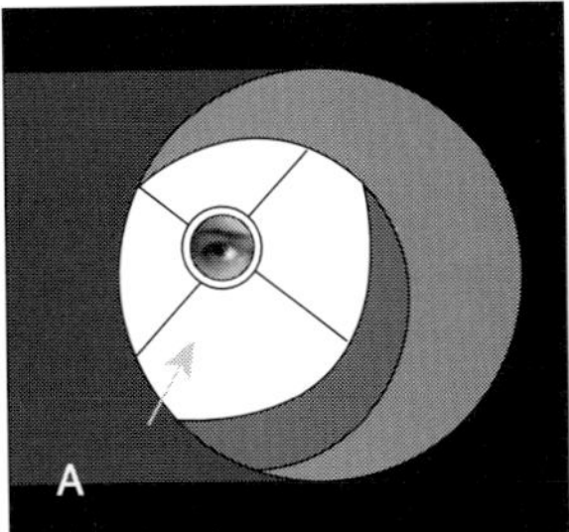

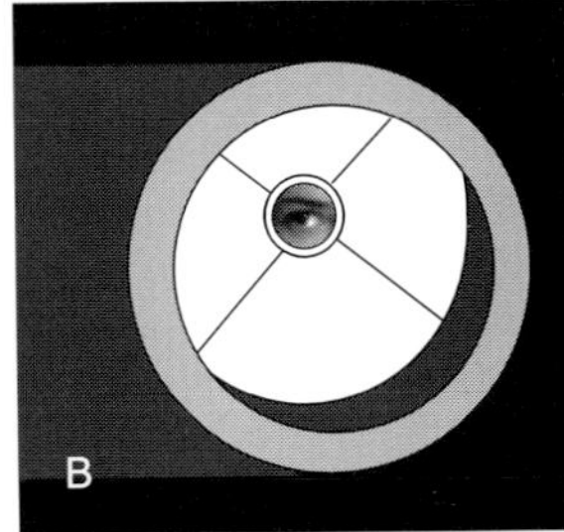

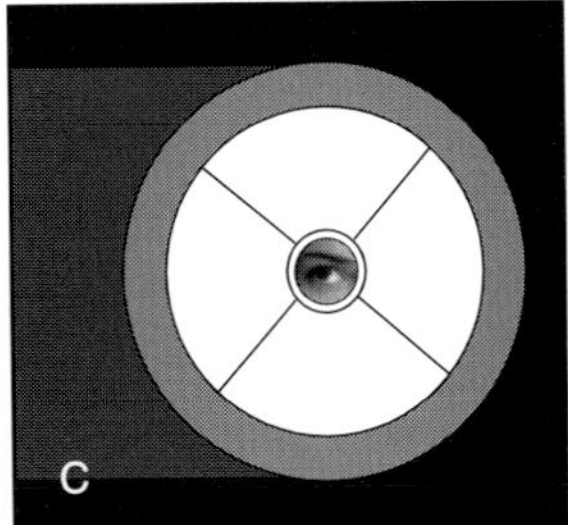

Fig. 27.1 Procedure for aligning the mirrors in a Newtonian telescope. **A.** The two mirrors are unaligned. **B.** The secondary mirror is aligned, but the primary is not. **C.** The two mirrors are aligned, and the reflection of the eye is centered in the reflection of the primary mirror.

ror is too small. In this case, the primary mirror is vignetted—diaphragmed by the flat mirror with a resulting loss of light.

Still looking through the eyepiece holder, observe the reflection of the spider's arms and the secondary's support within the reflection of the primary mirror. Center the image of the secondary by turning the primary mirror's adjustment screws located in the back of its cell. Usually only two of the three screws need to be adjusted to position the mirror (see Figure 27.1C on page 261).

When properly aligned, a reflection of your eye, seen through the eyepiece holder, should be centered in the reflection of the primary mirror. If you move your head back 20 or 30 centimeters from the eyepiece holder, the reflection should also be centered in the opening of the eyepiece holder.

27.1.1.2 Fine-Tuning the Alignment

At night point the telescope at a magnitude 1 or 2 star (Polaris, Deneb, etc.). Choose a star relatively high in the sky to ensure that the primary mirror is resting on all the contact points in its cell. Do not choose any of the planets, as they are not a true point source. Use an eyepiece that provides a magnification equal to the diameter of the primary mirror in millimeters. Adjust the eyepiece *outside* focus until you can clearly see the shadow of the secondary mirror inside the now blurred star image.

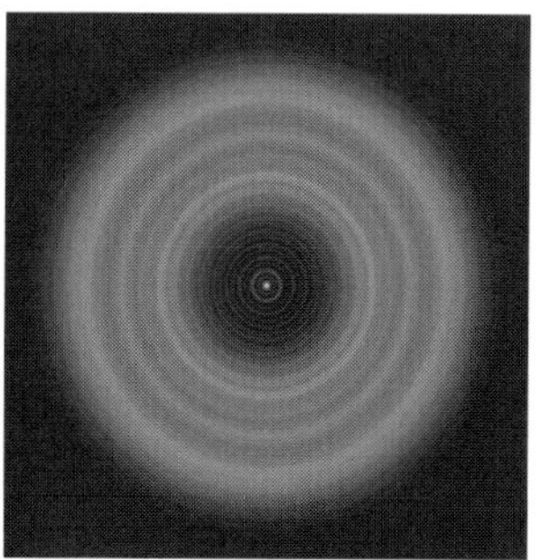

Fig. 27.2

Turn the primary mirror's alignment screws to approximately center the shadow of the secondary in the star image. Make these adjustments slowly, because the star will move in the field as the primary mirror is reoriented; you must be able to follow the star and re-point the telescope throughout the alignment procedure.

When the shadow is aligned properly, the defocused star centered in the eyepiece should look like Figure 27.2. Change the eyepiece to one that gives a magnification 2 to 2.5 times the diameter of the primary mirror and re-center the shadow.

Refine the alignment of the telescope by observing the Airy disk and the diffraction rings. If the star has a comatic shape, adjust the primary mirror to move the star in the direction opposite from the reinforced rings (Figure 27.3*A*). Alignment is perfect when the diffraction rings are centered and circular (Figure 27.3*B*).

27.1.2 Aligning a Cassegrain Telescope

As was done for the Newtonian, the first operation in aligning a Cassegrain telescope is to roughly align the mirrors during the day. If the telescope is a conventional two-mirror Cassegrain, centering the optics is done by observing from behind and through the primary mirror. The reflection of the primary should be just inside the border of the secondary. Backing away from the tube, you should

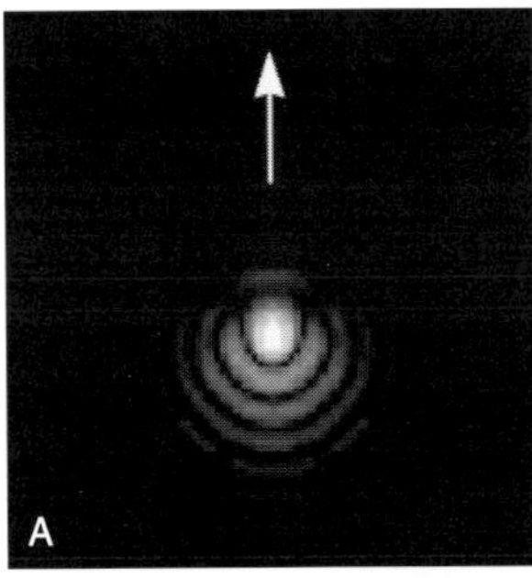

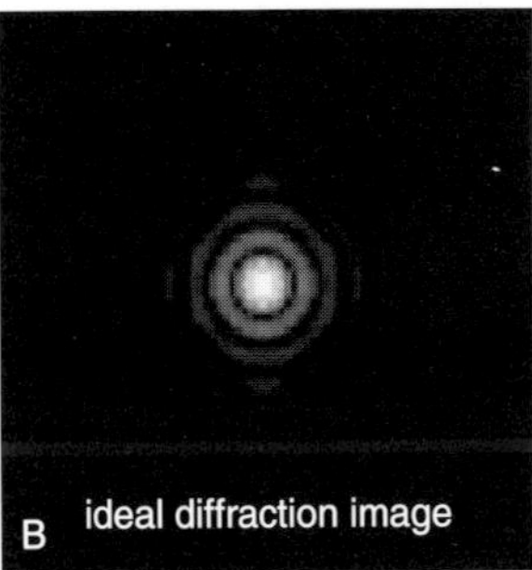

Fig. 27.3

see the secondary also exactly aligned in the hole of the eyepiece holder.

If you are aligning a three-mirror Nasmyth (as is the case with the 300-mm telescope design in Part 4 starting on page 203 of this book), you must first align the coudé (flat) mirror and center the image of the secondary within it. After this is done, center the image of the primary in the secondary without adjusting the coudé mirror.

Fine alignment of the primary mirror must be done at night, and, if possible, with calm stars (not twinkling). Use an eyepiece magnifying 2.5 to 3 times the primary diameter. Make the initial alignment and observe the movement of the star in the field as you turn the adjustment screws: the star should move in a direction opposite from that of the brighter rings (Figure 27.3*A*).

If there is a lot of turbulence or the telescope has a very large diameter, defocus the eyepiece slightly and center the star as in Figure 27.2.

The telescope window does not require any specific adjustment. Avoid setting the faces of the window exactly perpendicular to the optical axis of the telescope, as doing so can cause reflections inside the telescope tube during observation, especially when observing a faint object near a bright star or planet. The simplest solution is to set the window at a small angle equal to half the field at the instrument's focus, or about 20 minutes of arc for the 300-mm Cassegrain described in Part 4 beginning on page 203.

27.2 Polar-Aligning an Equatorial Telescope

After the optics are collimated, equatorial telescopes require another mechanical adjustment called polar alignment. This operation consists of aligning the principal rotation axis of the telescope (the right ascension axis or polar axis) with the axis of rotation of the Earth.

To observe visually or to make sketches, a cursory polar alignment is sufficient. However, a complete and exact polar alignment is necessary for astrophotography. The procedure takes three steps.

27.2.1 Step One: Quick Alignment

During the day, orient the axis of the mounting with the north-south meridian. Use a compass or a shadow from the sun at noon to determine the direction. At nightfall, verify optically that Polaris is located within the path of the right-ascension axis. Correct the orientation if necessary.

Transporting Your Telescope

Before transporting your telescope, the optics must be removed and protected. Automobile transport can vibrate the secondary mirror loose from its holder and it can fall onto the primary. The telescope must be partially disassembled: separate the tube from the fork and the fork from the mounting. In the case of an equatorial mount, do not separate the right-ascension axis from the fork. If you put the telescope in the back seat of a vehicle, don't forget to fasten the heavy parts in with seat belts in case you have to stop quickly.

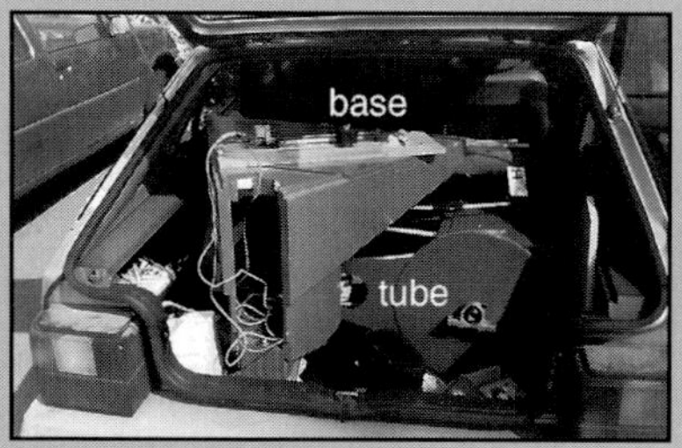

Fig. 27.4 The 250-mm telescope fits in the hatchback of a Citröen BX.

27.2.2 Step Two: Aligning on the Meridian

After nightfall, place a reticulated eyepiece in the holder. Point the telescope at a bright star on the meridian and close to the celestial equator, such as Altair, Procyon, α Aquarii, δ Orionis, or γ Virginis. Before turning on the clock drive, let the star drift in the field in order to determine the east-west axis (the star will drift to the west). Orient the reticle's cross-hairs along the east-west axis. Start the clock drive, and adjust the right-ascension drive rate. Let the telescope follow the star for a few dozen seconds and observe the movement of the star in reference to the reticle. If the star moves to the south along the north-south cross-hair (toward the top of the field if the image is inverted), move the south end of the base a little toward the east. If the star moves to the north (it descends in the field), move the south end of the base slightly to the west

27.2.3 Step Three: Latitude Adjustment

Aim at a star located 6 hours past the meridian, with a declination between 40° and 50° (Deneb, Capella, Algol, γ Andromedae). Observe the movement of the star in the field. If the star drifts to the north, lower the polar axis slightly (for example, place a shim under the south end of the base).

27.3 Observing

Once collimation and polar alignment are done, the telescope is ready for use. The exploration of the sky through the eyepiece is fascinating and unforgettable. Here are a few essential tips.

There are two basic types of observing, planetary and deep-sky. For the planets (and lunar), high magnification is necessary in order to see detail. Let D be the diameter of the telescope in millimeters. Jupiter and Venus have relatively low-contrast details and can be observed with magnifications from 1.5 to 1.8 times D.

Choosing an Observing Site

The choice of an observing site depends on what you want to observe.

Deep sky observations require a dark sky, away from the city or any other aggregation of lights, such as monuments, highways, or malls.

For example, you need to be about twenty miles from Paris in order to observe the Milky Way at the zenith and about sixty miles away to observe under truly dark skies. Industrial smog diffuses light (streetlights, car headlights, the Moon) and give the sky a milky appearance. High altitudes (> 3,000 feet) are required to escape humid and light-absorbing layers of the atmosphere. Above 8,000 feet, the air is usually very clear, and the sky becomes very dark.

Planetary observation is more affected by atmospheric turbulence than light pollution. Agitation of planetary images is more often tied to temperature and wind. In the Parisian basin, for example, the wind coming mostly from the west produces variations in telescopic images; a warm wind from the south often promises good images; and an eastern or northern wind brings cold air and a lot of turbulence.

If possible, avoid sites near the ocean, because the salt in the air can attack your mirror's coatings. Aluminized mirrors are also vulnerable in humid or polluted sites, within cities or close to factories.

Urban observers can benefit from a slight turbulence when there is a temperature inversion (a layer of cold air held at ground air by high air pressure during good weather), generally at the beginning of the evening during the first few days of an approaching high weather system.

Mars, Saturn, and the Moon have higher-contrast formations, and work well with magnifications from 2.5 to 3 times *D*. Of course, the "visual" resolution diminishes slightly (see Section 3.4.3 on page 29), but the principal details are large and easy to see, and their colors look better. Images of Mars and of Saturn's rings at 600x magnification in a 250-mm telescope are almost palpable. Mercury is often too low on the horizon to be seen without turbulence, even on a calm night. Other planets—Uranus, Neptune, and Pluto—are too small and dim to see detail, even under good conditions. In general, planetary observation requires little turbulence (little or no wind, stars do not twinkle) but not necessarily a very dark sky (many accomplished planetary and lunar observers are located in densely populated areas).

Deep-sky observing focuses on low-contrast objects (many of which are only 1 or 2% above the sky background) and requires a dark, transparent sky. To observe very dim stars you should seek equipupilar magnification. A magnification giving an exit pupil of 1.5 mm is ideal to begin to see detail in galaxies (their shape, arms, knots) and bright diffuse nebulae. Bright nebulae, planetary nebulae, and globular clusters are comfortably observed at up to 2 times *D*.

Antipollution filters (UHC, LPR, OIII, Hβ) can be used to lower the brightness of the sky and to increase the contrast of nebulae, especially inside large cities. These filters also give surprisingly good results at dark sites because the brightness of stars in the field is diminished, which helps make weaker objects more visible.

Using a telescope out under the stars requires some preparation. You must dress warmly (even in summer—bring clothes for the cool early morning hours before twilight) also bring a few snacks and a thermos full of a warm drink, and perhaps a portable radio or CD player with some music to listen to.

Fatigue makes observations long and difficult. You need to observe in a comfortable position (sitting down if possible). If the area is dry and dusty bring a blanket to cover the ground so you and the instrument don't get unnecessarily dirty. Another blanket wrapped around your legs and feet can help keep you warm.

Once you arrive at your observing site, let your eyes adapt to the darkness. You may have to wait 30 minutes to an hour before you see dim objects. In a Newtonian telescope, avoid breathing close to the secondary mirror so you don't fog it. If the mirrors get covered with dew, definitely *do not wipe them*. Bring the telescope into a warmer place and let the dew evaporate. You can also dry the surfaces with a fan, but make sure not to use hot air (as in a hair dryer).

27.4 Sketching at the Eyepiece

Sketching allows you to keep a record of your observations. In planetary observation, the eye is able to see details in a fraction of a second, while photography usually requires more exposure time and thus is more subject to blurring. Experienced observers always use the same eye to see fine details; the other eye, more sensitive to dim objects, is better used to detect "deeper" objects.

To make a drawing at the eyepiece, you'll need a wooden sketch pad (20 by

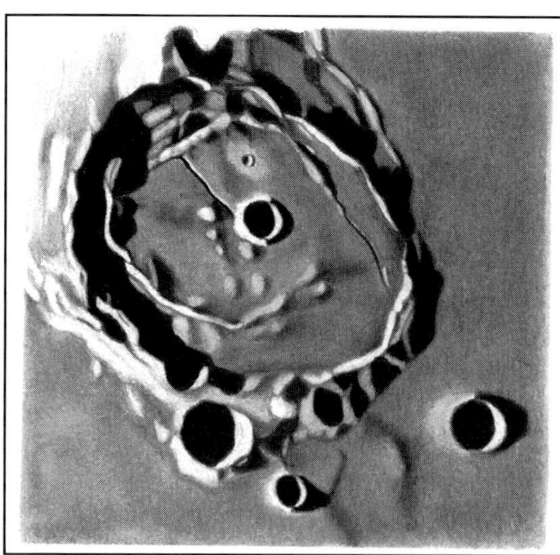

Fig. 27.5 The crater Poseidon, drawn by Nicolas Biver on May 7, 1992, between 20:00 and 22:00 Universal Time, using a 200-mm telescope with 300x magnification. There was slight turbulence at the beginning, and more at the end of the evening.

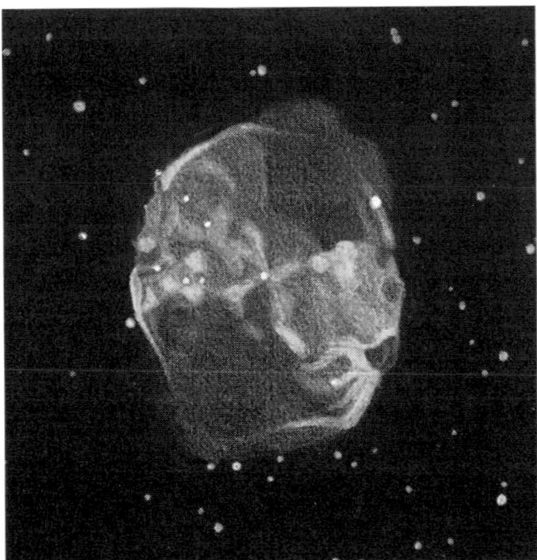

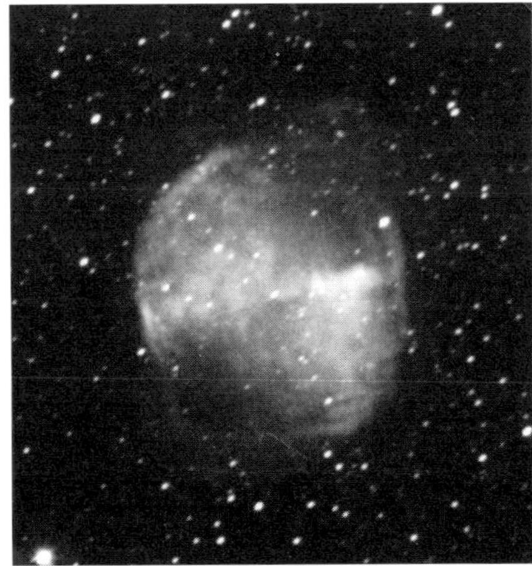

Fig. 27.6 M27 (the Dumbbell nebula, 8′ x 4′, magnitude 7.6) in the constellation Vulpecula. **Left:** a drawing made July 7, 1991, from 22:30 to 0:15 universal time by Frédérick Auchère at the eyepiece of a 250-mm *f*/8 telescope. Magnifications used: 400x and 800x. **Right:** a photograph made by Georges Beaume on July 3, 1992, at the focus of a 300-mm *f*/5.4 telescope. Thirty-minute exposure on hypersensitized Kodak Technical Pan 2415. Eyepiece drawing is a skill that requires experience and great eye endurance. But the results speak for themselves! © Frédéric Auchère.

30 cm), an HB or B pencil for planetary sketches, suitable paper in a water proof case (because of dew), an eraser, and a red diffusing lamp. The drawing should include information such as the diameter of the instrument, the magnification used, the date and time of the drawing (beginning and end, in Universal Time), and the state of the sky (transparency, turbulence). The drawing should be recopied neatly as soon as possible (so you don't forget anything).

Drawing can take from ten minutes to several hours at the eyepiece. It's important to use a black pencil on white paper. Use only a dim red lamp to light the picture as you're drawing it. To see fine planetary details, observe for several minutes to get your eye accustomed to the contrast and luminosity of the object. Always begin with the large details and progress on to the finer

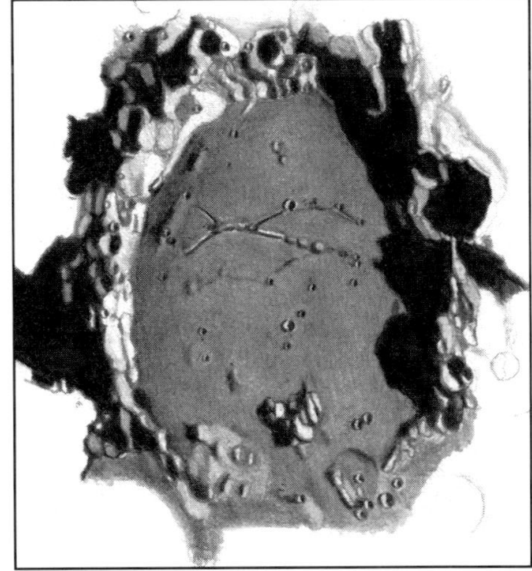

Fig. 27.7 The crater Fracastor drawn by Nicolas Biver on August 7, 1993 from 2:40 to 5:35 Universal TIme, using a 200-mm telescope. Magnification: 480x; little turbulence. © Nicolas Biver.

Using Colored Filters for Visual Observation and Astrophotography

For deep sky observing, antipollution filters can darken the sky background and increase the contrast of nebulae. In planetary observation, colored filters can help to bring out very fine detail.

- **Red filters,** useful for observing or photographing Jupiter or Mars, increase image contrast and let you penetrate the high-altitude fog of these planets. In deep-sky photography, a red filter can bring out hydrogen-containing nebulae (the Hα line at 0.656 μm) and gives a sky background that's very dark.
- **Yellow and green filters** are used to diminish the secondary spectrum visible in refractors. In astrophotography, a filter with a passband at 0.55 μm (yellow) increases the contrast and resolution of planetary images without significantly increasing exposure time. These filters also attenuate the chromatic dispersion of stars near the horizon caused by the atmosphere (equally valuable for reflectors and refractors).
- **Blue filters** bring out the high-altitude cloud formations on Mars along with the cirrus dust clouds over its large volcanic formations. They bring out the clouds, the giant red spot, and the equatorial bands of Jupiter as well as the equatorial storms of Saturn.
- **Violet filters** are used in observing Venus. In effect, the cloud formations on this planet are only visible in violet or ultraviolet light—and even then they're still very difficult to see. A violet filter strongly darkens images, since the eye is not very sensitive to the wavelengths it transmits, which is useful when observing bright Venus.

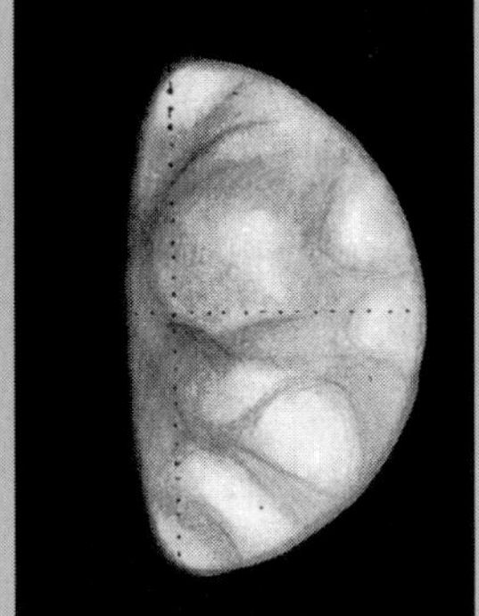

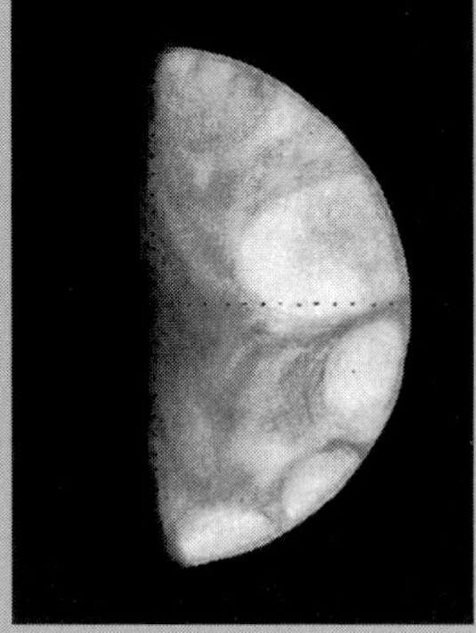

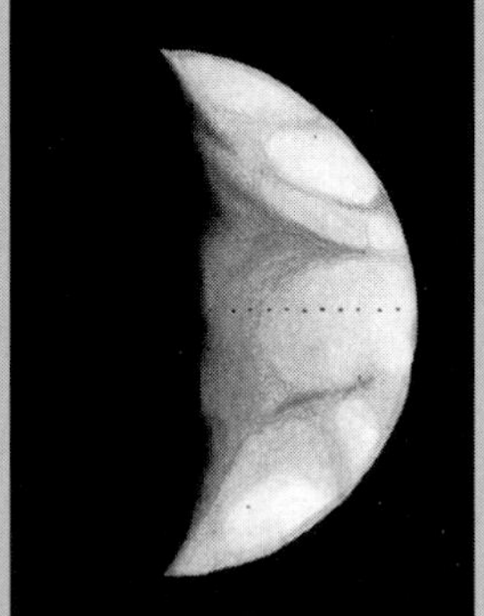

Fig. 27.8 Observations of Venus made by Frédéric Auchère at the Sorbonne Observatory in Paris, using a 250-mm telescope (*f*/7.9), magnification: 400x. A UG5 violet filter was used to bring out the clouds. From left to right: May 24, 1991, from 19:00 to 19:08 UT; June 4, 1991, from 18:21 to 18:28 UT; June 30, 1991, from 20:20 to 20:31 UT. © Frédéric Auchère.

ones. To maintain your objectivity, you should always see a detail two times before you draw it (three times if there is a lot of turbulence). To better distinguish the details of a bright object (a planet or nebula), keep your eyes open and shine a dim light on the telescope tube near the eyepiece mount. Observation of dim objects should be done using averted vision—that is, look slightly to the side of the object so that the more sensitive rods in your eye come into play.

27.5 Astrophotography

Photography at the telescope requires a manual "reflex" camera (not automatic) that allows timed exposures and is equipped, if possible, with a clear reticulated viewing screen. It should also have a flexible cable shutter release and several suitable camera-to-telescope adapters.

Deep-sky photography is usually done with the focus of the telescope projected directly on the film, while planetary photographs are often taken by projecting the image through an eyepiece and then on to the film.

For film, black-and-white is usually finer grained than color. The black-and-white film standard is Kodak's Technical Pan 2415, an exceptional fine-grained emulsion that gives high-contrast pictures. It is possible (and necessary for Technical Pan) to treat films with hydrogen before using them in order to increase their sensitivity (the process is called hypersensitizing).

27.5.1 Deep-Sky Photography

Deep-sky pictures are taken directly at the focus of the instrument, on a special

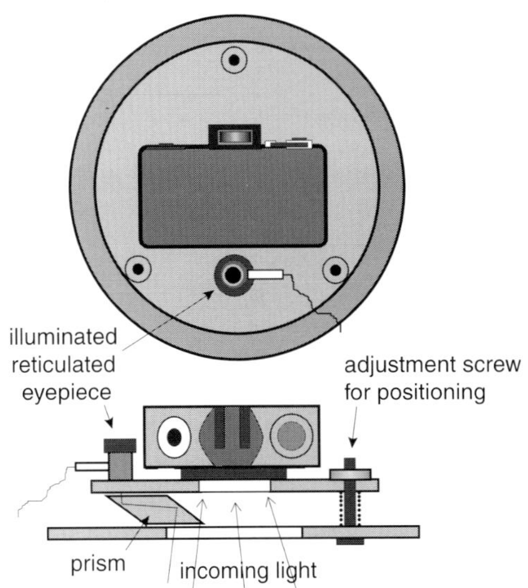

Fig. 27.9 Left: a photographic mounting at the focus of a telescope. The camera body is attached to a plate that's adjustable in orientation and height. The guiding eyepiece views an image of a star at the edge of the field from a small prism or mirror. **Above:** an example of a photographic mounting on a 200-mm telescope. © Sébastien Debruyne.

Eyepiece Projection Astrophotography

Astrophotgraphy is often done at the telescope's focus, but the images tend to be small, and there is little lunar or planetary detail. The following equation gives the dimension b (in millimeters) of an object at instrumental focus as a function of its apparent size in the sky in degrees, α, and the instrument's focal length (F in mm):

$$b = F \times \tan\alpha .$$

For example, an object ½° degree in diameter, like the Moon, would measure 8.73 mm at the focus of a telescope with a focal length of 1 meter.

To obtain larger images, install an eyepiece on the instrument and attach the camera (without its lens) behind it on a platform made for this purpose or on a plate purchased from a company that sells astronomy equipment. Measure the distance (T) that separates the last lens element of the eyepiece from the camera's film plane. Knowing the focal length (f) of the eyepiece, we can calculate the magnification (γ) of the image. The magnification is the ratio between the size of the image projected by the eyepiece on the negative and the initial size of the same image at the focus of the instrument. It is given by the formula:

$$\gamma = \frac{T}{f} - 1 .$$

For example, a distance (T) of 90 mm for an eyepiece with a focal length f of 12 mm gives a magnification of 6.5. The image on the film is thus 6.5 times larger than it is at the focus. The Moon from the previous example now measures 56.75 mm across and won't completely fit on a frame of 24 x 36 film.

Note: to obtain sharp photographs over a whole 24 x 36 negative frame, you should use eyepieces that give a magnification larger than 6x. This will reduce the field curvature.

We can calculate the final focal ratio (using the focal lengths of the eyepiece and the telescope and the distance, T), which lets us know the luminosity of the image by multiplying the focal ratio of the instrument by the magnification γ:

$$f/\#_{\text{final}} = f/\#_{\text{instrument}} \times \gamma .$$

For example, a 130-mm telescope with a focal length of 1200 mm has a focal ratio of 9.2. With a magnification (γ) of 6.5 the combination produces a final focal ratio of 60. The difference in brightness between two focal ratios (for example, 9.2 and 60) is equal to the ratio of the diameters of the projected images: a planet photographed at *f*/60 is twice as large as it is at *f*/30, but it is also 4 times less bright and requires 4 times as much exposure time to obtain the same density on film. A photo made at *f*/60 will take $(60/9.2)^2 = 42$ times longer than at *f*/9.2.

platform that allows you to adjust the orientation of the camera and its alignment (see Figure 27.1 on page 261). Guiding is checked using an eyepiece with a lighted reticle, installed on the photographic platform. Alignment is achieved, before inserting the film, by Foucault-testing using a bright star observed through the back of the camera. The beveled knife-edge slides on the camera's film guide-rails when it cuts into the light beams coming from the telescope. If the camera is placed exactly at the focus, a lighted disk of the mirror can be seen when the eye is directly behind the knife-edge. This disk will darken uniformly and a null as the knife is cut into the beam. If the camera is too close to the mirror, the knife-edge will intercept the light before the focus and the mirror will darken on the same side as the knife. If the camera is too far from the focus, the mirror will darken on the opposite side of the knife. The figures observed are similar to those in Figure 5.9 on page 55, but here the mirror is paraboloidal and the source is at infinity.

Fig. 27.10 Platform for photography at the focus of the telescope. View of the orientation and polar alignment system. The camera is installed on a T2 ring. © Sébastien Debruyne.

27.5.2 Planetary Photography

To record images of planets, install the camera on a support that allows an adjustment of about 50 to 250 mm in the distance between the film and the eyepiece. See

Fig. 27.11 (Above) Eyepiece projection platform used in planetary photography. Assembly by Sébastien Debruyne. **Right:** photographic platform made at the Sorbonne observatory for use with a 250-mm *f*/8 telescope. © Sébastien Debruyne.

Determining Exposure Time

It can take much time and effort, and a huge waste of film, to get a properly exposed photograph. Knowing all the parameters and the exposure time (t_1) of one good photograph taken with instrument (I_1), it is possible to calculate the exposure time (t_2) necessary to make a Foucaultogram on any instrument (I_2). The variable parameters to account for include: m, the magnitude of the star; D and f, the diameter and the focal length of instrument (I_1); τ its rate of transmission, τ_S, the rate of transmission of the separating lens; f', the focal length of the photographic objective (open to its maximum). The exposure time (t_2) is determined using the following equation:

$$t_2 = \left(\frac{f'_2 \times f_1}{f'_1 \times f_2}\right)^2 \times \left(\frac{\tau_1 \cdot \tau_{S1}}{\tau_2 \cdot \tau_{S2}}\right) \times t_1 \times 10^{-0.4 \times (m_1 - m_2)} .$$

The diameter (ϕ) of the image on the mirror is given by:

$$\phi = \frac{D \times f'}{f} .$$

The photographs for this exercise were taken on non-hypersensitized Kodak Technical Pan 2415, developed for 12 minutes at 20°C in HC 110 (1+9). The film transmitted 50% of the light, the objective used was 50 m set at $f/1.8$. The transmission rate of the different instruments was 0.8. The Foucaultgram on page 278 was exposed for 20 minutes on the star Polaris (magnitude 2); the photographs in Figure 28.4 on page 279 were exposed for 6 minutes on Vega (magnitude 0).

the box on page 270 for the necessary formulas, and see Figure 27.11 on page 271 for examples of how to adapt a camera body for planetary photography.

27.6 Reducing Turbulence

There are two kinds of turbulence: atmospheric and local. Atmospheric turbulence, which is dependent on the observation site and the weather, causes stars to twinkle and telescopic images to dance around. There is no solution for this kind of turbulence other than to wait for a calmer night or to move to a different site.

Local turbulence, which is caused by thermal exchanges in the area of or inside the instrument, causes irregular swirls in images, sudden changes in focus, or fuzzy images. There are a few ways to deal with local turbulence, depending on its cause.

- **The instrument.** Telescopes with open tubes are very sensitive to turbulence. The addition of a telescope window can significantly reduce air exchanges between the inside and the outside of the tube. A refractor with a window will give images as "stable" as those of a refractor of the same size. Don't paint the spider and the secondary support black.

Black paint will absorb infrared light from the sky, and the metallic pieces will heat up by a fraction of a degree, which will cause a variation in the index of the air and a visible dephasing of the light passing close to these pieces. This phenomenon, described by André Couder,[1] causes a directional flaw and reinforces the diffraction rings. The metallic pieces in the tube should be left shiny; brass or copper is better than aluminum.

At the beginning of the night, it takes several hours for the mirrors to stabilize in temperature. The center of the primary mirror will remain warmer than its periphery, which causes a spherical aberration that can sometimes be severe. A small fan at the back of the mirror will accelerate temperature equalization.

- **The observer.** Avoid moving your body directly under the tube opening or against the wind. To find the direction of a warm air vein, observe the visible swirls on the mirror while looking at a star directly at the focus without an eyepiece. Avoid breathing in the direction of the tube's top end. Cover truss tubes with reflective material, with the shiny side (reflecting the infrared radiation) turned towards the exterior.
- **The observing site.** Never observe through an open window. Avoid observing from a balcony, from under a warm roof, or near a chimney. On the ground, grass holds (and thus releases) less heat than concrete. In small observatories, the interior temperature should be the same as the exterior.

27.7 Eliminating Vibrations

Telescope vibration is a problem for photographs and planetary observation. Fast and slow vibrations have different causes.

- **Low frequency vibrations** (from 1 to 5 Hz) are caused if the moving parts of the telescope are not correctly balanced or if the instrument is not securely positioned on the ground. To remedy this problem: orient the telescope tube horizontally, with the mirrors in place but with no eyepiece or other accessories. Release the declination axis break. Carefully balance the telescope by re-positioning the weights symmetrically in relation to the central axis of the tube, so that you do not change its balance over its right-ascension axis. Turn the tube 180° and repeat the operation.

  Low frequency vibrations are also caused by a poor fit between the base and the fork or the fork and the tube. The right-ascension axis should be solidly attached to the fork, and its three points should rest on the base (the ball and the Teflon pads; see Figure 16.10 on page 187). The center of gravity for the tube should project down to the ground between the three feet of the base. The telescope's axes of rotation should be perpendicular.

  Make sure the counterweights are spaced correctly; the tube should be

[1] Couder, André, "Sur un effet thermique observé dans les télescopes à réflexion," *l'Astronomie*, Sep.-Oct. p. 253, 1949.

lighter than the mounting; make sure there is not too much weight on the tube.

The right ascension axis should be at least as long as the arms of the fork. It should not bend under the weight of the fork and the tube. The diameter of the axis should be greater than 40 mm for a 150-mm telescope, 60 mm for a 200-mm telescope, or 80 mm for a 300-mm telescope. If the axis becomes too heavy, you can make it hollow. The arms of the fork should be rigid and held together by a panel (see Figure 16.6 on page 184) so they don't vibrate like a tuning fork.

- **High frequency vibrations** (>10 Hz) follow a jolt or a nudge and are transmitted throughout the structure of the telescope. If they last longer than two seconds, they will be a major distraction. You can dampen high frequency vibrations by filling the hollow parts of the telescope (the fork, the sides of the truss tube) with polyurethane foam. The head of the 250-mm telescope can be isolated from vibrations using pieces of rubber (from a bicycle innertube) placed at the ends of the truss tubes.

28 Evaluating and Improving Your Instrument

28.1 Observing the Diffraction Disk

The quality of an image formed at the focus of a telescope depends on all of the optical elements in the instrument. Each element, mirror or lens, has its own defects and changes the light stream as it passes through the instrument. Any modification, however small, of the structure of the light that strikes the primary mirror will change the appearance of the diffraction spot. The same is true for the obstacles in the light's path: the spider and the secondary mirror, the eyepiece holder, mirror retainers, and so on. You can simulate these diffraction events with a computer and see the artificially flawed images. This allows you to appreciate the influence of each of these problems on an image.

By analyzing the diffraction image formed by an instrument at high magni-

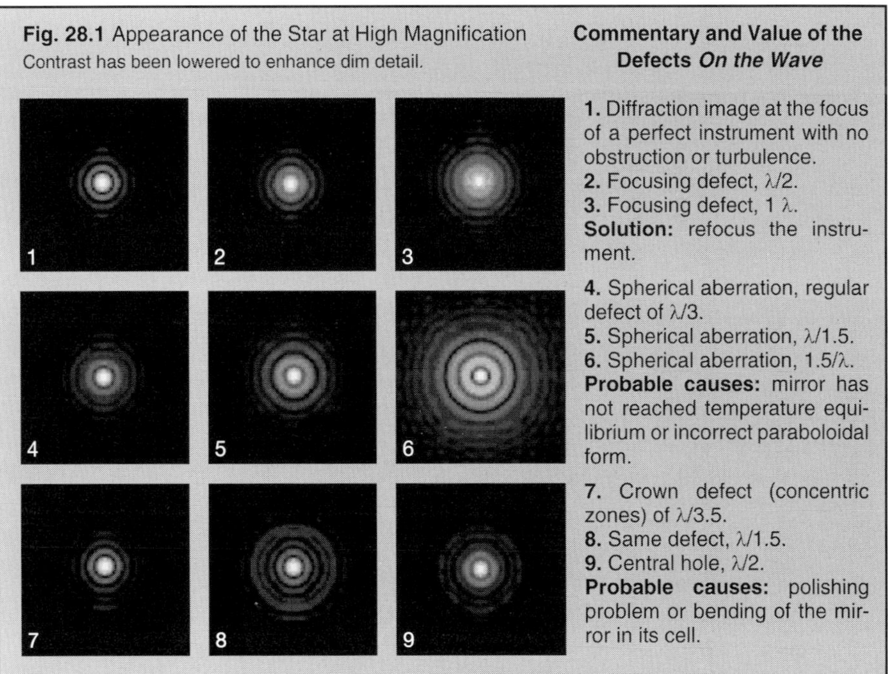

Fig. 28.1 Appearance of the Star at High Magnification
Contrast has been lowered to enhance dim detail.

Commentary and Value of the Defects *On the Wave*

1. Diffraction image at the focus of a perfect instrument with no obstruction or turbulence.
2. Focusing defect, $\lambda/2$.
3. Focusing defect, 1 λ.
Solution: refocus the instrument.

4. Spherical aberration, regular defect of $\lambda/3$.
5. Spherical aberration, $\lambda/1.5$.
6. Spherical aberration, $1.5/\lambda$.
Probable causes: mirror has not reached temperature equilibrium or incorrect paraboloidal form.

7. Crown defect (concentric zones) of $\lambda/3.5$.
8. Same defect, $\lambda/1.5$.
9. Central hole, $\lambda/2$.
Probable causes: polishing problem or bending of the mirror in its cell.

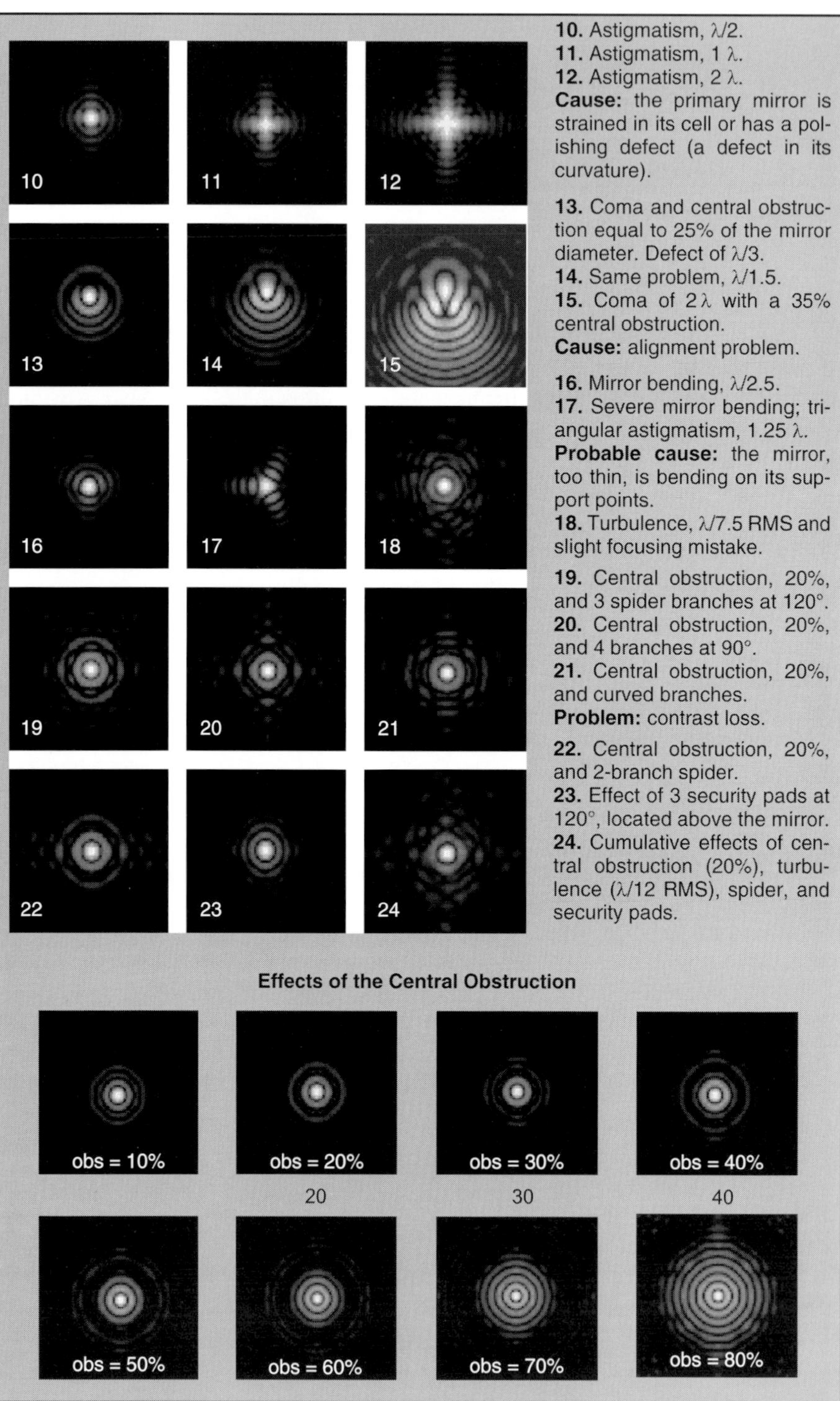

10. Astigmatism, $\lambda/2$.
11. Astigmatism, 1 λ.
12. Astigmatism, 2 λ.
Cause: the primary mirror is strained in its cell or has a polishing defect (a defect in its curvature).

13. Coma and central obstruction equal to 25% of the mirror diameter. Defect of $\lambda/3$.
14. Same problem, $\lambda/1.5$.
15. Coma of 2 λ with a 35% central obstruction.
Cause: alignment problem.

16. Mirror bending, $\lambda/2.5$.
17. Severe mirror bending; triangular astigmatism, 1.25 λ.
Probable cause: the mirror, too thin, is bending on its support points.
18. Turbulence, $\lambda/7.5$ RMS and slight focusing mistake.

19. Central obstruction, 20%, and 3 spider branches at 120°.
20. Central obstruction, 20%, and 4 branches at 90°.
21. Central obstruction, 20%, and curved branches.
Problem: contrast loss.

22. Central obstruction, 20%, and 2-branch spider.
23. Effect of 3 security pads at 120°, located above the mirror.
24. Cumulative effects of central obstruction (20%), turbulence ($\lambda/12$ RMS), spider, and security pads.

fication during a calm night, it is possible to recognize (in simple cases) the type of defect and to determine its cause.[1]

28.2 The Foucault Null Test on a Star

The Foucault test is generally used to test a concave mirror from its center of curvature. It is possible to extend the method to a finished telescope by testing the complete optical combination on a star.

The test can be performed visually. Point the telescope at a bright star (Vega, Sirius) at the center of the field and remove the eyepiece. The reflection of the primary mirror, brightened by the star, will be visible in the secondary mirror and will appear uniformly white (except if there is a lot of turbulence or visible dust). Next, place the beveled knife-edge at the focus of the instrument (resting on the tube of the eyepiece holder); advance the knife-edge progressively into the light path until the mirror is completely darkened. (Follow the procedure for positioning the camera found described in Figure 27.9 on page 269.) When the edge of the knife is exactly within the focal plane of the instrument and it "cuts" the diffraction disk, it becomes impossible to determine if the mirror darkens from the left or the right; it produces a null. Any defects in the optical assembly—a turned-down edge, a crown, ripples, astigmatism, coma, misalignment—will be visible; they will show up as features distinct from the gray background. This visual method is efficient, but it will only give you low-contrast images. It is also limited by atmospheric turbulence, which will obscure fine details.

To obtain more precise results, this test can be used in conjunction with photography. Place a camera, with its objective attached, behind the Foucault knife-edge right where your eye would go. Focus the camera on the primary mirror. The exposure will last several minutes, which will eliminate any turbulence effects.

This photographic method allows you to enlarge the image of the mirror and to increase the contrast of the image. Kodak's black-and-white Technical Pan 2415 works well for this purpose. You could also use a CCD camera, which is faster and easier to use than a conventional camera.

This method lets you test a complete instrument in real observing conditions and still test the form of a paraboloidal mirror as it is polished. Glass alone reflects 4% of incident light, $\tau = 0.04$. Knowing that the null is what we want to see on the mirror, it is no longer necessary (with experience) to use a Couder screen to make a paraboloidal mirror. The schedule for retouching depends on the weather.

To perform the photographic test on a star, you need to make a box containing the Foucault tester and a *beam splitter* (the procedure is described in Figure 28.3 on page 278). The knife-edge should be perpendicular to the perceived direction of the movement of the sky in the telescope. In an equatorial telescope, the "cutting" of the star is achieved either by the movement of the knife-

[1] For greater detail, see *Star Testing Astronomical Telescopes* by H.R. Suiter, Willmann-Bell, Inc. 1994.

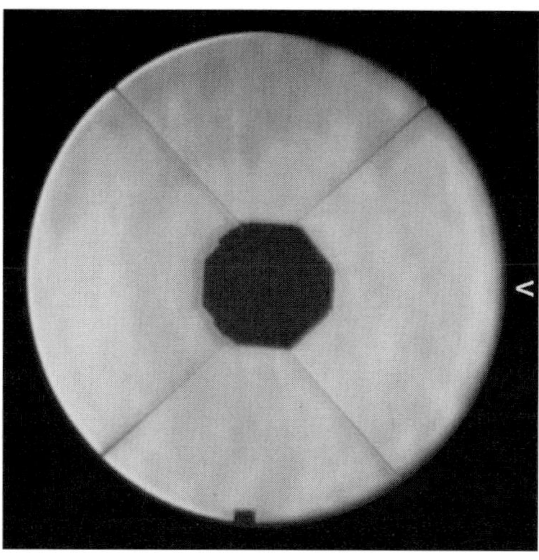

Fig. 28.2 Foucaultgram of a 600-mm mirror, focal length 2,764 mm, installed in a Newtonian telescope. The arrow at the right indicates the position of the knife-edge. The secondary mirror's support and the spider branches are visible in the reflection of the primary. The "relief" visible in the surface of the mirror is caused by the bending of the glass disk (40 mm thick) in its 27-point cell. The mirror has a small turned-down edge (visible to the left of the disk) and a slight astigmatism caused by a slight strain from the mirror cell. The defects are on the order of λ/25. The Couder and Rayleigh criteria (see Section 5.4.9 on page 72) are met. Optics made and photographed by David Vernet. © David Vernet.

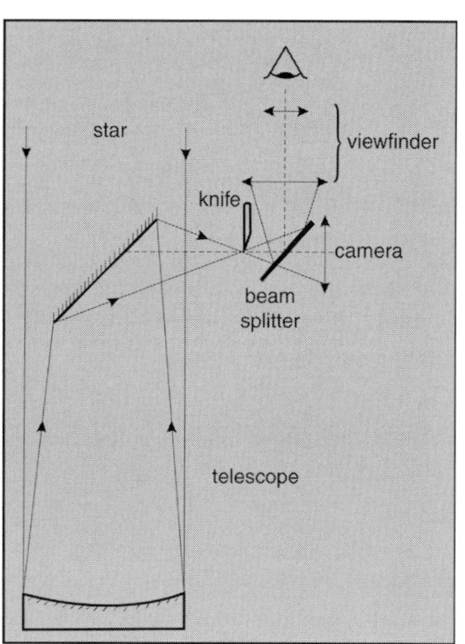

Fig. 28.3 The knife is positioned at the focus of the telescope in front of a separating lens. The light beam is divided into two parts: one traverses the separating lens and reaches the film or CCD plate; the other is diverted to the viewfinder that's used to guide the telescope during the photographic exposure.

In order to get a good photograph, the star should remain on the optical axis, at the center of the field. An equatorial mounting is recommended, but not necessary. Alt-azimuth telescopes can be pointed at Polaris and easily hand-guided for twenty minutes. Figure: © S. Debruyne.

edge or by the slow movement of the telescope in right ascension. The boxed Foucault tester is installed on an eyepiece holder with a barrel diameter of 50 mm. The camera is attached at the back of the Foucault box, with its objective attached. A 50-mm focal length sufficed for an *f*/4 or *f*/5 mirror; 135 or 200 mm of focal length are needed for an *f*/10 or *f*/12 mirror. For more information, see Figure 28.6 on page 281.

Qualitative analysis of these defects is not possible without making one or several "cuts" in the photographed surface. Circular zones are easy to recognize; they are caused by polishing irregularities and can be removed only by polishing.

Spherical aberration appears as lowered circular figures at the edge and at the

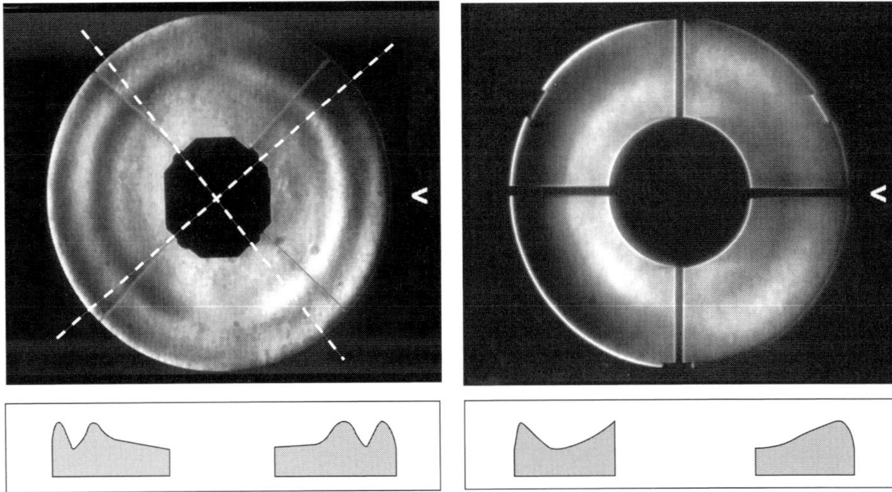

Fig. 28.4 Left: photograph of an aluminized mirror with a diameter of 600 mm and a focal length of 2,000 mm. The surface has strong defects: raised edge, circular zones, and astigmatism (the two dashed white lines on the photo represent the axes). **Right:** Foucaultgram of a Cassegrain telescope mirror, 200 mm diameter and 1800 mm focal length. The mirror has a raised edge, spherical aberration, and coma (seen in the asymmetrical lighting from right to left); these are possibly due to misalignment of the telescope or of the star on the optical axis. Note that the large central obstruction and the spider arms are framed by bright borders (perhaps caused by hot air veins). The corresponding horizontal profiles are shown below the photographs. © S. Debruyne and David Vernet.

center (see Form 6 on page 113). This appearance can be caused by a parabolization that is incomplete or carried too far, thermal inequality around the primary mirror (the center will be warmer than the edge at the beginning of a night), a mirror cell that does not hold the mirror correctly (i.e., it bends it), a mistake in the distancing of the primary and the secondary in a Cassegrain telescope, or simply the presence of a field corrector.

Astigmatism appears as a figure with two axes that are symmetric *but not a figure of revolution*. This defect happens if the mirror is in a "horseshoe" shape at the end of polishing (see Chapter 23 beginning on page 231). It is also caused if the lateral screws holding the mirror in its cell are too tight, or if the mirror is not resting on its supports correctly (for example, if it's resting on four points instead of three, or if there's a cloth underneath it). Mirrors can also bend under their own weight (this happens most frequently when the instrument is aimed near the horizon) and appear astigmatic. Astigmatism can also appear if the secondary mirror is slightly curved.

Coma is caused by misalignment of the optics, whether they are centered or oriented incorrectly. This defect forms an image that has only one axis of symmetry and *is not of revolution* (see Figure 28.4, right hand image). To detect coma in an instrument, eliminate any errors by making sure that each of the optical parts is perfectly aligned and that the star is centered exactly on the optical axis.

The photographic Foucault null test also lets you know exactly the number, the positions, and the size of all the obstructions in the instrument's light path.

Fig. 28.5 Foucault camera assembly with a guiding eyepiece. Construction and photograph: © S. Debruyne.

These are usually the central obstruction of the secondary mirror, the spider branches, and the security pads visible above the primary mirror.

28.2.1 Building the Foucault Camera

The object is to obtain, in the negative or with your eye, the image of the optical system that you're testing, *not* the image of the star. Figure 28.6 shows the position of the different mechanical and optical elements of the apparatus—the Foucault knife, the separating lens, the viewfinder, and the photographic objective.

28.2.1.1 Parts of the Foucault Camera

The Sleeve. This has a diameter of 50.8 mm and slides into the eyepiece holder of the telescope. Make it from PVC, Nylon, or aluminum in order to keep the tester light.

The Foucault knife-edge. If the instrument you're testing is on a non-motorized alt-azimuth mounting, you will need to use Polaris to test the optics because it is relatively fixed. To move the knife-edge perpendicular to the optical axis, mount the knife on an adjustable plate with no play and a 4- to 10-mm range of motion. The mobility of the knife will compensate for the absence of tracking in the instrument and the proper motion of Polaris. These platforms are rather expensive if purchased commercially; it is possible to make them much more cheaply.

If the tube is mounted on a motorized alt-azimuth or equatorial mounting, the moving platform is superfluous.

The beam splitter. This piece, inclined at 45° and placed as close to the knife edge as possible, allows you to see the same image in the visor that is formed on the photographic film (or CCD array).

Its index of reflection (R) determines the quality of the light received by your eye, and its index of transmission (T) the quantity of light received by the film (or

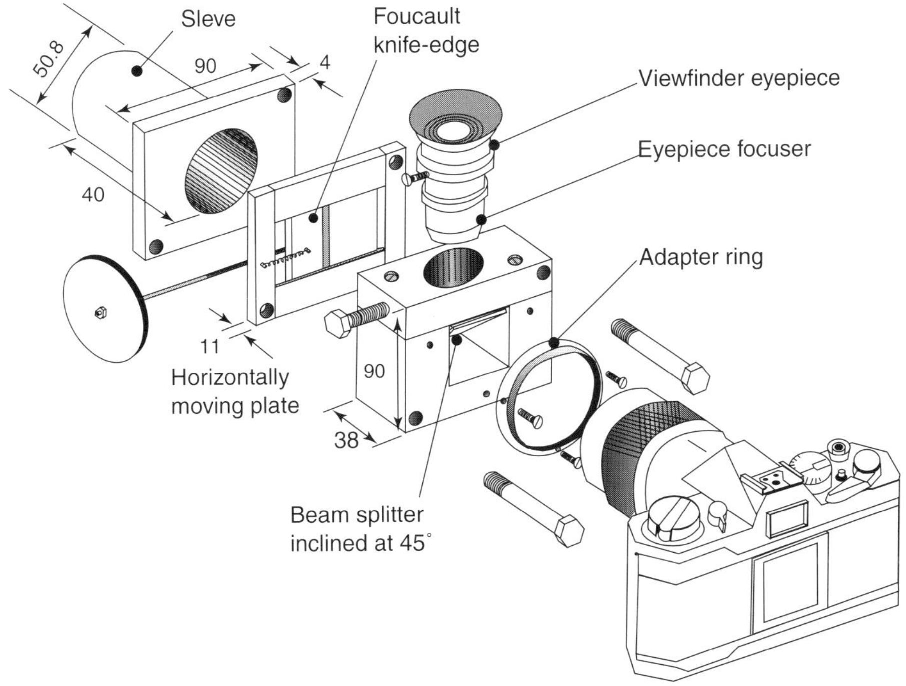

Fig. 28.6 Exploded view of the Foucault camera assembly. Dimensions, in millimeters, are indicative only. Figure: © S. Debruyne.

CCD). The choice of the lens is made, therefore, as a function of the type of camera used. Below are three examples of lenses sold by Edmund Scientific. They each measure 35 x 35 x 1 mm and are treated with antireflective coatings on their rear face. The quality of the optical glass is unimportant.

R/T	Reference	Price	Use
50/50	F 45325	$22	Universal
30/70	F 45324	$22	Photographic
70/30	F 45326	$28	CCD

The test viewfinder. This small refractor, similar to a low-power finder scope, consists of an objective and a non-reticulated eyepiece. It is placed as close as possible "above" the separating lens in order to recover all of the light coming from the star. The visor permits direct observation of the telescope mirror (or the objective, in the case of a refractor). Focusing and guiding are done not on the star but on the mirror.

In the place of the viewfinder you can use a 5 x 24 finder. If that's not convenient, you can make one using the following instructions.

For the objective, use a 20-mm achromatic doublet with a focal length of

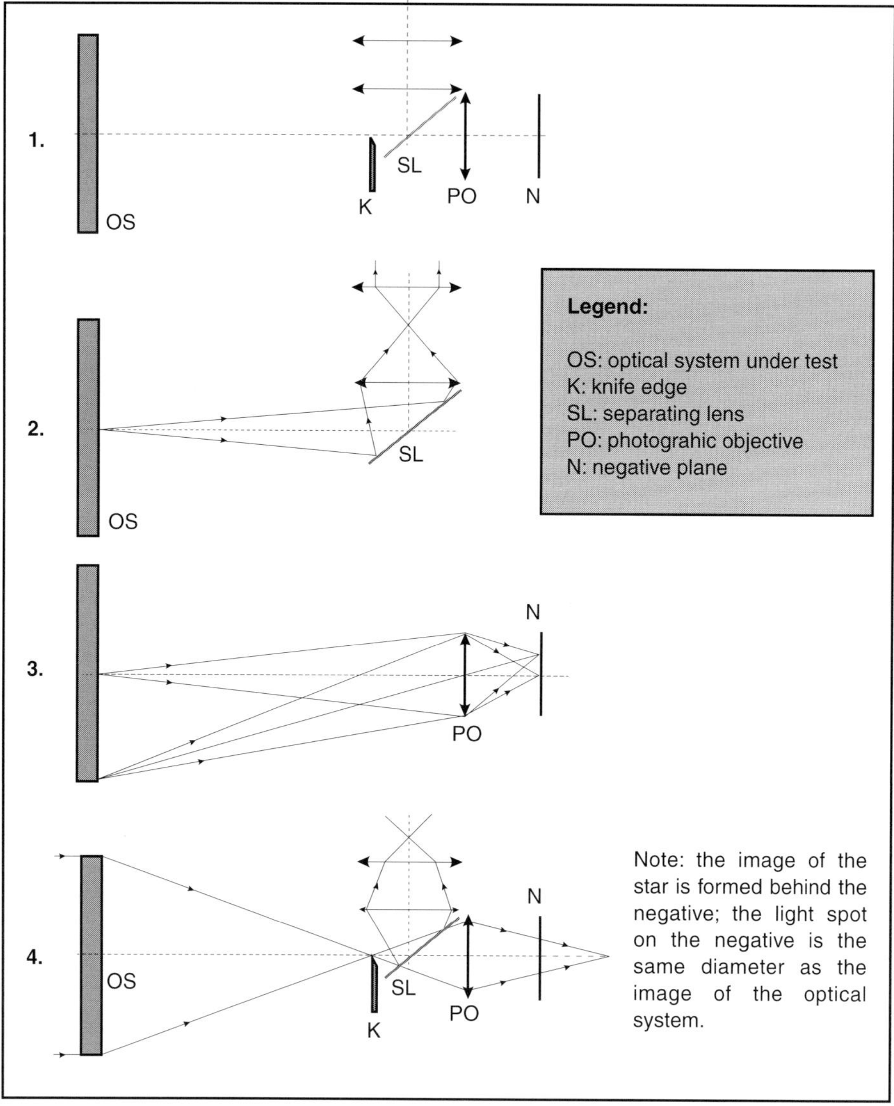

Fig. 28.7 Schematics for the Foucault camera.

1. The elements of the Foucault camera: the knife, separating lens, camera objective, and the eyepiece and objective of the viewfinder.

2. The path of the light beam coming from the optical system through the visor to the observer's eye.

3. The image formed by the optical system on the film or CCD plate.

4. The path of the light beam from the star through the Foucault camera. This schematic shows how to place the different elements in the optical system in order not to diaphragm the light path. The Foucault knife is positioned at the focus of the instrument.

Calculating Exposure Time and Image Diameter for the Foucault Camera Test

The values in the middle column are determined by experiments with a reference instrument. They can be used with the formulas on page 272 to calculate the exposure time and the dimension of the image on the negative for any other instrument. Note that in this example, the telescope's mirror is not aluminized ($\tau = 0.04$). These parameters are valid for Kodak Technical Pan 2415, not hypersensitized, developed for 12 minutes at 20°C in HC-110, dilution B (1 + 9). Parameters and formulas given by S. Debruyne.

Parameters	Reference Instrument (1)	Test instrument (2)
Focal length of the telescope	$f_1 = 2820$ mm	$f_2 = 10800$ mm
Optical diameter of the telescope	$D_1 = 600$ mm	$D_2 = 400$ mm
Focal length of the camera objective	$f'_1 = 50$ mm	$f'_2 = 200$ mm
Diameter of the image on the negative	$\phi_1 = 11$ mm	$\phi_2 = ? = 7$ mm
Index of transmission of the instrument	$\tau_1 = 0.8$	$\tau_2 = 0.04$
Index of transmission of the separating lens	$\tau_{Se1} = 0.3$	$\tau_{Se2} = 0.7$
Magnitude of the star	$m_1 = 2$	$m_2 = -1.4$
Exposure time	$t_1 = 25$ min.	$t_2 = ? = 10$ min.

about 40 mm. The Edmund Scientific catalog has the following references:

Diameter	Focal Length	Coating	Quality	Reference	Price
20 mm	40 mm	antireflective	TechSpec	F 32722	$40
20 mm	55 mm	antireflective	commercial	F 94818	$26

For the eyepiece, use standard eyepieces with 24.5-mm or 31.75-mm barrel diameters. A focal length of 25 mm works well for testing *f*/3 instruments; use shorter focal-length eyepieces for higher *f*-numbers.

The camera adapter. To attach the camera to the Foucault apparatus, you can use a ready-made camera adapter ring (where *P* is followed by the diameter of the objective in millimeters). Its large collar is attached to the Foucault camera using several screws.

The photographic objective. It is preferable to use a fixed-focal-length objective that's brighter and of a better quality than a zoom lens. The focal length (f') of the objective is determined using the following equation:

$$f' = d \times \frac{f}{D}$$

where D is the diameter of the telescope's primary and f is its focal length; d is the desired dimension of the image on the negative. The objective, situated as close as possible to the separating lens, will always be used wide open.

28.2.2 Alignment

Daytime alignment. Insert the test apparatus in the eyepiece holder and focus the Foucault camera directly on the primary mirror as you look through the tester's viewfinder and the camera's viewfinder. If you are testing a catadioptric telescope like a Schmidt-Cassegrain, do not insert the tester in the eyepiece holder, and focus it on a piece of white paper placed at a distance equal to the focal length of the instrument you're testing.

Nighttime alignment.

- With the test apparatus still uninstalled, choose a bright star, depending on the season and the type of instrument you're testing (Polaris if its a Dobsonian).
- Using high magnification, center the star precisely in the field of the perfectly collimated telescope.
- Replace the eyepiece with the Foucault camera, already focused (from the daylight alignment).
- Orient the tester so that the knife-edge is perpendicular to the movement in right ascension, since it is generally on this axis that slight movement will occur. Declination corrections are useless because they will just move the star along the knife-edge, which would not help the test any.
- Observe through the Foucault tester. Using the right-ascension movement or moving the knife-edge on its plate, occult the light beam coming from the star.
- Place the knife-edge at the focus of the instrument to be tested by turning the focus knob on the eyepiece holder. This will move the entire test apparatus. The star should stay positioned on the knife-edge.

Focusing is done by finding the null in the viewfinder (in practice, turbulence generally prevents you from seeing a full null even if the mirror is perfect). By "opening" or "closing" the knife or turning the right ascension knob, the direction the shadow comes from should not be detectable.

Notes: When you are testing, guiding constraints are less strict than they are for photography. Guiding affects only the contrast and the visibility of the defects.

It is necessary to occult the star 50%, which translates to a null by equal presence of dark and light turbulence swirls.

28.3 Cleaning Your Telescope's Mirrors

Over time, the optics in your telescope will accumulate several types of dirt: dust, gray or blue spots, fingerprints, etc. As this dirt builds up, it increases light scattering off the mirrors causing a loss of contrast. How often you will have to clean your optics depends on how often you use your telescope and how it is stored. Mirrors must be cleaned very carefully in order to avoid harming their metal coatings and it is best to err on the side of waiting longer to do this job than to do it too frequently.

The method described here requires a large bucket or tub, demineralized water, dishwasher detergent, acetone (be careful—this product is flammable and must not be inhaled), a new artificial ("cellulose") sponge[2], and soft, worn cloths (very fine 100% cotton).

Before beginning, make sure that the coating is not damaged or worn, because cleaning can aggravate these problems. If the coating is damaged or worn, you can follow this procedure, but avoid touching the mirror surface with anything other than the cleaning liquids (if you do, the mirror will probably spot slightly). Take the same precautions if the surface is silvered and has brown circles or cracks in the film (don't use a sponge).

28.3.1 Procedure

Fill the bucket or tub with lukewarm tap water and a few drops of dishwashing liquid. Dunk the mirror in the water, holding it from its back. Agitate the mirror to remove the larger dust particles and dirt from its surface. Wipe your fingers lightly, without pressing, over the submerged mirror's surface to dislodge and remove any particles stuck to its surface (see Figure 27.7A). The texture of your skin, associated with the soapy water, will not scratch the surface.

Empty and rinse the basin, and fill it with the demineralized water. Rinse the mirror with clean water before submerging it again in the tub. Dunk the sponge in the water and wipe it softly over the mirror surface (*B*). When the mirror is clean, take it out of the water and pass the sponge very lightly over the surface to remove any drops of water (*C*). With it lying on a flat surface, pour a little acetone in the center of the mirror and spread it over the surface using the cotton cloth (*D*). Always wipe from the center to the edge. When the acetone is almost completely evaporated, breathe hot air on the mirror to create a little fog (*E*). Keep the cloth damp with acetone; a dry cloth will scratch the metallized surface. Clean the mirror

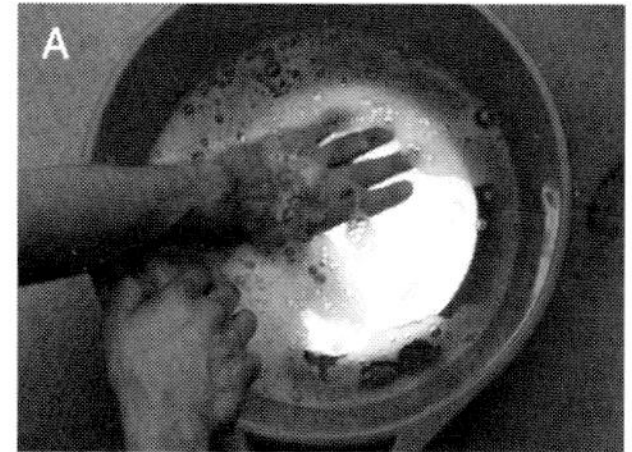

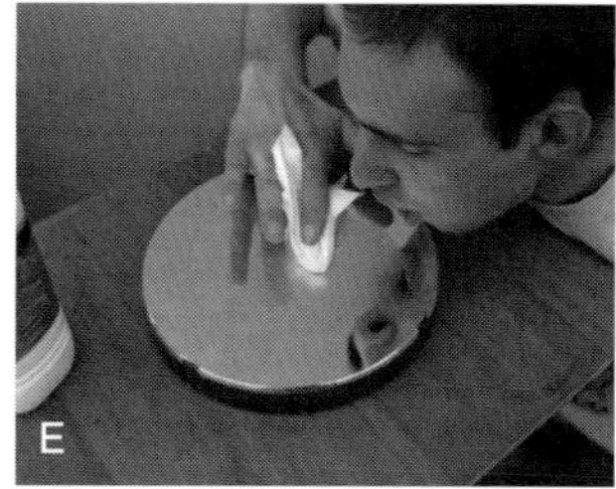

Fig. 28.8 Cleaning steps for a 250-mm mirror. Photos: © Karine Lecleire

[2] Natural sponges may contain grit.

this way until all the grease spots (blue spots) have disappeared. Erase any acetone spots by breathing on the mirror and wiping away any spots that appear.

28.3.2 Durability of Optical Coatings

In general, a vacuum silvering or aluminizing protected by a thin silicon oxide layer is highly resistant to abrasion. Outside of a city and under normal use, metal coatings that have been overcoated last about ten years. In the city, or in a polluted or humid site, your coatings can deteriorate in one or two years. If you look at a lamp from behind the mirror, a worn-out surface will see small pinpoints of light coming through the coating. The coating on such a mirror can be removed chemically and a new coating applied (these steps are usually done by the same company that recoats your optics).

Metal coatings last longer if you put a fan inside the telescope tube to circulate the air. A telescope window also helps to reduce moisture in the tube; it cuts down on the dust deposits. Finally, you may find that a desiccant placed within the tube, and renewed periodically, is beneficial.

Appendix A

Clubs and Associations/ Annual Conferences on Telescope Making

A.1 Conferences and Star Parties

Midwest Astrofest: Held near Kankakee, IL, in late September. Camping or bunk beds available in dormitory cabins. Lots to see here with many innovative telescopes shown every year. Large swap-meet. About 800 attendees. Details at http://www.chicagoastro.org.

RTMC Astronomy Expo: Held near Big Bear City, CA, on Memorial Day weekend, elevation 7,500 feet. Best turnout of commercial exhibitors anywhere; has fantastic astronomical equipment swap-meet. About 1,500 attendees. Details at: http://www.rtmc-inc.org.

Stellafane Convention: Held annually on a newmoon weekend in late July or early August on Breezy Hill, near Springfield, VT. No commercial exhibitors. Limited primitive camping at site; other campgrounds and motels nearby. About 1,500 attendees. Details at http://www.stellafane.com.

The Texas Star Party: Held in West Texas, where the sky is really dark, in late May or early June. Commercial exhibitors. Camping or bunk beds available in dormitories; about 750 attendees. Details at http://www.metronet.com/~tsp.

The Winter Star Party: Held in the Florida Keys in late January or early February. Good show of telescopes, although primary focus is observing with telescopes. About 600 attendees. Details at http://www.scas.org/wsp.html.

Every year there are many other local and regional star parties, featuring telescope exhibits or telescope design competition. Watch for announcements in the astronomical media.

A.2 Clubs and Associations

The Astronomical League is composed of over two hundred and forty local amateur astronomical societies from all across the United States. The local societies provide a good way to get to know other amateur astronomers in your area. The Astronomical League can be found on the Web at http://www.astroleague.org.

Appendix B

Suppliers

For the most current information concerning suppliers the reader should consult *Sky and Telescope* and *ASTRONOMY* magazines. Suppliers are constantly entering or leaving the market and therefore any list stands a high probability of becoming dated rather rapidly. The list presented below is representative and generally reflects businesses which have served the amateur telescope maker for a number of years. Neither the inclusion of those listed below or the exclusion of others should be considered approval or disapproval by the author, translator or the publisher.

Telescope Parts

University Optics
Box 1205
2122 E. Delhi Rd.
Ann Arbor, MI 48106
www.universityoptics.com

Meade Instruments Co.
6001 Oak Canyon
Irvine, CA 92620
www.meade.com

Celestron International
2835 Columbia St.
Torrance, CA 90503
www.celestron.com

Finished Diagonal Mirrors

Hardin Optical Company
P.O. Box 219
Bandon, OR 97411
http://www.hardin-optical.com/

Coatings

Clausing, Inc.
8038 Monticello Ave.
Skokie, IL 60076
www.clausing.com

QSP Optical Technology
1712-F Newport Circle
Santa Anna, CA 92705
www.qsptech.com

Majestic Optical Coatings
152 Willow Way
Clark, NJ 07066
www.majestic-coatings.com

Finished Mirrors

ASTROTELESCOPE S.A.R.L.
Jean-Marc Lecleire
"LeWalter"
1 promenade du Belvédère
F-77200 TORCY, France
Internet: www.astrotelescope.com

Mirror Making Kits

ASTROTELESCOPE S.A.R.L.
Jean-Marc Lecleire
"LeWalter"
1 promenade du Belvédère
F-77200 TORCY, France
Internet: www.astrotelescope.com

Newport Glass
P.O. Box 127
10564 Fern Avenue
Stanton, CA 90680

Willmann-Bell, Inc.
Box 35025
Richmond, VA 23235
www.willbell.com

Software

Foucault data reduction program, written in C: http://etienne.deforas.free.fr/foucault/foucault.htm

Bibliography

Amateur Telescope Construction

Berry, Richard and David Kriege. *The Dobsonian Telescope: A Practical Manual for Building Large Aperture Telescopes* (Richmond, VA: Willmann-Bell, Inc., 1997). This book provides complete, detailed and clear instructions on building a state-of-the-art Dobsonian telescope—with an aperture of up to 40 inches—using readily available materials and supplies. Anyone building a large aperture telescope or an altazimuth mount should have this book before beginning.

Berry, Richard. *Build Your Own Telescope* (Richmond, VA: Willmann-Bell, Inc., 1994). This book features complete directions for four reflectors ranging from 4 inches to 10 inches and a 6-inch refractor. It is recommended for any telescope maker's book shelf.

Berry, Richard. *How to Build a Dobsonian Telescope*—2nd Edition (Milwaukee, WI: AstroMedia Corp., 1982). This pamphlet is based upon articles that appeared in *Telescope Making* magazine.

Berry, Richard. *How to Build Your Own Observatory* (Milwaukee, WI: AstroMedia Corp., 1981). Based upon articles that appeared in *Telescope Making* magazine.

Brown, Sam. *All About Telescopes*—2nd Edition. (Barrington, NJ: Edmund Scientific Co., 1975). This book was written originally to feature various Edmund products (primarily surplus); however, the author's ability to draw and his knowledge of telescopes and optics make this a valuable addition to any collection of telescope making books.

Howard, Neale E. *Standard Handbook for Telescope Making—Revised* (New York: Harper and Row, 1984). This book is based on the author's experience teaching telescope making at a private New England school. It has a very good section on making a fiberglass tube.

Ingalls, Albert G. *Amateur Telescope Making, Books 1, 2,* and *3* (Richmond, VA: Willmann-Bell, Inc., 1996). A rearrangement of the original three-volume work, based on articles published in *Scientific American* from the 1920s to the mid-1950s, this book covers every aspect of telescope making. Most of the information in these books is as relevant today as when it was written, and it makes for fascinating reading.

Macintosh, Allan. *Advanced Telescope Making Techniques, Volume 1: Optics, Volume 2: Mechanical* (Richmond, VA: Willmann-Bell, 1986). Features selected articles from the Maksutov Circulars, which were published for about 20 years starting in 1957.

Orford, Henry. *Lens-Work for Amateurs* (London: Pitman & Sons, 1949). Parts of this book are interesting, particularly the sections on working small lenses.

Rutten, Harrie and Martin A.M. van Venrooij. *Telescope Optics: Evaluation and Design* (Richmond, VA: Willmann-Bell, Inc., 1988). Provides a comprehensive overview of telescope optics for the amateur astronomer.

Suiter, Harold Richard. *Star Testing Astronomical Telescopes, A Manual for Optical Evaluation and Adjustment* (Richmond, VA: Willmann-Bell, Inc., 1994). The best (and only) must-read for the testing of finished telescopes.

Texereau, Jean. *How to Make a Telescope*, Second English Edition (Richmond, VA: Willmann-Bell, 1984). The bible for amateur astronomers who wish to grind and polish a high-quality telescope mirror.

Thompson, Allyn J. *Making Your Own Telescope* (Cambridge, MA: Sky Publishing Corp., 1947). A standard work that has stood the test of time.

Trueblood, Mark and Russell M. Genet. *Telescope Control* (Richmond, VA: Willmann-Bell, Inc., 1997). This is the second edition of *Microcomputer Control of Telescopes* (1984), and deals with every aspect associated with computerizing a telescope.

Books and Articles About Telescopes

Barlow, Boris. *Astronomical Telescope* (London: Wykeham Publications Ltd., 1975). Describes theory of optics and large professional telescopes but written for the intelligent lay-person.

Bell, Louis. *The Telescope*, (New York: Dover Publications Inc., 1981). This is a republication of the original 1922 edition. Good treatment of antique refractors and general history of the telescope.

Bureau of Naval Personnel. *Basic Optics and Optical Instruments (*New York: Dover Publications Inc., 1969). Covers in detail, and with a clarity found only in military training manuals, optical principles and repair of binoculars and other optical instruments found in the US Navy.

Couder, André. *Recherches sur les déformations des grandes mirroirs employés aux observatoires astronomiques. (*Thesis, 1932)

Couder, André. "Sur un effet thermique observé dans les télescopes à reflexion." *l'Astronomie* (September–October, 1949): 253.

Danjon, A., and André Couder. *Lunettes et Télescopes: théorie, conditions d'emploi, description, réglage histoire. (*Paris: Editions de la Revue d'Optique theorique et instrumentale, 1935).

Dévé, Charles. *Optical Workshop Principles* (London: Adam Hilger, 1945).

Dimitroff, George and James G. Baker. *Telescopes and Accessories* (Philadelphia, PA: The Blakiston Company, 1945). Interesting review of telescopes in use during the first half of the 20th century.

Françon, M. *Vision dans un instrument entache d'aberration spherique* (Vision in an instrument with spherical aberration). (Thesis, 1945). *Editions de la revue d'optique.*

Frank, Charles. *Frank's Book of the Telescope (*Glasgow: Charles Frank, 1955). Introductory book for the prospective telescope purchaser.

Hoskin, Michael. *The Cambridge Illustrated History of Astronomy.* (Cambridge, MA: Cambridge University Press, 1997). A readable text by a renowned science historian with excellent illustrations traces the history of astronomy from its ancient roots into the twentieth century.

Johnson, Benjamin K. *Optics and Optical Instruments (*New York: Dover Publications, Inc., 1960) reprint of 1947 book titled *Practical Optics.* Covers basic optical principles and instruments.

King, Henry C. *The History of the Telescope* (New York: Dover Publications, Inc., 1979). Starts from the beginning and is complete through mid-1950s. The best book of its kind.

Learner, Richard. *Astronomy Through the Telescope* (New York: Van Nostrand Reinhold, 1981). Covers the instruments, inventors and their discoveries during the past 500 years. Intended for the general reader interested in science.

Moore, Patrick. *Astronomical Telescopes and Observatories for Amateurs* (New York: W.W. Norton, Inc., 1973). Provides a good overview with emphasis on English instruments.

Maréchal, André "Étude des Effets Combinés de la Diffraction et des Aberrations Géométriques sur L'image d'Un Point Lumineux" *Reveue d'Optique*, Vol 26(9) pp. 257–277, 1947.

Maréchal, A., and M. Françon. *Diffraction, structure des images, influence de la coherence*. (Masson, 1970).

NASA. *The Space Telescope*, NASA SP-392 (Washington, DC: Government Printing Office, 1976). Provides details of the instrument and its intended work.

Page, Thornton and Lou Page. *Telescopes: How to Make Them and Use Them (*New York: The Macmillan Company, 1966). Largely a collection of various "Gleanings for ATM's" articles from *Sky and Telescope* Magazine.

Platzeck, R., and E. Gaviola. *Journal of the Optical Society of America* (JOSA) 29 (November, 1939).

Paul, Henry E. *Telescopes for Skygazing*, 3rd Edition (Garden City, NY: Amphoto, 1979). Besides telescopes, this book describes binoculars.

Emphasis is upon commercial equipment of the 1950s and '60s, but it does contain many pictures of amateur and antique instruments.

Sidgwick, J.B. *Amateur Astronomers Handbook.* 3rd Edition (New York: Dover, 1980). Republication of the original 3rd Edition by Faber and Faber, London, 1971. 4th Edition, revised by James Muirden, Hillside, NJ: Enslow, 1980. A great deal of useful information on telescopes, mountings, and their accessories.

Texereau, Jean. "Les principaux défauts réels des surfaces optiques engendrés par differentes techniques de polissage." *Ciel et Terre* 3–4 (March–April, 1950).

Van Helden, Albert. *Siderius Nuncius, or The Sidereal Messenger.* Translation from the original written in 1610 by Galileo Galilei. (Chicago: University of Chicago Press, 1989). Learn the "new astronomy" from the astronomer who initiated telescopic astronomy.

Van Helden, Albert. *Invention of the Telescope* (Philadelphia, PA: The American Philosophical Society, 1977). Excellent study of the very beginnings of the telescope.

Woodbury, David O. *Glass Giant of Palomar* (New York: Dodd, Mead & Co., 1946). Starting in 1939 this book was reprinted 7 times by 1946. This is the best popular account of the building of the 200-inch Hale Telescope.

Warner & Swasey Co. *The McDonald Telescope* (Cleveland, OH: Warner & Swasey Co., OH, 1939). This little book describes the McDonald Telescope of the University of Texas.

Books About Telescope Makers

Land, Barbara. *Telescope Makers: From Galileo to the Space Age* (New York: Thomas Y. Crowell Co., 1968). Includes Galileo, Kepler, Newton, Herschel, Fraunhofer, Rosse, Hale, Schmidt, Reber and Friedman.

Scaife, Ed. *John A. Brashear—An Autobiography* (New York: The American Society of Mechanical Engineers, 1924).

Warner, Deborah J. and Robert B. Ariail, *Alvan Clark & Sons: Artists in Optics* (Richmond, VA: Willmann-Bell, Inc., 1996). Tells the story of the greatest lensmakers in American history. In this, the only definitive work published about the smaller instruments, are catalogued over 600 optical instruments made by Clark and his successors.

Willard, Berton C. *Russell W. Porter—Arctic Explorer, Artist, Telescope Maker* (Cambridge, MA: Sky Publishing Corp., 1976). The story of the man who popularized telescope making from the 1920s through the mid-1940s.

Books About Optical Design

Conrady, Alexander E. *Applied Optics and Optical Design* (2 vols., New York: Dover Publications Inc., 1960). Part 1 is a reprint of the 1929 edition. Part 2 covers the author's complete system of optical design and was first published in 1992. Covers the fundamental concepts of geometrical and physical optics.

Geary, Joseph M. *Introduction to Lens Design With Practical ZEMAX® Examples* (Richmond, Willmann-Bell, 2002). A textbook designed for home study. Concepts are presented mathematically. Worked examples, problems with answers. Hundreds of illustrations. Practically any lens design software can be used with this book.

Kingslake, Rudolf. *Lens Design Fundamentals* (New York: Academic Press, 1978). Kingslake, student of Conrady and recognized expert in his own right intends this book to be a modernization of Conrady's *Applied Optics and Optical Design*, Part 1.

Malacara, Daniel, ed. *Optical Shop Testing.* 2nd Edition (New York: John Wiley and Sons, 1992).

Smith, Gregory Hallock. *Practical Computer-Aided Lens Design* (Richmond: Willmann-Bell, 1998). A book that recognizes that modern computers take most of the drudgery out of lens design and consequently concentrates on concepts and practical examples. Very little math. Beautifully written.

Smith, Warren J. *Modern Optical Engineering* (New York: McGraw-Hill Book Company, 1966). This is a good reference for anybody needing a broad knowledge of current-day optical engineering and design.

Southall, James P. C. *Mirrors, Prisms, and Lenses* (New York: Dover Publications Inc., 1966). This is a reprint of the 1933 (3rd) edition. Provides

Index